in the United States and Canada

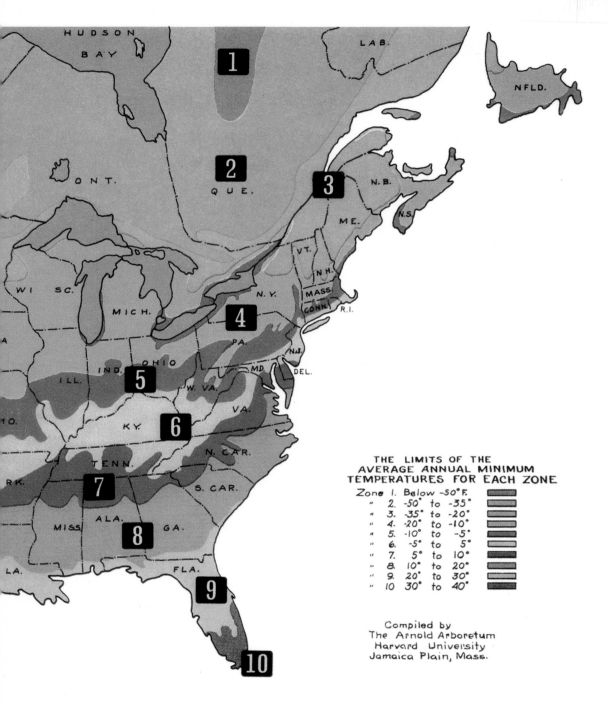

THE LIMITS OF THE
AVERAGE ANNUAL MINIMUM
TEMPERATURES FOR EACH ZONE

Zone 1.	Below	-50°F.	
" 2.	-50°	to	-35°
" 3.	-35°	to	-20°
" 4.	-20°	to	-10°
" 5.	-10°	to	-5°
" 6.	-5°	to	5°
" 7.	5°	to	10°
" 8.	10°	to	20°
" 9.	20°	to	30°
" 10	30°	to	40°

Compiled by
The Arnold Arboretum
Harvard University
Jamaica Plain, Mass.

*Complete
Manual
of
Perennial
Ground Covers*

Complete
Manual
of
Perennial
Ground Covers

David S. MacKenzie

Professional Horticulturist

PRENTICE HALL, ENGLEWOOD CLIFFS, NEW JERSEY 07632

Library of Congress Cataloging-in-Publication Data

MacKenzie, David S.
    Complete manual of perennial ground covers.

    Bibliography: p.
    Includes indexes.
    1. Ground cover plants—Handbooks, manuals, etc.
2. Ground cover plants—United States—Handbooks,
manuals, etc. 3. Ground cover plants—Canada—
Handbooks, manuals, etc. I. Title.
SB432.M33   1989   635.9 '64   88-9989
    ISBN 0-13-162058-4

Editorial/production supervision and
    interior design: Ed Jones
Cover design: Diane Saxe
Manufacturing buyer: Bob Anderson
Page layout: Steve Sopkia
Endpapers: Map of hardiness zones supplied
    and used with the permission of the Arnold Arboretum
    of Jamaica Plain, Mass. (Harvard University).

Illustrations by Marlan Winter Cotner
Photography by David S. MacKenzie

©1989 by Prentice-Hall, Inc.
A Division of Simon & Schuster
Englewood Cliffs, New Jersey 07632

ISBN 0-13-162058-4

Printed in the United States of America

10 9 8 7 6 5 4 3 2 1

Prentice-Hall International (UK) Limited, London
Prentice-Hall of Australia Pty. Limited, Sydney
Prentice-Hall Canada Inc., Toronto
Prentice-Hall Hispanoamericana, S.A., Mexico
Prentice-Hall of India Private Limited, New Delhi
Prentice-Hall of Japan, Inc., Tokyo
Simon & Schuster Asia Pte. Ltd., Singapore
Editora Prentice-Hall do Brasil, Ltda., Rio de Janeiro

In Dedication

The writing of this book is dedicated to the following people
who have been influential in my life:

Dr. Donald and Marjory MacKenzie, my parents,
for all that they have done for me through the years.
To them are my thanks and admiration.

Harold and Marguerite Swartz, my grandparents,
for their exemplary lifestyles.
Everyone should be so fortunate as to have
such a companion as my grandfather.

Ralph Shugert, friend and horticultural mentor,
for all of the encouragement and guidance he has given me.
Since I began in the field of horticulture,
Ralph has been a true inspiration.
The same can be said by countless others.

Dr. Anthony Kooiker, Dr. Larry Lup, and John Huisken,
whose friendships I had the pleasure of making while in college.
Their guidance and support is much appreciated.

David Heiss, friend, former employer, and accomplished agronomist,
for the countless hours spent instructing and advising me
in the areas of business management
and responsible and ethical horticultural practices.
Dave has encouraged me greatly and I am most grateful.

The late Donald Miller, an excellent gardener and friend,
for the enthusiastic and unselfish spirit
in which he shared his experiences, knowledge, and ideas.

Contents

Preface

This is not the first book to be written on the landscape use of perennial ground covers; it is preceded by those that encompass plants from a broad geographic range, yet contain only superficial information. Others are limited to plants of a small geographic region and supply a good deal of information. It is hoped that this book, which is a compilation of the author's experience and research, in addition to a survey of existing literature and the contributions of many professional and amateur horticulturists, will satisfy the need for a reference that addresses ground covers of a vast geographic range (North America), while providing thorough, useful, and interesting information about each genus. It is intended to fill the need for a standard North American reference on ground covers for the professional and serious amateur, landscaper, nursery owner, horticulturist, and gardener. It will also function more than adequately as a supplemental text for students of landscaping and horticulture. While written primarily for use in the United States and Canada, much of the information herein is adaptable to Europe, Australia, and New Zealand.

David S. MacKenzie

Acknowledgments

My thanks are due to the following people who participated in the production of this book.

Editors Ed Moura, Paul Corey, and Ed Jones for their advice and hard work. Marlan Winter Cotner (artist) for her dedication, creativity, cheerfulness, and encouragement.

Friends and colleagues in the horticulture profession who unselfishly shared their knowledge: Mary Walters of Walters Gardens, Tom Kimmel of Twixwood Nursery, Scott Farnsworth of Perry's Nurseries, Marvin Berson of Mapleleaf Growers, Leo Blanchette of Blanchette Gardens, Fred Devries of Ottawa Kent Perennial Gardens, Dr. Kent Kurtz of California Polytechnic College, Kathy Musial of Huntington Botanic Gardens, Jerry Colley of Siskiyou Rare Plant Nursery, Peter Orum of Midwest Ground Covers, Reiner Kruger and Roger Duer of Monrovia Nursery Company (California), Nancy Vandervere of Leaves-N-Weaves Landscaping, Wilbur Mull of Classic Ground Covers, Harlan Hamernik of Bluebird Nursery, Dr. Donald Huttleston of Longwood Gardens, Ralph Shugert of Zelinka Evergreen Nursery, David Heiss of Turf Services, Joan Defato of Los Angeles State and County Arboretum, Suzanne Nelson of the Desert Botanic Garden.

Nurseries who freely donated specimens or slides which Marlan used for illustration: Walters Gardens, Twixwood Nursery, Greentree Nurseries (Phoenix), Perry's Nursery (California), Midwest Ground Covers, Monrovia Nursery Company (California), Great Lakes Greenhouse, Bluebird Nursery, Weston Nursery, Tip Top Nursery (Phoenix), Springbrook Gardens.

Professors and authors who helped guide me in the preparation of the manuscript: Dr. Steven Still, Dr. Harrison Flint, and Dr. Michael Dirr.

Those who helped type the manuscript: Chris Payne, Connie Bronson, and Nancy Benefiel.

Nancy and Janet MacKenzie and Steven Sloan, who compiled the indexes of plant names.

Roy Putnam and Brett Brolick, who assisted with photography.

The reviewers who donated their assistance to read the manuscript: Ralph Shugert, Leo Blanchette, and Dr. Steven Still.

The reviewer of the proofs: Dr. Anthony Kooiker.

The Hamlyn Publishing Group Limited, which gave permission to reproduce definitions of plant genera.

Special thanks are due to the following individuals: Joel Kammeraad (my business partner) for his advice and for his help and flexibility in dealing with the many inconveniences which the writing of this book created. And to Rita Hassert of the Morton Arboretum for being indispensable by supplying me with a tremendous amount of botanical literature, information, support, and humor.

Uses and Culture of Ground Covers

1

Exactly what is a ground cover? With respect to landscaping, a ground cover is any material that is used to cover the soil. Commonly used ground covers include nonliving materials such as wood chips, bark, stone, concrete, husks of various fruit, plastic sheeting, woven plastic, and many others.

Living ground covers are plants that generally and naturally (without pruning) range in height from less than 1 inch to around 4 feet tall. They may be annual, biennial, or perennial and woody, succulent, or herbaceous. Ground covers may spread broadly, spread little, spread fast or slowly, clump, trail, twine, vine, creep, or hummuck. They may be deciduous, semievergreen, or evergreen. Conifers and angiosperms, along with primitive spore-forming plants, qualify. In short, ground covers are a very diverse group, yet one trait unites them more than any other: the ability (when properly cultured and established) to densely cover the soil in a manner that excludes weeds.

PHYSICAL USES OF GROUND COVERS

The physical landscape uses of ground covers are numerous. Let's examine them individually.

Facing

Facing simply refers to any material applied as a covering or screen in front of something. Facing usually enhances appearance through masking or ornamentation. Inanimate facings commonly are latticelike and made of iron or wood. The use of living ground covers for facing generally pertains to facing building foundations (*foundation planting*) or facing trees, shrubs, and nonliving ornaments within the landscape.

Figure 1–1 Foundation facing of blue sargent juniper (*Juniperus sargentii var. glauca*) in a raised bed.

Foundation Planting (Facing Building Foundations).   Following construction, it is common for a few inches to a foot or so of the concrete foundation of homes and commercial buildings to be left exposed. A foundation planting will mask such unsightliness and help blend in the building with other elements in the landscape (Figures 1–1 and 1–2). Planting close to buildings is often a job for which standard shrubs and turf are inadequate. Many living ground covers, on the other hand, are particularly well adapted for this purpose.

In the north, because the first 3 or 4 feet out from the base of buildings is subject to bombardment by falling ice and heavy snow accumulation (effective agents

Figure 1–2 A simple but effective use of japanese spurge (*Pachysandra terminalis*) to conceal building foundation.

of plant destruction), facing with upright or broad shrubs often invites disastrous consequences. Branches crack, causing plants to become weakened and physiologically distressed. Turf, too, fails. Often turf grasses are too low or simply uninteresting. Herbaceous ground covers (which often die back anyway) and horizontally spreading woody ground covers, however, have few problems with snow or ice damage and usually come through winter in good shape.

The area closest to building foundations is not only subject to inordinate accumulations of snow and ice, but also to torrents of rainwater as it pours from the roof above. Erosion of soil is unsightly; but more important, if left unchecked the rain will form a gutter that can effectively channel water down to the subterranean foundation, where it can do major damage. The answer is to use ground covers to stabilize the soil. Fibrous and densely rooted ground covers will easily bind the soil and thereby check erosion. When preparing for foundation planting, keep in mind that the soil near the foundation is usually the poorest on the site. Therefore, sample first and amend accordingly.

In addition to allowing damage from snow, ice, and rain, planting close to the foundation can result in other undesirable situations. Planting too close to a wall prevents a shrub from spreading naturally. Moreover, the reduced light exposure on the side nearest the foundation will cause thin leggy growth should a particular shrub be intolerant to shade. The use of ground covers (many of which are shade tolerant) as facing near the foundation will allow shrubs to be sited far enough away to receive adequate sun and thus develop in a natural manner. Planting shrubs away from the foundation also allows easy access to utility meters, valves, and outlets. Finally, by siting shrubbery away from the foundation, the opportunity for its sometimes destructive, larger roots to damage foundation and utility lines and pipes is eliminated.

Facing Trees, Shrubs, Benches, and Ornaments. Like building foundations, trees and shrubs may be faced with ground covers for the purpose of concealment (Figures 1–3 and 1–4). In this case, trunks, stems and bases are the objects to mask. Also, trees and shrubs, along with benches and ornaments, may be surrounded by ground covers to provide an attractive setting or to link them with other landscape elements.

Figure 1–3 This beech and oak are effectively linked with each other, and brought into harmony with surrounding turf through the use of facing with *Pachysandra terminalis.*

Figure 1–4 The contrasting white color and vertical orientation of this statue are linked with surrounding greens and horizontal elements by facing with *Euonymus fortunei* 'Longwood' (Courtesy of Wavecrest Nursery and Landscaping Company.)

Other Uses for Ground Covers

Fire Protection. Certain ground covers, notably those with succulent stems and leaves, curtail the spread of flame. Such ground covers are useful when planted around foundations where they act as fire breaks. The use of such plants is of primary interest in areas where very dry, hot periods occur.

Setting for Bulbs and Annuals. The interplanting of spring flowering bulbs or annuals with certain ground covers can be very attractive (Figure 1–5) and is practiced too infrequently. They complement each other, and the bulbs or annuals sup-

Figure 1–5 Here, the showy flowers of pansies (annual) heighten interest when combined with this variegated selection of liriope.

ply color and interest in early spring, a time when ground covers are often least attractive. Later the bulbs die back without leaving a patch of bare soil or ragged leaves, as they do when planted alone. With annuals, color may be provided throughout the season. The main rule for interplanting is to match species so that the foliage of the ground cover is no taller than one-third to one-half the height of the foliage of the bulbs or annuals.

Living Mulch.   A mulch is any material used to cover the soil to prevent weed growth. The most common mulches are turf grass, stone, bark, wood chips, and concrete. Ground covers also are frequently used to cover the soil as "weed substitutes" (Figure 1–6). Some ground covers even make good substitutes for turf grass where foot traffic is not too heavy (Figure 1–7). The benefits derived from mulching (such as erosion control, air conditioning, and esthetics) are discussed later under the topic Merits of Ground Covers.

Figure 1–6   This planting of dwarf ophiopogon serves as a living mulch and lawn substitute.

Figure 1–7   Along this sidewalk a cultivar of *Juniperus horizontalis* is far more interesting and much easier to maintain than turf.

Accent Planting.   Accent planting refers to the use of plants in such a manner that their presence emphasizes or helps to draw attention to particular elements in the landscape. Ground covers may lend accent to other plants (Figure 1–8), or they may be used alone or with other plants to accent benches, steps, entryways, ornaments, and other garden structures (Figure 1–9). Accent may be conveyed subtly, moderately, or boldly, using such elements as foliage, bark, stems, and blossoms. Collectively and in various proportions, these elements determine the degree of accent. The cool, green, erect, fine-textured sweet woodruff (*Galium odoratum*) lends moderate accent to the curve of the gray-barked stems of the downy serviceberry, (*Amelanchier canadensis*). When spring brings a profusion of tiny white stars to sweet woodruff, the serviceberry is accented all the more.

Hedging.   Similar to shrubs and tall ornamental grasses, ground covers may be used for hedging (Figure 1–10). Being lower, they differ in that they provide no

Figure 1–8   Strikingly, this fall-flowering planting of Stonecrop (*Sedum* 'Rosy Glow') draws attention to the rich foliage of a corner planting of Japanese yew.

Figure 1–9 More subtly, the southern maidenhair fern (*Adiantum capillus veneris*) accents this rugged rock ledge.

Figure 1–10 The hedge of crimson pygmy barberry to the left of this photograph effectively delimits the back boundary of this rockery planting. (Courtesy of Twixwood Nursery)

Figure 1–11 An edging of intermediate height, loosely growing variegated liriope is all that is needed to ease the transition from a jungle-like tropical border to the formality of the precisely manicured turf in the foreground.

screening effect. Neither do they defend against invasion (except a few thorny ones) from the neighbor's pets and children. They can, however, delimit and identify property, garden, and border periphery in a manner that is both interesting and attractive.

Transition Planting. Transition planting refers to the use of plants (usually ground covers, but sometimes taller shrubs) for the purpose of unifying different elements or textures in the landscape. Transition plants ease one's perception of change from one character to another (Figure 1–11). Ground covers are commonly used in

Figure 1–12   Strikingly beautiful, the outstanding foliage of *Agapanthus orientalis* 'Variegatus' (variegated lily of the Nile) allows this fine ground cover to stand alone as a specimen in this woodland border.

this regard to aid one's visual transition from the vertical orientation of a tree trunk or building wall to the horizontal orientation of the ground below. Ground covers also function well in unifying corner and entryway accent plantings. Surrounding or facing such plantings with ground covers achieves unity with the remainder of the landscape.

Specimen Planting.   Although less common as specimens than are trees and shrubs, unique ground covers contribute nicely when planted alone as specimens (Figure 1–12). Specimen planting of ground covers should be conducted in locations to which the eye naturally falls. Planting on hillsides and berms and along banks that parallel ascending walks or steps will expose special ground covers to the attention they deserve.

MERITS OF GROUND COVERS

The merits of ground covers can be divided into (1) the material and (2) the esthetic, emotional, or psychological. Each is important to our enjoyment of the landscape.

Material Merits

Soil Benefits.   Two common uses of ground covers are mulching and facing other ornamentals. Soil stabilization and erosion and runoff control are obvious benefits of such use. Other ornamentals (trees and shrubs) benefit as well. To be specific, the processes of root growth and the death of the more shallowly rooted ground covers increase the porosity of the soil. Oxygen and water then more easily penetrate to the deeper roots of trees and shrubs. Additionally, a ground cover facing can prevent foot and lawn-mower traffic and thereby eliminate unnatural causes of compaction. The result is reduced stress and better health and vigor of trees and shrubs.

Direct Monetary Benefits. A case can be made for the monetary benefits of all the merits previously discussed. However, some monetary benefits are more direct and should be factors considered in landscape planning and design. Ground covers provide financial advantages in a number of ways, the primary of which is reduced labor cost derived through lower maintenance requirements. As a mulch, turf grass is less expensive initially, but considering its intense maintenance needs (frequent mowing, edging, fertilization, irrigation, disease and weed control, leaf removal, etc.), it is usually the least economic over time.

Using ground covers to face buildings, trees, and shrubs is a good alternative to turf grass in that it eliminates the intricate labor practices of mowing and trimming around and underneath them. Scratched faces and arms and expensive plant loss due to the voracious string trimmer become only a memory. Similarly, planting ground covers in areas of difficult access, such as patios, courts, narrow strips paralleling walks and fences, underneath ornaments, and upon severely sloping or undulating terrain, eliminates much intense and often dangerous work.

Some ground covers can also eliminate much annual cleanup work. In nature, a cycle exists by which organic matter is returned to the earth. Leaves, flowers, fruit, and animals, virtually every component of all living matter, are reduced to humus by microorganisms. Rich in nutrients, the humus supplies fertility and assures the continuation of life. Today, people clear land, erect buildings, and surround them with concrete, stone, and manicured turf. Then countless hours are spent befuddling nature's attempts to deposit humus on our creations. Memories of past weekend afternoons spent raking leaves and grass clippings, or picking up, sweeping, and vacuuming fallen fruits, nuts, flowers, twigs, and whatever else, are not always cherished. Using ground covers, many of which benefit from and effectively conceal fallen debris, underneath and around trees and shrubs can eliminate much cleanup work, save money, and allow one to spend more time doing things that are more enjoyable.

Not to be overlooked among the material benefits of ground cover use are the environmental influences (other than those related to soil) that ground covers exert on their immediate surroundings. In winter in the North, ground covers are often appreciated as they stabilize snow and reduce drifting. In summer, especially in the South, they benefit trees and shrubs that have large water needs. These larger plants often experience water stress; to relieve this, ground covers, which usually consume less water than would evaporate from exposed soil, can be used as a conservation measure. Other plants, as well as people and animals, benefit as ground covers help to keep the surrounding air cool and well supplied with oxygen.

Esthetic, Emotional, and Psychological Benefits

Alleviation of Dissonance. Psychologists refer to the phenomenon whereby we become anxious over differences between our values or expectations and the reality that confronts us as dissonance. Usually, the words confusion or frustration apply in everyday life, and we may experience this when confronted with a disharmonious landscape. Fortunately, in a number of ways, ground covers, the heroes of this book, can help prevent such unnecessary mental stress.

First, ground covers unify unrelated elements in the landscape. For example, a coarse, symmetrical, brown, two-story brick house and a winding, clear, tranquil, blue, cool brook have little in common. Yet creative use of ground covers (plants that do this better than any other) can join the two in harmony with each other and the remaining components in the landscape. Similarly, ground covers can soften the sharp edges and angles of unnatural objects, such as walks, drives, benches, and fences, and make them at home among the living elements of the landscape. To illustrate, consider the stark and isolated appearance of many trees when suspended in a sea of immaculately groomed turf. Linking trees with a common facing of ground covers will unite them and ease the transition from coarse bark to fine blade. In the same way, edging a winding path of stone or brick with a ground cover helps to combine it with its surroundings of turf or woodland.

Creating an Impression. For landscaping purposes, ground covers are indispensable for creating a rich, warm, pleasant atmosphere. Edging walkways with broad sweeping beds of loosely arranged ground cover will direct traffic and provide a cool, welcoming appearance that communicates an invitation to visit.

Modifying Perception. Ground covers also function in altering our perceptions. Sharp grade changes can be made to appear moderate by covering with graceful, shrubby covers. Planting richly textured ground covers in broad, curving borders gives the perception of spaciousness, while coarsely textured ground covers may seem to reduce space and enhance intimacy. Bright colors or fine textures can lighten an area and elevate one's mood. Conversely, coarse textures in subdued shades of blue, gray, or green reduce brightness and enhance feelings of tranquility and peacefulness.

Relieving Monotony. The endless variety of ground covers is a key element in relieving monotony in the landscape and our lives. Used for accent, they inject interest into such mundane fixtures as steps, entryways, decking, and shrubs. They can amaze us with colorful and sometimes exfoliating bark, intricate branch patterns, and colorfully painted foliage. Evergreen ground covers bring year-round beauty, and many types grace the landscape with colorful and sometimes heavenly scented flowers. Flowers and fruit alike attract many kinds of songbirds and wildlife.

GENERAL TIPS FOR THE SUCCESSFUL USE OF GROUND COVERS

The number of uses for ground covers in the landscape is limited only by one's imagination. Even so, three general guidelines should be followed:

1. Avoid using too many different types of ground covers in the same landscape. It is usually more pleasing to see a few types used to cover greater areas than to risk creating a busy, cluttered appearance by breaking up the beds and increasing the varieties.
2. In general, plant large-foliaged ground covers when the scale is large and small-foliaged ground covers when the scale is small.
3. Adhere to companion planting between ground covers and other plants in the landscape. For example, avoid planting a ground cover near a tree or shrub of very different environmental adaptations. This only creates excess work, since each will have unique cultural demands. Moreover, the mixing of culturally dissimilar plants often creates a chaotic appearance that disrupts harmony in the landscape.

PLANTING AND CARING FOR GROUND COVERS

When to Plant

The best time to plant ground covers varies with the local climatic conditions. Always, though, plants should be acclimatized to existing weather conditions.

In the North, planting from spring until the ground freezes in the fall should involve little risk of loss. Spring is ideal; but for the most part, planting in mid-summer is quite acceptable, even during hot weather and lack of rain. Irrigation needs (and in most northern areas, water is plentiful) will be only moderate and will involve little expense. Extending the planting season until the ground freezes or snow falls is also fine, provided that the plants are mulched to prevent the heaving that sometimes occurs with exposed soil. If the plants have had adequate time to establish good root growth (generally, to the extent that a firm tug will not uproot them) prior to the ground freezing, the danger of heaving will be minimized and mulching is not necessary.

Warm or hot climates necessitate planting in fall, winter, and spring, unless water is readily available and relatively inexpensive. If penetrating frost is expected, mulching should be used as it is in the North.

Where to Plant

Ground covers are available that will grow in virtually any terrain or naturally oc-curring soil conditions. Prior to planting, the soil should be tested to assess fertility, pH, physical composition, and, most importantly, the ability to drain excess water. Knowing these factors allows the selection of proper plant material and will deter-mine what amendments are needed to make the soil suitable for the desired plant. Other factors, such as exposure to drying winds and prevailing light conditions, should be considered and are as important as soil type.

How to Plant

Bed Preparation. Prior to planting, all weeds should be killed, including the roots, and all debris removed. All amendments are then incorporated. For broad-spreading, woody ground covers, only the area in the vicinity of the root zone (usually the top 8 to 12 inches of soil in an area about 3 feet in diameter) need be amended. But for plants that stolonize, layer, or spread by division or rhizomes, or those that are planted very closely together, the entire bed must be amended. Once weeds are removed and amendments made, the soil should be worked well and final grading completed. Holes or trenches should be dug with a spade, trowel, or hoe and the material planted to the same depth as it was originally growing. For large jobs, mechanical transplanters may be feasible, depending on plant type.

Spacing. Triangular spacing (parallel rows with plants staggered) places all plants at an equal distance apart. Square planting spaces plants an equal distance from their neighbors placed above, below, or across, but not diagonally. Each method is acceptable and creates a very organized and professional appearance. Random spac-ing looks less organized, but sometimes takes less planting time; and for plants that form a solid mat, such as pachysandra or aegopodium, the end appearance is the same. To calculate the number of plants needed, use the following formula:

$$\text{number of plants needed} = \frac{\text{area to be covered}}{(\text{on center distance between plants})^2}$$

Or consult the accompanying chart (continued on next page).

	Plants Needed Per:			Area per Plant
Spacing*	Square Foot	100 Square Feet	1000 Square Feet	(in square feet)
3″	16	1,600	16,000	0.06
4″	9	900	9,000	0.11
5″	5.75	575	5,750	0.17
6″	4	400	4,000	0.25

＊Plants same distance apart in rows as between rows.

Spacing	Plants Needed Per:			Area per Plant (in square feet)
	Square Foot	100 Square Feet	1000 Square Feet	
7″	2.93	293	2,930	0.34
8″	2.25	225	2,250	0.44
9″	1.78	178	1,780	0.56
10″	1.44	144	1,440	0.69
12″	1	100	1,000	1.00
14″	0.73	73	730	1.37
16″	0.56	56	560	1.78
18″	0.44	44	440	2.27
20″	0.36	36	360	2.77
24″	0.25	25	250	4.00
30″	0.16	16	160	6.25
36″	0.11	11	110	9.09
4′	0.06	6	60	16.7
6′	0.028	2.8	28	35.7
8′	0.015	1.5	15	66.7
10′	0.01	1	10	100.

General Maintenance

One of the greatest selling points of ground covers is their relatively low maintenance requirements. It is, however, a misconception to believe that no maintenance is involved in the culture of ground covers; as with other plants, they may need periodic irrigation, weeding (at least until established), soil amendment, mulching, pruning, disease and insect control, litter removal, and occasional division and mowing.

Irrigation. The first item on the maintenance schedule for all ground covers is to thoroughly water them in immediately after planting. This will help alleviate transplant shock and eliminate large air pockets in the soil. Thereafter, new plantings should be watered frequently enough to prevent wilting until the roots become well established; a period of about three months is usually adequate. From then on, during the first year, it is wise to watch closely for signs of wilting and to supply water as needed. Always water to the extent that water percolates through the entire root zone. Never water unless there is a genuine need. After the first year, many ground covers will need little or no supplemental irrigation. If needed, it is usually only during warm, dry summer days. On the other hand, even though a particular ground cover can survive a stretch of drought, it will often appear more esthetically pleasing if an occasional deep watering is exercised. Regardless of plant type, in areas of freezing winter temperatures, it is wise to water thoroughly in late fall before the ground becomes solid. Plants that enter winter without drought stress have greater winter survival rates.

Weed Control. Weed control demands are usually most intense during the first year following planting and then will taper off as plants begin to fill in and cover exposed soil. Once filled in, little or no weed control measures will be called for. To control weeds and minimize maintenance demands, the most critical step is the preplant procedure. Completely eliminating all living weeds and their roots will in large part reduce the amount of weed control effort and expense thereafter. Following planting, weed control can be exercised in three ways. The first is manual and consists of hoeing, tilling, or hand pulling; for most people, it is least desirable. The second is chemical. Today, many preemergent herbicides are labeled for use on ground covers of various types. For the most part, preemergent herbicides are relatively inexpensive and simple to use. Postemergent herbicides, on the other hand, involve a little more labor and, if drifting occurs, may damage desirable plants. Selective postemergent herbicides (those which kill developed weeds selectively) are increasing in popularity and for the purpose of killing weed grass in broad-leaved ground covers, are often the only solution for severely overrun plantings. Finally, mulching around ground covers with organic materials such as wood chips or bark is often the preferred method. Like preemergent chemicals, mulching prevents weeds from becoming established. It does not, however, greatly risk damaging the environment in the process, as many chemicals do. Moreover, organic mulches eventually decompose to supplement the soil. Nondegradable mulches, such as stone, can also be used around ground covers that thrive in infertile, rocky soil conditions.

Pruning.   Most ground covers benefit from some type of annual pruning. Pruning ground covers is generally undertaken to make them look neater, to open them up to increase light penetration and air circulation, or often to stimulate new growth and branching.

The tools used in pruning ground covers vary, but the most useful are hand shears, hedge shears, and a modified lawn mower. Converting a lawn mower for pruning ground covers involves adding extensions to the wheels to increase the elevation (Figure 1–13). General guidelines for pruning evergreen herbs, subshrubs, and woody shrubs are as follows:

1. Prune spring flowering plants immediately after bloom.
2. Prune summer and fall flowering plants in spring.
   Removal of more than one-third of the length of branches is never recommended. For herbaceous plants that die back, the dead stems and leaves can be removed either in fall or spring. In all cases, clippings should be cleaned up and disposed of away from the pruning location. This will help to prevent the accumulation and spread of pathological organisms. If a mower is used, it should have an attached bag to catch the clippings.

Figure 1–13   Elevated lawn mower, modified to prune ground covers.

Fertilizer. Most ground covers will benefit from annual fertilization. In general, a granular product in which both nitrogen and potassium are in a slow or controlled release form is desirable. Phosphorous need not be in this form since it is relatively insoluble to begin with. The NPK ratio (ratio of the percentages of elemental nitrogen, phosphorus as P_2O_5, and potassium as K_2O in a fertilizer) should be 2–1–2, as a 10–5–10 formulation, and it should be applied at a rate of 1 1/2 pounds of actual nitrogen per 1000 square feet in mid-spring, followed by an application of 1 pound actual nitrogen per 1000 square feet in early fall. More properly, the soil should be sampled in early spring and fertilization (both product and rate) carried out in accord with the individual needs of the particular plant. At any rate, granular fertilizer should only be applied when leaves are dry. Following application, any fertilizer that has adhered to or become trapped in the foliage should immediately be washed off with a fast stream of water.

Amending Soil. Where a ground cover grows in soil that was originally amended with organic matter, periodic supplemental amendment will probably be needed. Garden compost, the most frequently used organic amendment, is usually acceptable. Instead of incorporating with a rototiller or spade, organic amendment to densely growing ground covers is best accomplished by simply top dressing with occasional thin layers. Wash the amending material from stems and leaves; then let the worms and other soil organisms do the muscle work of mixing it with existing soil. If spagnum moss is used, it will have to be dug in as it may become dry and prevent water from penetrating.

Pathology. Although all plants are susceptible to various pathogens, susceptibility does not mean that a certain plant is without value or will inevitably succumb to a disease or pest. Proper siting and attention to cultural needs are the best defenses against pathology and will ensure that very little or no maintenance will have to be dedicated toward control measures. Preventive practices, other than proper cultural maintenance, are rarely required. If preventive chemical applications are constantly necessary to keep a plant alive, then cultural problems exist or nature has deemed that plant unfit for life in that particular environment. Applying preventive applications to plants in the landscape is expensive and all too often harmful to the environment.

Division. Certain ground covers (usually those that spread by crown division) sometimes grow together so tightly that they become overcrowded. The result is stunted growth, reduced floral display, and greater susceptibility to pathology. To alleviate the crowded condition, plants can be removed intermittently and the gaps filled with soil, or all plants can be removed, split apart, and replanted. These practices are referred to as thinning and division, respectively.

Rodent Control. If rodents decide a particular patch of ground cover is suitable for food, it is time to consult a pesticide distributor and get rid of the problem. Rodents will generally prefer poisoned grain to ground covers and eradication usually proves to be quite simple. Mechanical trapping may also be practiced should it be feasible.

Leaf Cleanup. In the case of low, dense-growing ground covers where patches of the entire planting can be covered up by leaves, they should be raked up or vacuumed. On the other hand, if the ground cover is large enough so that fallen leaves slip through its canopy and are either obscured or are not unsightly, they can be left to decompose and augment the soil. If leaves are suspected of containing toxic substances (as has been said of the walnuts) or of altering the pH adversely, removal is in order.

Dusting.   Occasionally, in times of low rainfall, dust will accumulate on the foliage of ground covers. Although not directly dangerous, dust accumulation is a stressor and is unsightly. Wash with a fast stream of water and gain the added benefit of removing a few insects should any be present.

Using This Book

2

This chapter provides an explanation of the topics discussed in the ground cover encyclopedia. To derive the greatest educational benefit (and return on your educational dollar), you should be thoroughly familiar with this material.

The topics are arranged in the following order:

Genus (pronunciation of genus name): Definition: description of genus.
FAMILY:
Species (pronunciation of species name): Definition.
COMMON NAME:
NATIVE HABITAT:
HARDINESS:
HABIT:
SIZE:
RATE: Spacing, recommended container sizes.
LANDSCAPE VALUE:
FOLIAGE:
STEM:
INFLORESCENCE:
FRUIT:
OF SPECIAL INTEREST:
SELECTED CULTIVARS AND VARIETIES:
OTHER SPECIES: Pertinent information regarding other species.
CULTURE: Soil:
  Moisture:
  Light:

PATHOLOGY: Diseases:
  Pests:
MAINTENANCE:
PROPAGATION: Cuttings:
  Division:
  Layering:
  Seed:

NOMENCLATURE

Simply stated, nomenclature is a system of naming. The accurate use of nomenclature in botany and horticulture enables us to communicate effectively. In the case of horticulture, each plant has only one recognized, valid scientific name (sometimes, however, nurserypeople forget this and create confusion through erroneous identification or careless use of plant names). Horticultural plant names are under the control of the International Association of Plant Taxonomy.

Carolus Linnaeus (1707–1778), who is generally recognized as the father of modern botany, is credited with the initial development of the botanical classification system that we use today. Linné, as he was called in his native country, Sweden, based his classification on the reproductive structures of individual plants. The original system was published in his book *Species Plantarum* in 1753. The system has undergone some changes through the years, but the diligent work of Carolus Linnaeus will long be recognized as paramount in the science of botany.

Nomenclature Terms

For the purpose of this text, we take into consideration identification by family, genus, species, occasionally subspecies, and variety or cultivar. From the general to the more specific, these terms allow us to effectively classify the plants that we will study. The terms are defined as follows:

Family. Families are a way of grouping plants that are broadly related. For example, common families are grasses, orchids, and legumes.

Genus. This is the next more precise group following family. Literally, the word means general. The genus is part of the binomial name (meaning two names) that is made up of the genus and species. Genera (plural of genus) names are usually composed of a combination of Latin or Greek roots. The genus is always capitalized. For instance, with *Sedum kamtschaticum, Sedum* is the genus name.

Species. Following the genus is the species name (literally meaning specific), which is not capitalized. The species name is also commonly derived from Greek or Latin. In the botanical name *Hedera helix, helix* is the species name.

Subspecies. A subspecies name is infrequently used as a designation of lower taxonomic standing than species. A subspecies consists of a biotype of which more than 90 percent of the population is distinct from the species. The difference is often subtle and is usually attributed to regional influences. Subspecies are rarely discussed in horticultural texts. Subspecies names are not capitalized. An example of a subspecies is ssp. *petiolaris*, which is represented in the botanical name *Hydrangea anomala* ssp. *petiolaris*.

Botanical variety. A botanical variety is similar to a subspecies, but is considered to be a variant that has *arisen in nature*, with the difference that it *may or may not have a clear geographical distribution*. Botanical varieties are not capitalized and are preceded by the abbreviation var. One commonly cultivated variety of ground cover is *Juniperus chinensis* var. *sargentii*.

Cultivar. A cultivar is a distinct strain that has *arisen in cultivation.* In effect, it is a cultivated variety, with the term cultivar being a compression of these two words. Cultivar names are written in the common language of the place of origin. Cultivar names are capitalized and placed between single quotation marks. They are frequently descriptive in a manner that emphasizes the plants uniqueness and enhances consumer appeal. An example is *Pachysandra terminalis* 'Green Carpet.'

Other taxonomic terms that are less frequently used include the following:

Clone. Plants originating from a single plant that are propagated *asexually* and are genetically identical.

Form. Naturally arising, morphologically distinct groups that vary in an outstanding manner such as flower color, leaf color, or growth habit. Most frequently, a form is thought of as a means of distinguishing varients that have arisen from subspecies or varieties.

Hybrid. Hybrids are usually designated by a multiplication sign (×) between the generic and specific epithets, that is, between genus and species names. An example of a hybrid is *Abelia* × *grandiflorum*, which resulted from crossing *A. chinensis* with *A. uniflora.*

Finally, pronounciations and the meanings for most of the generic and specific epithets have been given. Pay close attention to these as they are indispensable, academically for memory retention, economically for presenting yourself authoritatively, and idealistically as items to satisfy your yearning for knowledge.

Common names are given where possible.

PLANT AND GROWTH CHARACTERISTICS

Native Habitat. Under this heading, the areas of natural distribution of a specific species are given.

Hardiness. Hardiness ratings in this book are based on the map of the Arnold Arboretum (see four-color end papers at front or back of book). In many cases, only the northernmost zone that a plant may be expected to survive in is listed. In these cases survival of the plant may most often be expected in many warmer zones, provided suitable cultural requirements (moisture, light, soil types, etc.) are met, either naturally or artificially. If a southernmost hardiness limit is known, a range of hardiness from north to south is listed.

Limitations of hardiness zones are inherent and should be understood. Hardiness ratings are based on average annual minimum temperatures and are therefore overly simplistic for use as the sole means of determining whether a particular plant should be used. Hardiness zones are to be used only as a guide to plant selection and are not intended as an absolute rule for determining the survivability of a particular plant in the zone listed.

Factors besides minimum temperature that affect a plant's hardiness or ability to thrive include drought, exposure to sun and wind, soil conditions, altitude, length of growing season, amount of rainfall, soil characteristics, snow cover, humidity, frost occurrence, nearness to water, pollution, and others. With this in mind, it is easy to see that a plant's hardiness is governed by the particular microclimate it is subjected to, either naturally or with help from people. For example, nearness to a body of water tends to increase hardiness by moderating temperature and, essential to some varieties, by increasing humidity. Or winter dessication (burning, which may limit the use of some broadleaved evergreens where snow cover is unreliable) may be overcome by planting in a microclimate of reliable snow cover, mulching, or planting in a sun-sheltered location.

Finally, to summarize, hardiness ratings are designed to serve only as a general guide for determining the expectations of a plant's behavior in a particular geographic region. Growing in areas farther north will in many cases result in poor growth, winter injury, or even death. Growing farther south will in most cases prove satisfactory, but southern limits do exist for some varieties, and will be better understood and more available in the future. If a particular plant is grown north of the northernmost recommended hardiness zone (or south of its southernmost zone, if such information is available), the planting site and cultural practices should be such that they ameliorate the adversity that normally prevents healthy growth in that particular zone.

Habit. Refers to the general form and mode of growth that a ground cover displays.

Size. Mature size is a difficult subject to address. Height and spread are both stated for many plants. For others, only the approximate mature height is given. In either case, sizes may vary with climate, insect and disease attacks, cultural practices, plant location with respect to wind and shading, fertility, sun exposure, and so on. In short, many variables exist, and actual sizes will vary somewhat. Moreover, most ground covers, like trees and shrubs, may be kept lower or narrower than average if maintained so by mowing or shearing.

Rate. Rate refers to the relative speed with which a ground cover performs its duty of blanketing the ground *when its cultural requirements are met.* Terms such as relatively slow, moderate, relatively fast, fast, or invasive are used to describe the rate of spread. Although not empirical, they generally suffice for our purposes.

Spacing: Suggested ranges, of recommended distance between plants, are somewhat subjective. The time for establishment will vary with individual plants based on their growth rate and planting distance. Choosing a closer spacing is generally advised when using relatively small material. Thus, as plant size increases, planting distance should increase. Very loosely speaking, spacing as recommended will result in relatively complete cover in approximately one to two growing seasons when cultural needs are optimally met and reasonably sized plants are used. Slow-maturing plants or broad-spreading woody shrubs will take longer, but generally an attractive and easy to maintain bed is realized in two or three growing seasons.

Container sizes: Container sizes and types will vary with regional preferences and availability. Sizes listed are generally common in the trade. Containers are recommended either by volume (that is, pint, quart, gallon, etc.) or by diameter. Container diameters are given for plants that are generally shallowly rooted, and containers for them need not be more than a couple of inches deep. If bareroot material is used, its soil/root mass should correspond to the size and volume of the recommended container.

Landscape Value. This is a discussion of customary uses of a particular genus or species for landscape purposes. The uses listed are by no means conclusive, and you are encouraged to be creative and use ground covers in new and innovative manners.

Foliage. Foliage is discussed with respect to persistence, morphology, and coloration. Various foliage types are illustrated in Figures 2–1 through 2–8. A definition of morphological terms can be found in the Glossary.

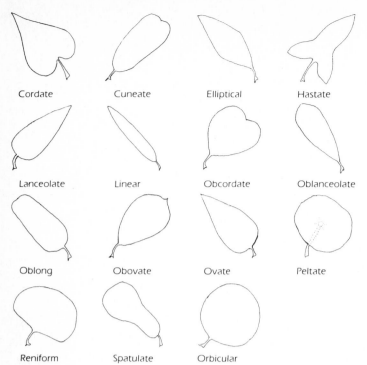

Cordate   Cuneate   Elliptical   Hastate

Lanceolate   Linear   Obcordate   Oblanceolate

Oblong   Obovate   Ovate   Peltate

Reniform   Spatulate   Orbicular

Figure 2–1 Simple Leaf Shapes.

Figure 2–2 Simple Leaf Arrangements.   Alternate   Opposite   Subopposite   Whorled

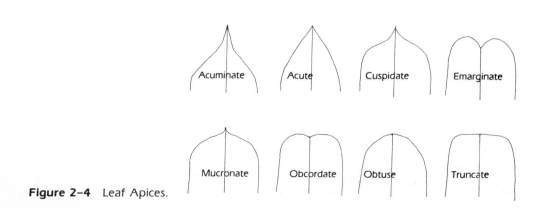

Even Pinnate   Odd Pinnate   Palmate   Bipinnate   Trifoliate

Figure 2–3 Compound Leaf Arrangement.

Acuminate   Acute   Cuspidate   Emarginate

Mucronate   Obcordate   Obtuse   Truncate

Figure 2–4 Leaf Apices.

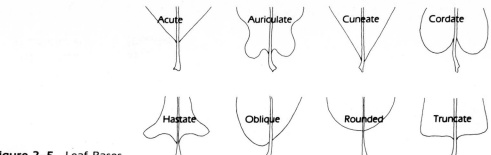

Figure 2–5 Leaf Bases.

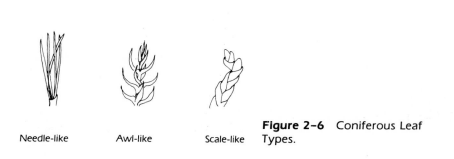

Needle-like   Awl-like

Scale-like

Figure 2–6 Coniferous Leaf Types.

Figure 2–7 Leaf Junctions.   Sessile   Perfoliate   Petioled

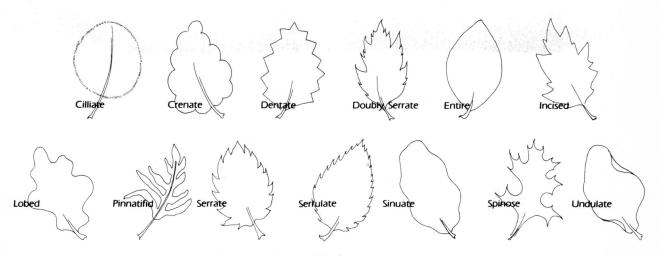

Figure 2–8 Leaf Margins.

Texture.   Texture refers to the degree of coarseness displayed by a particular ground cover. Texture may vary from season to season for deciduous plants, and the texture of foliage may differ from bark or stems. If discussed, this topic is limited to foliage texture.

Bark/Stem.   A comprehensive discussion of this topic has been avoided. Generally, it is restricted to those species that have relatively interesting branching patterns or unusual bark. Bark and stem are usually not considered the most outstanding features of ground covers.

Inflorescences.   Inflorescences are the structures upon which flowers are arranged. Inflorescences are termed either determinate, in which the oldest flower terminates the main axis and the progression of blooming is downward or outward, or indeterminate, in which the youngest flower is terminal or central and the progression of bloom is upward or inward. In reality, most plants do not strictly follow these descriptions. These terms are to be used loosely and interpreted thusly in the following descriptions of the more common types of floral arrangement. Moreover, the morphological descriptions that follow are not always adequate to fully describe an individual inflorescence; nature is never content to fit perfectly into our organizational sytems. Typical flower parts are shown in Figure 2–9, and Figure 2–10 shows the types of inflorescences.

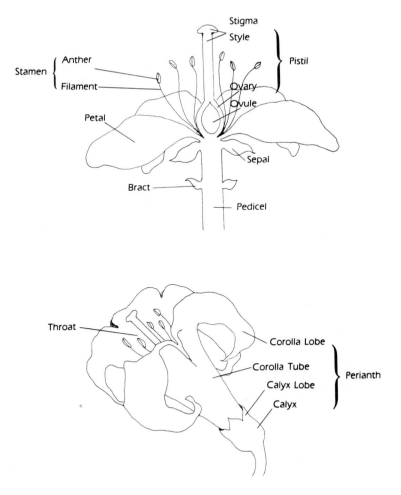

Figure 2–9 Typical Flower.

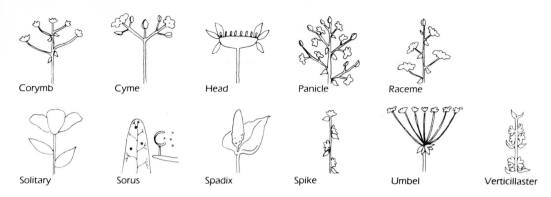

Figure 2–10 Inflorescences.

Corymb: Indeterminate; this inflorescence is composed of individual flowers held on lateral stalks of varying lengths, the longest at the base and shortest terminally. The result is usually a flattish-topped cluster. Outer flowers open first.

Cyme: Determinate; this inflorescence is flat-topped or convex with central flowers opening first.

Head: Sometimes called a capitulum, a head is the typical arrangement of flowers in the family Compositae (sunflower family). A single head is composed of several sessile flowers in a rounded or flat-topped cluster. Progression of bloom is usually from periphery to center. Sterile, ray flowers are generally more ornamental and are found to surround the fertile disc flowers that are arranged atop of a receptacle. Heads may exist solitarily or in various arrangements such as spikes, corymbs, panicles, and racemes.

Panicle: Indeterminate, this pattern is simply a more or less elongated inflorescence with branches from a central axis, which themselves in turn are branched. Congested, somewhat cylindrical panicles are called a *thyrse* and may be referred to as thyrsoid.

Catkin: Sometimes referred to as an *ament,* a catkin is a loose spike or raceme composed of many, most frequently apetalous, sessile flowers. Catkins may be erect or pendant and are particularly well adapted to make use of the wind for pollination.

Spadix: This type of inflorescence is most simply described as a spike with a thickened fleshy axis. If is often surrounded by a conspicuous, hoodlike bract called a spathe. Flowers are usually very small and in some cases exude copious amounts of viscous, insect-attracting compounds.

Raceme: Indeterminate, but frequently variable in blooming. A raceme is an elongated inflorescence that has a central axis with simple pedicels of about equal length.

Spike: Indeterminate; a spike is similar to a raceme, but flowers are sessile.

Umbel: Indeterminate; an umbel is an inflorescence in which the pedicels arise at and radiate from about the same place on the end of the peduncle. Generally, the flowers open from the periphery inward. Sometimes the peduncles branch and end in secondary umbelletes, in which case the inflorescence is termed a compound umbel.

Solitary flower: A solitary flower exists separately from others of the same plant.

Fruit. Listed for fruit are type, size, color, time of effectiveness, and ornamental worth. Figure 2–11 shows the various fruit forms.

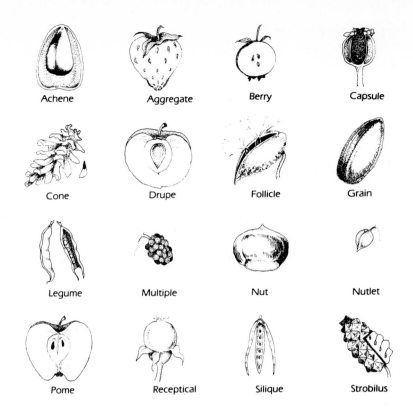

Figure 2–11 Common Fruit Types.

Selected Cultivars and Varieties. Those cultivars and varieties that are frequently cultivated and generally deemed worthy of use are listed. Most of these can be obtained readily in the trade, but others may take considerable effort to locate.

Of Special Interest. This category is reserved for discussion of outstanding or unique features of the genus or species in question.

Culture. Of primary importance to success with ground covers (as well as with other plants) is attention to cultural needs. Soil (type, pH), moisture (actual moisture requirements or range of suitable moisture level), drought tolerance, drainage, and other topics that might be of worth are covered. Finally, light requirements are listed. The types of light conditions are as follows:

Full sun: The area is exposed to sunlight all day long.
Light shade: The area is exposed to a few hours of direct, but mainly partially obscured sun. The usual filter is a tree with very open canopy, such as a honey locust *(Gleditsia).*
Moderate shade: Little direct sunlight is available. Reflected light makes up a great deal of that which is available. The floor of an open hardwood forest is typical of this condition.
Dense shade: Almost no or no direct sunlight is available, and reflected light accounts for almost all the light that is available. The ground along a north-facing wall, the floor of a dense forest, or under dense trees such as Canadian hemlock typify areas of dense shade.

Of course, these descriptions of shade are inexact, but given that few ground covers are unadaptable to ranges of available light, they serve the purpose adequately.

Pathology. Listed under this heading are the pathological elements that are most frequently encountered with a given host plant. Physiological, environmental, and biological pathology are listed. Chemical treatments are rarely discussed, as chemical labels and products change very frequently. For control measures, several texts and bulletins are available and are updated continually.

Maintenance. This is a discussion of the practices that ought to be adhered to to keep ground covers optimally healthy and attractive. All ground covers can benefit from moderate soil fertility maintenance, but in most cases specific recommendations for fertilization are not given unless a particular plant is very unusual in its needs. For most cases, the program discussed in Chapter 1 is adequate.

Propagation. This category includes means of propagation that are both commonly implemented and have proved to be successful. For the most part, the information in this section is easy to understand. One commonly used abbreviation, however, is IBA, which stands for indole-3-butyric acid. Expressed in parts per million concentration (ppm), IBA is the most frequently used chemical for the stimulation of root growth in stem cuttings. Cuttings treated with the proper concentration of this invaluable growth regulator root more quickly and in greater number than untreated stem cuttings. IBA is to be used on *stem* cuttings only. On root cuttings it usually prevents bud formation.

Ground Covers from A to Z

3

GROUND COVER QUICK REFERENCE CHART

The following chart is useful in determining which particular plants will perform under the specific conditions of a particular climate and site. It also summarizes the various attributes of each species. Abbreviations for plant types are as follows:

  B = bamboo
  F = fern
  G = grass
  H = herb
  P = partially so; for example, under the heading **Evergreen**, P would mean that the particular plant is semievergreen.
  S = shrub
 SU = succulent
  V = Vine
  ? = information presently unknown

If a combination is used (for example, H/S), it indicates that the plant has characteristics of both groups.. In this example, the plant would have characteristics of both a herb and shrub.

Using This Chart

This chart can be of great help in determining which ground covers will function in a particular site for a given function. Simply match the categories with the conditions presented by the particular landscape setting. If certain aspects do not match, you will know that amendments will need to be made (for instance, raising the soil pH or planting a tree to provide more shade) or that a different plant should be considered.

PLANT NAME	Type of Plant	Northern Hardiness Limit	Southern Hardiness Limit	Less Than 6 Inches Tall	6 Inches to 1 Foot Tall	1 to 2 Feet Tall	More than 2 Feet Tall	Full Sun	Light Shade	Moderate Shade	Dense Shade	Large Areas	Moderate Areas	Small Areas	Tolerates Foot Traffic	Erosion Control	Fire Retardant	Drought Tolerant	Spring Bloom	Summer Bloom	Fall Bloom	Winter Bloom	Evergreen	Alkaline Soil	Acidic Soil	Neutral Soil	Salt Tolerant	Tolerates Saturated Soils
Abelia × grandiflorum	S	6	9			×	×		×	×	×	×					×		×	×	×		P	×				
Acacia redolens	S	9	10			×	×	×	×	×	×	×				×	×		×	×	×		×	P		×		
Acaena microphylla	H	6	7	×			×						×											×	×			×
Achillea ageratifolia	H	3	9		×		×	×				×	×	P	×	×		×	×		×		×	×				
A. filapendula	H	3	9			×	×	×	×	×	×	×	×			×	×	×	×			×	×	×				
A. millefolium	H	2	9			×	×	×	×	×	×	×				×	×	×	×			×	×	×				
A. tomentosa	H	2	9	×			×	×	×	×	×	×	×	P		×	×	×	×	×		×	×	×				
Adiantum pedatum	F	3	8				×		×	×	×	×				×								×	×			
Aegopodium podagraria	H	3	9	×	×		×	×	×	×	×	×	×		×				×	×	×		×	×	×			
Aethionema grandiflorum	S	6	?				×						×		×									×	×			
Agapanthus africanus	H	9	10				×	×	×	×	×	×			×	×	×	×	×	×	×	P	P	×	×			
A. inapertus	H	9	10				×	×	×	×	×	×			×		×	×	×		P		×	×				
A. orientalis	H	9	10				×	×	×	×	×	×	×		×		×	×	×		P	P	×	×				
Ajuga genevensis	H	3	9	×				×	×	×	×		×				×		×	×			×	×	×			
A. pyramidalis	H	5	9	×				×	×	×	×	×	×				×		×			×	×	×				
A. reptans	H	4	9	×				×	×	×	×	×	×			×	×		×	P		×	×	×				
Akebia quinata	V	4	8		×		×	×	×	×	×	×	×				×			×	×		×	×	×	P			
Alchemilla vulgaris	H	3	8		×		×	×	×	×	×	×	×			×	×			×	×		P	×	×			
Ampelopsis aconitifolia	V	4	?		×		×	×	×	×			×	×			×							×	×			
A. brevipedunculata	V	4	8		×		×	×	×	×	×	×	×							×	×			×	×			
Andromeda polifolia	S	2	7	×			×	×	×			×		×				×	×		×		×	×	×		×	
Anemone hupehensis	H	5	8		×	×	×	×	×	×	×	×							×	×	×			×	×			

27

PLANT NAME	Type of Plant	Northern Hardiness Limit	Southern Hardiness Limit	Less Than 6 Inches Tall	6 Inches to 1 Foot Tall	1 to 2 Feet Tall	More than 2 Feet Tall	Full Sun	Light Shade	Moderate Shade	Dense Shade	Large Areas	Moderate Areas	Small Areas	Tolerates Foot Traffic	Erosion Control	Fire Retardant	Drought Tolerant	Spring Bloom	Summer Bloom	Fall Bloom	Winter Bloom	Evergreen	Alkaline Soil	Acidic Soil	Neutral Soil	Salt Tolerant	Tolerates Saturated Soils
A. pulsatilla	H	5	8		x			x					x					x						x	x			
Antennaria dioica	H	3	9	x				x					x			x	x	x					?	x	x			
A. plantaginifolia	H	3	?	x				x					x			x	x	x					?	x	x			
A. rosea	H	4	?		x			x									x	x					?	x	x			
Arabis alpina	H	4	8		x			x							x	x	x	x					x	P	x			
A. caucasica	H	3	?	x				x							x	x	x	x					x	P	x			
A. procurrens	H	4	?	x				x							x	x	x	x					P	P	x			
Arctostaphylos edmundsii	S	7	?	x				x				x	x			x	x	x			x			x	x			
A. hookeri	S	7	?			x		x		x		x	x			x	x	x			x	x		x	x			
A. pumila	S	7	?	x				x		x		x	x			x	x	x			x	x		x	x			
A. uva-ursi	S	2	7	x				x	x	x		x			x	x	x			x	x		x	x	x		
Arenaria balearica	H	6	?	x						x			x			x						P	x	x			
A. laricifolia	H	5	?		x			x				x				x				P	P	x	x			
A. montana	H	4	?	x				x	x			x	x			x	x			x	x	x	x			
A. verna	H	3	9/10	x				x							x	x			x	P	x	x			
Armeria juniperifolia	H	3	?	x				x				x			x	x	x			x	P	x	x	x		
A. maritima	H	3	8/9		x			x		x		x	x			x	x			x	P	x	x	x		
Aronia melanocarpa	S	4	9				x	x		x		x				x						x	x			
Artemisia abrotanum 'Nana'	H	4	8			x		P				x			x	x				P	x	x	x		
A. caucasica	H	4	?	x				P					x		x	x			x		P	x	x	x		
A. frigida	H	2	?		x			P				x			x	x			x		P	x	x	x		

Species																										
A. schmidtiana	H	3	9		×	×	×		P	P		×	×	×		×	×	P	×		×	×	×	×	P	×
A. stellerana	H	2/3	?		×	×	×		P	P		×	×	×	×	×	×	P	×		×	×	×	×	P	×
Arundinaria humulis	B	6/7	?			×	×	×					×			×	×	×			×	×	×		×	×
A. pumila	B	8	?		×	×	×		×	×		×	×	×	×	×	×	×	×	×	×	×	×	×	×	×
A. pygmaea	B	6/7	?		×	×	×	×	×	×		×	×	×	×	×	×	×	×	×	×	×	×	×	×	×
A. variegata	B	6/7	?		×	×	×	×	×	×		×	×	×	×	×	×	×	×	×	×	×	×	×	×	×
A. viridistriata	B	7	?		×	×	×	×	×	×		×	×	×	×	×	×	×	×	×	×	×	×	×	×	×
Asarum canadense	H	3	9	P			×		×	×	×	×	×	×	×	×	×	×	×	×	×	×	×		×	×
A. caudatum	H	4	7		×		×	×	×	×	×	×	×	×	×	×	×	×	×	×	×	×	×	×	×	×
A. europeum	H	4	8		×		×	×	×	×	×	×	×	×	×	×	×	×	×	×	×	×	×	×	×	×
A. shuttleworthii	H	6	?				×	×	P	P		×	×	×	×	×	×	×	×	×	×	×	×	×	×	×
Asparagus densiflorus 'Sprenger'	H	8	10		×	×	×	×	×	×	P			×		×	×	×	×	×	×	×	×	×	×	×
Asplenium bulbiferum	F	10	10		×	×	×	×	P	P				×	×	×	×	×	×	×	×	×	×	×	×	×
A. platyneuron	F	3	?		×	×	×	×	P	P		×	×	×	×	×	×	×	×	×	×	×	×	×	×	×
A. risiliens	F	6	?		×	P	×	×	×	×		×	×	×	×	×	×	×	×	×	×	×	×	×	×	×
A. trichomanes	F	3	?		×	×	×	×	×	P		×	×	×	×	×	×	×	×	×	×	×	×	×	×	×
Astilbe arendsii	H	4/5	7/8				×	×	×	×	×		×	×	×	×	×	×	×	×	×	×	×	×	×	×
A. astilboides	H	4/5	?		×	×	×	×	×	×	×		×	×	×	×	×	×	×	×	×	×	×	×	×	×
A. chinensis 'Pumila'	H	4/5	?		×	×	×	×	×	×	×		×	×	×	×	×	×	×	×	×	×	×	×	×	×
A. × crispa	H	4/5	?		×		×	×	×	×	×	×	×	×	×	×	×	×	×	×	×	×	×	×	×	×
A. japonica	H	4/5	?		×	×	×	×	×	×	×	×	×	×	×	×	×	×	×	×	×	×	×	×	×	×
A. × rosea	H	4/5	?		×	×	×	×	×	×	×	×	×	×	×	×	×	×	×	×	×	×	×	×	×	×
Athyrium filix-femina	F	3	3		×	×	×	×	×	×	×		×			×	×	×	×	×	×	×	×	×	×	×
Atriplex semibaccata	H	8	?		×	×	×	×	×	×	×	×	×	×	×	×	×	×	×	×	×	×	×	×	×	×
Aubrieta deltoidea	H	4	9		×	×	P	×	P	P		×	×	×	×	×	×	×	×	×	×	×	×	P	×	×
Aurinia saxatilis	H	3	?		×	×	×	×	×	×	P	×	×	×	×	×	×	×	×	×	×	×	×	×	×	×

29

PLANT NAME	Type of Plant	Northern Hardiness Limit	Southern Hardiness Limit	Less than 6 Inches Tall	6 Inches to 1 Foot Tall	1 to 2 Feet Tall	More than 2 Feet Tall	Full Sun	Light Shade	Moderate Shade	Dense Shade	Large Areas	Moderate Areas	Small Areas	Tolerates Foot Traffic	Erosion Control	Fire Retardant	Drought Tolerant	Spring Bloom	Summer Bloom	Fall Bloom	Winter Bloom	Evergreen	Alkaline Soil	Acidic Soil	Neutral Soil	Salt Tolerant	Tolerates Saturated Soils
Baccharis pilularis	S	7	?			×	×	×			×			×	×	×	×	×	×			×	P	×	×			
Bellis perennis	H	6	10	×				P	P				×					×	P	P			?	×	×			
Berberis thunbergii cv's	S	4	8			×	×	×	×	×		×				×	×	×	×					×				
B. stenophylla 'Irwinii'	S	5	?		×	×	×	×	×	×		×				×	×	×	×			×		×				
Bergenia cordifolia	H	2/3	8/9	×		×		×	×	×		×						×	×			P	P	×				
B. crassifolia	H	2/3	?	×		×		×	×	×		×						×		×		P	P	×				
Brunnera macrophylla	H	3	9	×		×		×		×	×	×						×	×				×	×				
Calluna vulgaris	S	4	6/7	×		×		×				×				×	×	×	×			×			×			
Campanula carpatica	H	3	9	×	×			×				×	×					×	×				P	P	×			
C. elatines var. garganica	H	6	?	×	×			×				×	×	×				×	×			P	P	P	×			
C. portenschlagiana	H	4	?	×	×			×		×		×	×	×				×	P	P		×	P	P	×			
C. poscharskyana	H	3	?	×	×			×	×		×	×	×	×				×	×			P	P	P	×			
Carissa grandiflora	S	7	?		×	×	×	×	×	×							×	×	×		×	×	×	×	×		
Carpobrotus chilensis	SU	9	10	×				×		×				×	×			×	×			×	×	×	×			
C. edulis	SU	9	10	×				×		×				×	×			×	×			×	×	×	×			
Ceanothus gloriosus	S	7	?		×	×	×	×		×						×	×	×	×			×	×	×	×		
C. griseus var. horizontalis	S	8	?		×	×	×	×		×				×	×	×	×	×	×			×	×	×	×		
C. prostratus	S	7	?	×	×	×		×		×						×	×	×	×			×	×	×	×		
C. thyrsiflorus var. repens	S	8	?	×	×	×		×		×		×				×	×	×	×		×	×	×	×	×		
Cephalophyllum alstonii	SU	9	10	×				×		×		×			×			×	×			×	×	×	×		
Cerastium tomentosum	H	3	10	×	×			×		×			×		×			×	×			×	×	×	×	×		

Ceratostigma plumbaginoides	H	5/6	9		×			×	×	×			×			×					×		P	×	×	×		
Chamaemelum nobile	H	4	10	×	×			×	×			×	×	×				×		×			×	×	?	P		
Cistus crispus	S	7	?			×		×					×			×	×	×	×	×			×	×		×	×	
C. salvifolius	S	7	?			×		×					×			×	×	×	×	×			×	×		×	×	
Comptonia peregrina	S	2	8				×	×				×				×		×		×					×		×	P
Convallaria majalis	H	3	7		×				×	×	×	×	×	×		×			×					P	×	×		
Coprosma × kirkii	S	8	9				×	×	×			×				×		×					×	×	×	×	×	×
C. petriei	S	7	?				×	×	×			×				×		×					×	×	×	×	×	×
Coreopsis auriculata	H	4/5	9		×	×		×	×				×	×				P	×	×	×		×	×	×	×		
Cornus canadensis	H	2	7	×					×	×			×	×					×	×					×			P
C. sericea 'Kelseyi'	S	2	7			×	×	×	×			×	×						×	×					×	×		×
Coronilla varia	H	3	10		×	×		×	×			×			P	×		×		×	×				×	×		
Correa pulchella	S	9	10				×	×	×			×				×		×	×		×	×	×	×	×	×		
Cotoneaster adpressa	S	4	7	×				×	×			×	×	×				×		×				P	×	×		
C. apiculatus	S	4	7				×	×	×			×	×					×	×	×				P	×	×		
C. congestus	S	5	9				×	×	×				×	×				×	×				×	P	×	×		
C. conspicuus decorus	S	6	8			×		×	×					×				×	×				×	P	×	×		
C. dammeri	S	5	8			×		×	×			×	×					×	×				P	P	×	×		
C. horizontalis	S	4/5	7		×	×		×	×			×	×					×	×	×			P	P	×	×		
C. microphyllus	S	5	9				×	×	×			×	×					×	×	×			×	P	×	×		
C. salicifolius cv's	S	5	9			×		×	×			×	×					×	×	×			×	P	×	×		
Cotula squalida	H	8	?	×				×	×				×	×	P					×			×	?	×	?		
Crassula multicava	SU	9	10		×			×	×				×	×					×			×	×	×	×	×		
Cyclamen hederaefolium	H	5	?	×					×	×			×	×						×	×			×	×	×		
Cymbalaria aequitriloba	H	8	10	×						×	×			×						×			×	P	×	×		
C. muralis	H	5	9	×						×	×			×		×				×	×		×	P	×	×		

PLANT NAME	Type of Plant	Northern Hardiness Limit	Southern Hardiness Limit	Less Than 6 Inches Tall	6 Inches to 1 Foot Tall	1 to 2 Feet Tall	More than 2 Feet Tall	Full Sun	Light Shade	Moderate Shade	Dense Shade	Large Areas	Moderate Areas	Small Areas	Tolerates Foot Traffic	Erosion Control	Fire Retardant	Drought Tolerant	Spring Bloom	Summer Bloom	Fall Bloom	Winter Bloom	Evergreen	Alkaline Soil	Acidic Soil	Neutral Soil	Salt Tolerant	Tolerates Saturated Soils
Cyrtomium falcatum	F	10	10	×					×	×		×	×									×	×	×				
Cystopteris bulbifera	F	3	?		×			×	×	×		×											×	×				
Cytisus decumbens	S	5	9	×		×	×	×				×				×	×	×	P				×	×	×			
Daboecia cantabrica	S	5	8	×			×	×			×	×				×	×	×	×				×		×	×		
Dalea greggii	S	9	10	×			×	×			×	×	×			×	×	×	×	×			×	×	×			
Delosperma cooperi	SU	7	10	×			×	×			×	×	×			×			×	×			×	×	×			
D. nubigenum	SU	4/5	10	×			×	×			×	×	×		×	×			×	×			×	×	×			×
Dennstaedtia punctilobula	F	3	8			×		×	×			×	×		×	×												
Dianthus arenarius	H	4	7	×				×				×	×					×	×	×			×	×	×			
D. deltoides	H	3/4	7	×	×			×	×			×	×					×	×	P			×	×	×			
D. plumarius	H	3/4	7	×	×			×	×			×	×					×	×				×	×	×			
Dicentra exima	H	3	9		×			×	×	×			×	×				P	×	P				×	×			
D. formosa	H	2/3	?	×				×	×	×			×	×				P	×	P				×	×			
Dichondra micrantha	H	10	10	×				×	×	×	×			×	×	×							×	×	×			
Drosanthemum floribundum	SU	9	10			×						×				×	×	×	×	×			×	×	×			
D. hispidum	SU	9	10		×	×		×	×	×	×			×	×	×	×	×					×	×	×			
Dryopteris intermedia	F	3	?						×	×	×											P	×	×				
D. filix-mas	F	3	?		×				×	×	×											P	×	×				
D. marginalis	F	3	?		×				×	×	×											×	×					
Duchesnea indica	H	5	8	×				×	×	×			×	P	×			×	×	×		×	×	×	×			

Epimedium alpinum	H	4	8		×	×			×	×		×	×	×					×						×			
E. grandiflorum	H	3	8		×				×	×		×	×	×					×						×			
E. pinnatum	H	5	8		×				×	×		×	×	×					×						×			
Erica carnea	S	4/5	7/8	×	×			×	×				×	×		×		P	×			×	×		×		×	
E. cinerea	S	6/7	8			×		×	×			×	×	×		×		P	×	×	×		×		×		×	
E. × darleyensis	S	6	?				×	×	×			×	×			×		P	×		×	×	×		×		×	
E. tetralix	S	4	7		×			×	×			×	×	×		×		P		×	×		×		×		×	
E. vagans	S	5	8		×			×	×			×	×			×		P		×			×		×		×	
Erigeron karvinskiianus	H	9	10			×		×					×	×				P	×	×	×		×	P	P	×	×	
Erodium chamaedryoides	H	7	10	×				×	×					×					×	×	×		×	×	×	×		
Euonymus fortunei cv's	V	5/6	9/10	×	×	×	×	×	×	×	×	×	×	×	P	×							×		×			
Festuca ovina var. *glauca*	G	4	9		×			×							P			×		×			×		×	×		
Ficus pumila	V	9	10	×				×	×					×									×		×			
Filipendula vulgaris	H	3	?			×		×	×	×			×	×						×								
Forsythia × intermedia 'Arnold Dwarf'	S	5	8				×	×	×	×		×	×	×		×			×						×		P	
F. viridissima 'Bronxensis'	S	5	8				×	×	×	×		×	×	×		×			×						×		P	
F. suspensa var. *sieboldii*	S	5	8				×	×	×	×		×	×			×			×						×		P	
Fragaria chiloensis	H	4/5	10	×			×	×					×	×	P	×			×				×		×			
F. vesca	H	5/6	10		×		×	×					×	×		×			×	×	×		×		×			
F. virginiana	H	5/6	?	×	×		×	×					×	×		×			×	×			×		×			
Fuchsia procumbens	H	10	10	×	P			×	×					×						×	×		×	×	×	×		
Galium odoratum	H	4	8		×					×	×		×	×					×				P		×			
Gardenia jasminoides 'Prostrata'	S	9	10			×		×	×	×			×	×					×	×	×		×		×			
Gaultheria hispidula	S	3	?	×					×	×		×	×	×	P				×				×		×			P
G. humifusa	S	4	?	×					×	×		×	×	×	P					×			×		×			P

PLANT NAME	Type of Plant	Northern Hardiness Limit	Southern Hardiness Limit	Less Than 6 Inches Tall	6 Inches to 1 Foot Tall	1 to 2 Feet Tall	More than 2 Feet Tall	Full Sun	Light Shade	Moderate Shade	Dense Shade	Large Areas	Moderate Areas	Small Areas	Tolerates Foot Traffic	Erosion Control	Fire Retardant	Drought Tolerant	Spring Bloom	Summer Bloom	Fall Bloom	Winter Bloom	Evergreen	Alkaline Soil	Acidic Soil	Neutral Soil	Salt Tolerant	Tolerates Saturated Soils
G. miquelina	S	5	?	×				×	×		×	×						×				×		×			P	
G. ovatifolium	S	5	?	×				×	×		×	×										×		×			P	
G. procumbens	S	3	?		×			×	×				×	P				×		×		×		×			P	
Gaylussacia brachycera	S	5	7	×				×	×		×	×										×		×				
Gazania linearis	H	8	10	×			×	×				×	×	×			P	×	×			×	×	×	×			
G. rigens	H	8	10	×			×	×				×	×	×			P	×	×			×	×	×	×			
Gelsemium sempervirens	V	7	9			×		×		×								×	P	×			×	×	×			
Genista germanica	S	5	7	×			×	×				×	×	×					×				×	P	×			
G. hispanica	S	7	9	×			×	×				×	×	×				×	×				×	P	×			
G. horrida	S	7	?		×		×	×				×	×	×						×			×	P	×			
G. lydia	S	7	?		×		×	×				×	×	×						×			×	P	×			
G. pilosa	S	5	8	×			×	×				×	×	×					×				×	P	×			
G. sagittalis	S	3	8	×			×	×				×	×	×					×				×	P	×			
G. sylvestris	S	6	?				×	×				×	×	×				×	×				×	P	×			
G. tinctoria	S	5	7				×	×				×	×	×						×			×	P	×			
Geranium cinereum	H	5/6	?		×		×	×				×	×	×					×				P	P	×			
G. endressii	H	5/6	?				×	×				×	×	×					×				P	P	×			
G. himalayense	H	4	?	×	×		×	×				×	×	×				×	×				P	P	×			
G. incanum	H	9	10	×			×	×				×	×	×						×	P		P	P	×			
G. macrorrhizum	H	3	?		×		×	×				×	×	×					×				P	P	×			
G. pylzowianum	H	7	?	×			×	×				×	×	×									P	P	×			
G. sanguineum	H	3	8	×			×	×				×	×	×				×	×				P	P	×			

	G. wallichianum	Glechoma hederaceae	Gypsophila paniculata	G. repens	Hebe chathamica	H. decumbens	H. pinguifolia 'Pagei'	Hedera canariensis	H. colchica	H. helix	Hedyotis caerulea	Helianthemum nummularium	Hemerocallis aurantiaca	H. fulva	H. lilioasphodelus	H. middendorffii	Herniaria glabra	Heuchera sanguinea	Hippocrepis comosa	Hosta decorata	H. fortunei	H. lancifolia	H. plantaginea	H. sieboldiana	H. undulata	H. ventricosa
											×		P	P	P	P				P	P	P	P	P	P	P
	×	×	×	×	×	×	×	×	×	×		×	×	×	×	×	×	×	×							
	P	×	×	×	×	×	×	×	×	×		×	×	×	×	×	×	×	×		×	×	×	×	×	×
		P						P	P	P																
	×			×				×	×	×	×			×	×	×	×							×	×	×
		P						P	P	P							P									
	×	×	×	×		×		×	×	×	×			×	×	×	×	×				×			×	×
		P											P	×	P	P			P							
	×	×	×	×			×				×		×	×	×	×	×	×	×		×	×	×	×	×	×
			×	×	×		×					×	×	×	×	×				×	×	×	×	×	×	×
					×	×	×						×		×			×		×	×	×	×	×	×	×
	×	×	×	×		×		×	×	×	×	×	×	×	×	×	×			×	×	×	×	×	×	×
	×	×	×	×	×	×	×	×	×	×	×	×	×	×	×	×	×	×	×	×	×	×	×	×	×	×
	×	×	×	×	×	×	×	×	×	×	×		×	×	×	×	×	×	×	P	P	P	P	P	P	P
			×			×					×		×	×	×	×									×	×
				×		×								×		×		×	×	×	×	×		×		
	×			×			×	×		×	×		×			×			×				×		×	×
	×	?	×	×	10	?	10	9/10	9	9/10	?	10	?	?	?	?	10	9	10	9	9	9	9	9	9	9
	6	3	3	3	8/9	5	8/9	7	5	5-6/7	3	5/6	6	2/3	3	3	5	3	6	3	3	3	3	3	3	3
	H	H	H	H	S	S	S	V	V	V	H	S	H	H	H	H	H	H	H	H	H	H	H	H	H	H

Plant reference chart (page 36). Plants are listed as rows; attribute categories as columns. "×" = applies; "P" = partial/possible; "?" = uncertain value.

PLANT NAME	Type of Plant	Northern Hardiness Limit	Southern Hardiness Limit	Less Than 6 Inches Tall	6 Inches to 1 Foot Tall	1 to 2 Feet Tall	More than 2 Feet Tall	Full Sun	Light Shade	Moderate Shade	Dense Shade	Large Areas	Moderate Areas	Small Areas	Tolerates Foot Traffic	Erosion Control	Fire Retardant	Drought Tolerant	Spring Bloom	Summer Bloom	Fall Bloom	Winter Bloom	Evergreen	Alkaline Soil	Acidic Soil	Neutral Soil	Salt Tolerant	Tolerates Saturated Soils
Hydrangea anomala ssp. petiolaris	V	4	7/8	×			×	×	×	×	×	×							×					×				
Hypericum buckleyi	S	5	8		×			×	×	×	×	×							×					×				
H. calycinum	S	6	10			×		×	×	×	×	×						×		×		P		×				
H. ellipticum	S	3	?				×	×	×	×	×	×								×		P		×				
H. olympicum	S	6	?		×			×	×	×	×	×							×					×				
H. patulum	S	6	?				×	×	×	×	×	×								×		P		×				
H. reptans	S	7	?	×				×	×	×	×			×					×					×				
Iberis gibraltarica	H	6	9		×			×					×					×	×	×			P	×	×			
I. saxatilis	H	3	7/8	×				×	×	×			×					×	×	×			P	×	×			
I. sempervirens	H	3	8		×			×	×	×			×			×		×	×				P	×	×			
Ilex cornuta cv's	S	7	9				×	×	×	×	×		×					×					×	×	×			
I. crenata cv's	S	6	8			×		×	×	×			×					×					×	×	×			
I. vomitoria	S	7	10				×	×	×				×					×					×	×	×			
Indigofera incarnata	S	5	?			×		×						×						×		P	P	×	×			
I. kirilowii	S	4	7			×		×	×	×				×						×		P	P	×	×			
Iris cristata	H	3	9	×				×	×	×	×			×			×	×				P	×	×		P	
I. pumila	H	4	?	×				×	×					×			×	×				P	×	×		P	
I. tectorum	H	5	9		×			×	×	×				×				×				P	×	×		P	
Jasminum floridum	S	8	10			×		×	×					×			P		×		P	P	×	×			
J. mesnyi	V	8	?				×	×	×					×			P	×	×			×	P	×	×			
J. nudiflorum	V	5	10			×		×	×					×			P	×				×		×	×			

Juniperus chinensis cv's	S	4	9			×	×	×				×	×			×		×					×		×	×		
J. communis cv's	S	2	6	×		×		×				×	×			×		×					×		×	×	×	
J. conferta	S	6	8		×	×		×				×	×			×		×					×	P	×	×	×	
J. × *davurica*	S	5	9				×	×				×				×		×					×		×	×		
J. horizontalis	S	3	9	×	×	×		×				×	×	×		×		×					×		×		×	
J. procumbens	S	4	9		×			×					×	×		×		×					×		×			
J. sabina cv's	S	4	7			×	×	×				×	×			×		×					×	P	×	×		
J. squamata cv's	S	4	8		×	×		×					×			×		×					×		×	×		
J. virginiana cv's	S	2	9			×		×					×			×		×					×	P	×	×	×	
Lamiastrum galeobdolon	H	4	9		×	×		×	×	×		×	×			×			×				×	?	×	×		P
Lamium maculatum	H	4	8/9	×	×			×	×	×				×					×				×	?	×	×		
Lampranthus aurantiacus	SU	9	10		×	×		×				×	×			×	×	×	×				×	×	×	×		
L. filicaulis	SU	9	10	×				×						×		×	×	×	×				×	×	×			
L. productus	SU	9	10			×		×				×	×			×	×	×	×			×	×	×	×	×		
L. spectabilis	SU	9	10			×		×				×	×			×	×	×	×				×	×	×	×		
Lantana montevidensis	S	9	10				×	×	×			×	×			×		×	×	P	P	P	×	×	×	×		
L. × *hybridus*	S	9	10			×	×	×	×			×	×			×		×	×	P	P	P	×	×	×	×		
Lathyrus latifolius	H	3	10	×				×	×			×	×			×		×		×	×			P	P	×		
Laurentia fluviatilis	H	8	10	×				×	×					×					×	P			×	×	×	×		
Lavandula angustifolia	H	5/6	9			×	×		×				×	×				P		×			P	P	×	×		
Liriope muscari	H	6	10		×	×		×	×	×		×	×	×				×		×			×		×			
L. spicata	H	4	10		×			×	×	×		×	×	×				×		×			×		×			
Lithodora diffusa	S	6	7		×			×	×				×						×	×			×		×			
Lonicera alpigena 'Nana'	S	5	?				×	×	×				×					P	×					P	×	×		
L. × *brownii*	V	3	9		×			×	×			×	×			×		P	×	×				P	×	×		
L. caprifolium	V	5	?		×			×	×			×	×			×		P		×	×			P	×	×		

PLANT NAME	Type of Plant	Northern Hardiness Limit	Southern Hardiness Limit	Less Than 6 Inches Tall	6 Inches to 1 Foot Tall	1 to 2 Feet Tall	More than 2 Feet Tall	Full Sun	Light Shade	Moderate Shade	Dense Shade	Large Areas	Moderate Areas	Small Areas	Tolerates Foot Traffic	Erosion Control	Fire Retardant	Drought Tolerant	Spring Bloom	Summer Bloom	Fall Bloom	Winter Bloom	Evergreen	Alkaline Soil	Acidic Soil	Neutral Soil	Salt Tolerant	Tolerates Saturated Soils
L. etrusca	V	7	?	×			×	×		×	×	×			×		P		×			P	P	×	×			
L. flava	V	6	9/10	×			×	×		×	×	×					P	×	×				P	×	×			
L. × heckrottii	V	5	9	×			×	×	×	×	×	×					P	×	×	×		P	P	×	×			
L. japonica	V	4	10		×		×	×	×	×	×	×	×				P	×	×	×		P	P	×	×	×		
L. pileata	S	5	10	×			×	×	×	×	×	×	×				P	×	×			P	P	×	×			
Lotus berthelotii	H	10	10		×		×	×									P	×				×	×	×	×			
L. corniculatus	H	5	10	×				×			×	×	×		×		P		×				×	×	×			
Lysimachia nummularia	H	3	9	×			×	×	×				×	P					×		P		×	×		×			
Mahonia nervosa	S	5	?				×	×				×	×					×				×	×		×		P		
M. repens	S	5	9	×	×	P		×	×		×	×	×					×	×			P	P	×	×			
Mazus pumilio	H	7	?	×			×	×	×				×	×										×	×			
M. reptans	H	3	9	×	×			×		×	×		×	P				×	×			P	P	×	×			
Mentha × piperita	H	3	?				×	×							×									×	×			
M. pulegium	H	7	?	×				×			×	×	×	P										×	×			
M. requienii	H	6	10	×				×			×		×	×										×	×			
M. spicata	H	3	?	×	×		×	×				×							×					×	×			
Microbiota decussata	S	2	9					×					×				P						×	?	×			×	
Mitchella repens	H	3	9	×	×			×			×	×						×	×	×		×	×	×	×		×	
Mitella diphylla	H	3	?	×				×		×	×	×	×		×										×		×	
Muehlenbeckia axillaris	V	6	10	×	×		×	×		×	×	×					×		×		?	×	×	×		×	
M. complexa	V	6	10	×	×		×	×		×	×	×					×		?	×	×	×	×	×	×		
Myoporum parvifolium	S	9	10	×			×	×		×	×	×				×		×				×	×	×	×		

Myosotis scorpioides	H	3	9
Nandina domestica cv's	S	7	9
Nepeta × faassenii	H	3	?
Nierembergia repens	H	7/8	10
Omphalodes verna	H	6/7	9
Onoclea sensibilis	F	3	?
Ophiopogon jaburan	H	7	?
O. japonicus	H	7	10
O.planiscapus'Nigrescens'	H	6	10
Origanum dictamnus	H	9	10
O. vulgare 'Compacta Nana'	H	5	?
Osmunda claytoniana	F	3	?
Osteospermum fruticosum	SU	9	10
Oxalis oregana	H	9	10
Pachysandra procumbens	H	4	9
P. terminalis	S	5	9
Parthenocissus quinquifolia	V	3	9
Paxistima canbyi	S	4/5	6/7
P. myrsinites	S	5	7
Pelargonium peltatum	H	9	10
Pernettya mucronata	S	6/7	?
P. tasmanica	S	7	?
Phalaris arundinacea var. picta	S	3	?
Phlox divaricata	H	3	?
P. nivalis	H	6	9

PLANT NAME	Type of Plant	Northern Hardiness Limit	Southern Hardiness Limit	Less Than 6 Inches Tall	6 Inches to 1 Foot Tall	1 to 2 Feet Tall	More than 2 Feet Tall	Full Sun	Light Shade	Moderate Shade	Dense Shade	Large Areas	Moderate Areas	Small Areas	Tolerates Foot Traffic	Erosion Control	Fire Retardant	Drought Tolerant	Spring Bloom	Summer Bloom	Fall Bloom	Winter Bloom	Evergreen	Alkaline Soil	Acidic Soil	Neutral Soil	Salt Tolerant	Tolerates Saturated Soils
P. × procumbens	H	5	?	×				×				×						×					?	×	?			
P. stolonifera	H	3	?	×				×				×						×						×	×			
P. subulata	H	2/3	9				×	×				×	×	P				×					×	×	×			
Phyla nodiflora	H	9	10			×	×	×	×			×	×	×	×			×		×			×	×		×	×	
Picea abies cv's	S	2	7		×		×				×	×					P			×				×	×	×		
P. pungens cv's	S	2	7		×		×				×	×					P							×		×		
Pinus banksiana cv's	S	2	6/7				×	P				×					P						P	×	×			
P. densiflora cv's	S	4	7	×	×		×	P				×					P							×	×			
P. flexilis	S	2	7		×	×	×	P				×					P						P	×	×			
P. mugo	S	2	7		×		×	P		×	×	×					P							×	×	×		
P. nigra cv's	S	4	7		×	×	×	P				×					P							×	×	×		
P. parviflora 'Bergman'	S	4/5	7		×	×	×	P	×			×					P						P	×	×			
P. strobus cv's	S	3	8		×	×	×	×		×	×	×					P							×	×	×		
P. sylvestris cv's	S	2	8	×	×		×	P			×	×					P							×				
Pittosporum tobira 'Wheeler's Dwarf'	S	7/8	9				×	×				×	×	P			P	×					P	×	×	×		
Polygonum affine	H	3	?		×		×	×				×	×	P			P	×		×		P	P	×	×			
P. capitatum	H	7	10			×	×	×					×	P			P	P	P			P	P	×	×			
P. cuspidatum var. compactum	H	3	10		×	×		×				×					P	×		×			P	×	×	×		
Polypodium aureum	F	10	10					×	×	×		×										×		×			P	
P. virginianum	F	3	?	×				×	×	×		×										×		×	×		P	

Species	Habit			1	2	3	4	5	6	7	8	9	10	11	12	13	14	15	16	17	18
Polystichum acrostichoides	F	3	9				×						×			×	×	×	×	×	
P. braunii	F	3	?				P						×			×	×	×	×	×	
P. lonchitis	F	3	?				×		×				×		×	×	×	×	×		
Potentilla cinerea	H	3	9			P	×	P		×	×			×		×			×		
P. nepalensis	H	5	9			P	P	×	×		×			×			×		×	×	×
P. tabernaemontani	H	5	9			P	P	P	×		×			×	×	×	×		×	×	×
P. × tonguei	H	6	9			P	P	×	×		×			×	×	×	×		×	×	×
P. tridentata	H	2	?			P	P	×	P		×			×	×	×	×		×	×	×
Pratia angulata	H	7	10	×		×	×	P				×		×	×		×	×	×	×	×
Primula × polyantha	H	3	8	×		×	×		×	×		×		×	×		×			×	
Pulmonaria angustifolia	H	2/3	9	×	×	×	×		×				×	×	×	×		×	×		×
P. saccharata	H	2/3	9	×	×	×	×		×				×	×	×	×		×	×	×	×
Pyracantha koidzumii	S	8	10	P	×	×	×	×		×				×	×	×	×	×	×		×
Ranunculus repens	H	3	9	×	P	×	×	×	×	×				×	×	×	×	×	×		×
Rhus aromatica	S	3	9	×	×	×	×	×	×		×	×				×	×	×	×		×
Rosa banksiae	S	7	10	×	×		×	×	×		×	×				×	×	×	×		×
R. wichuraiana	S	5/6	10	×	P	×	×	×	×		×	×				×	×	×	×		×
Rosmarinus officinalis	H/S	6	9	×	P	×							×			×	×	×	×		×
Sagina subulata	H	5	10	×	?			×	×	×						×			×		×
Salix purpurea 'Nana'	S	4	9	P	P	×	×	×	×		×	×				×	×	×	×	×	×
S. repens	S	4	9	P	P	×	×	×	×		×	×				×	×	×	×	×	×
S. tristis	S	2	9	P	P	×	×	×	×		×	×				×	×	×	×	×	×
Santolina chamaecyparissus	H/S	6	9	×	×	P	×		×					×	×			×	×	×	×
S. virens	H/S	7	9	×	×	P	×		×		×		×	×	×		×	×	×	×	×
Saponaria ocymoides	H	2	8	×	×	×	×	P			×	×			×	×	×	×	×	×	×

PLANT NAME	Type of Plant	Northern Hardiness Limit	Southern Hardiness Limit	Less Than 6 Inches Tall	6 Inches to 1 Foot Tall	1 to 2 Feet Tall	More than 2 Feet Tall	Full Sun	Light Shade	Moderate Shade	Dense Shade	Large Areas	Moderate Areas	Small Areas	Tolerates Foot Traffic	Erosion Control	Fire Retardant	Drought Tolerant	Spring Bloom	Summer Bloom	Fall Bloom	Winter Bloom	Evergreen	Alkaline Soil	Acidic Soil	Neutral Soil	Salt Tolerant	Tolerates Saturated Soils
Sarcococca hookerana var. humulis	S	5	8			×		P	P	×	×	×		×						×		×	×	×				
Satureja douglasii	H	4	9		×			×	×	×			×	×				×	×	×			?	×	×			
Saxifraga stolonifera	H	7	10	×					×	×	×		×	×						×		×		P	×	×			
Sedum acre	SU	3	9			×		×				×	×	×	×	×	×	×		×	×		P	×	×			
S. album	SU	3	?			×		×				×	P	×	×	×	×	×		×	×		P	×	×			
S. anglicum	SU	3	?			×		×				×	P	×		×		×		×	×		P	×	×			
S. brevifolium	SU	5	?			×		×				×		×		×		×		×			×	×	×			
S. cauticolum	SU	3	?			×		×				×		×			×			×	×		P	×	×			
S. confusum	SU	7/8	10	×				×				×		×		×				×			P	×	×			
S. dasyphyllum	SU	5/6	?			×		×				×		×		×	×	×		×	×		P	×	×			
S. ellacombianum	SU	3	?			×		×				×	P	×		×	×		×	×			×	×	×			
S. kamtschaticum	SU	3	?			×		×	×			×		×	×	×	×		×	×	×		P	×	×			
S. lineare	SU	7	?	×				×				×		×		×				×	×		P	×	×			
S. lydium	SU	3	?			×		×	×			×	P	×		×		×		×	×		P	×	×			
S. middendorfianum	SU	3	?			×		×	×			×		×		×				×	P	P	P	×	×			
S. moranense	SU	7	?			×		×	×			×		×		×		×		×	×		P	×	×			
S. oaxacanum	SU	6	?			×		×	×			×	P	×		×	×		×	×		P	×	×			
S. populifolium	SU	3	?	×				×		×		×		×		×				×			P	×	×			
S. reflexum	SU	3	?	×				×		×		×		×		×		×		×	×		P	×	×			
S. rubrotinctum	SU	9	10	×				×		×		×		×		×				×			P	×	×			
S. rupestre	SU	7	?	×				×		×		×		×		×	×		×	×	×		P	×	×			

Species	Type			Mark grid (× = present, P = partial)
S. sarmentosum	SU	3	?	× × × × P × × P × × ×
S. sexangulare	SU	2/3	?	× × × × P × × P × × ×
S. sieboldii	SU	2/3	?	× × × × P × × P × × ×
S. spathulifolium	SU	5	?	× × × × P × × P × × ×
S. spectabile	SU	3	9	× × × × P × × P × × ×
S. spurium	SU	3	?	× × × × P P × × × × ×
Sempervivum arachnoideum	SU	5	10	× × × × × × × × × × ×
S. arenarium	SU	5	10	× × × × × × × × × × ×
S. × barbulatum	SU	5	10	× × × × × × × × × × ×
S. × fauconnettii	SU	5	10	× × × × × × × × × × ×
S. montanum	SU	5	10	× × × × × × × × × × ×
S. ruthenicum	SU	5	10	× × × × × × × × × × ×
S. soboliferum	SU	5	10	× × × × × × × × × × ×
S. tectorum	SU	4	10	× × × × × × × × × × ×
Soleirolia soleirolii	H	10	10	× × × × ? × P P
Spiraea japonica cv's	H	5	9	× × × × × ×
Stachys byzantina	H	4	?	× × × × P P P ×
Stephanandra incisa 'Crispa'	S	3	8	× × × × ×
Symphoricarpos × chenaultii	S	4	8	× × × × × P ×
S. orbicularis	S	2/3	6/7	× × × × × P × ×
Tanacetum parthenium	H	6	10	× × × × ×
Taxus baccata cv's	S	5/6	7	× × × × × P × ×
Teucrium chamaedrys	S	5	8	× × × P × × ×
Thymus × citriodorus	H	3	?	× × × × P × × ×
T. herba-barona	H	4	?	× × × × P × × ×

PLANT NAME	Type of Plant	Northern Hardiness Limit	Southern Hardiness Limit	Less Than 6 Inches Tall	6 Inches to 1 Foot Tall	1 to 2 Feet Tall	More than 2 Feet Tall	Full Sun	Light Shade	Moderate Shade	Dense Shade	Large Areas	Moderate Areas	Small Areas	Tolerates Foot Traffic	Erosion Control	Fire Retardant	Drought Tolerant	Spring Bloom	Summer Bloom	Fall Bloom	Winter Bloom	Evergreen	Alkaline Soil	Acidic Soil	Neutral Soil	Salt Tolerant	Tolerates Saturated Soils
T. lanicaulis	H	3	?	X				X					X				X			X			X	P	X	X			
T. serpyllum	H	3	?	X				X	X				X				X			X			X	P	X	X			
T. vulgaris	H	5	?	X				X	X				X				X			X			X	P	X	X			
Tiarella cordifolia	H	3	?	X				X	X	X		X	X					X						X	X			
Tolmiea menziesii	H	7	10	X				X	X	X		X	X										P	X	X		X	
Trachelospermum asiaticum	V	7	9				X	X	X	X	X	X			X							X		X				
T. jasminoides	V	9	10	X			X	X	X	X	X	X			X							X		X	X		X	
Tradescantia fluminensis	SU	8	10	X				X				X	X						X	X		P		X			X	
Vaccinium angustifolium	S	2	6	X			X	X				X				X	X		X	X				X			X	
V. vitis-idaea	S	2	6	X			X	X	X			X				X	X		X		X	X		X	X		X	
Vancouveria hexandra	H	5	8	X					X	X		X				X			X	X		X		X	X			
V. planipetala	H	6	8	X					X	X		X				X			X			X		X	X			
Verbena bipinnatifida	H	3	?	X			X	X				X		X	X	X	X	X	X			P	X	P	X			

Species	Type																				
V. canadensis	H	5	10	x			x		x	x		x	x	x	x		x	P	x	P	x
Veronica chamaedrys	H	3	10		x		x	x	x	x		x	x	x			x	P	x	P	x
V. incana	H	3	?		x	x	x	x	x	x		x	x		x		x	P	x	x	x
V. latifolia 'Prostrata'	H	3	?	x	x	x	x	x	x	x		x	x		x		x	P	x	P	x
V. officinalis	H	3	?	x		x	x	x	x	x		x	x		x		x	x	x	x	x
V. prostrata	H	5	?	x			x	x	x	x		x	x		x		x	P	x	P	x
V. repens	H	5	?		x		x	x	x		x	x	x		x		x	P	x	P	x
Viburnum davidii	S	7	9			x	x	x	x	x		x	x	x			x	x	x	x	x
Vinca major	V	7	9	x	x	x	x	x	x	x	P	x		P	x	x		P	x	x	x
V. minor	V	4	7/8		x	x	x	x	x		P	x	x	P	x	x		P	x	x	x
Viola hederaceae	H	9	10	x				x	x	x		x	x		x		x	?	x	x	x
V. odorata	H	6	10	x			x	x	x	x		x	x	x	x		x	?	x	x	x
V. pedata	H	4	?	x	x	x	x	x	x	x		x		x	x		x	?	x	x	x
Waldsteinia fragarioides	H	4	7	x	x		x	x			P		x	x	x		x	x	x	x	x
W. ternata	H	4	?	x			x	x	x	x		x	x		x		x	x	x	x	x
Xanthorhiza simplicissima	S	3	8		x	x	x	x	x	x	x	x	x	x	x		x	x	x	x	x
Zoysia tenuifolia	H	9	10	x		x	x	x	x	x	x	x	x	x	x	x	x	x	x	x	x

ENCYCLOPEDIA OF PERENNIAL GROUND COVERS

Abelia (ȧ-bē′li-ȧ or ȧ-bēl′yȧ) Named for Dr. Clark Abel, physician and author of China, who discovered *A. chinensis* in 1816–1817. This genus is composed of about 30 species of shrubs from Asia and Mexico. Cultivars of the hybrid group *A. × grandiflora* are suitable for ground cover use.

FAMILY: CAPRIFOLIACEAE: Honeysuckle family.

A. × grandiflora (grand-di-flō′rȧ) Large flowered. (Figure 3–1).

COMMON NAME: Glossy abelia.

HYBRID ORIGIN: (*A. chinensis* × *A. uniflora*) Location and date unknown.

HARDINESS: Zones 6 to 9. Considered to be the most hardy and abundant flowering of the abelias.

HABIT: Usually a dense, rounded, semievergreen, spreading, many stemmed shrub with arching branches. If damaged by cold in zones 6 to 7, it regrows quickly into a dense, small shrub.

SIZE: 3 to 6 feet tall by 3 to 6 feet across (usually toward the lower sizes in the north).

RATE: Relatively slow growing; space 2 to 3 feet apart from gallon-sized containers in the north and 3 to 4 feet apart in the south and coastal areas.

LANDSCAPE VALUE: Useful en masse as a general cover in large areas. Often used on freeway banks and slopes as a soil binder. Also excellent for low, dense hedges. Not tolerant of foot traffic.

FOLIAGE: Simple, ovate, opposite, 3/5 to 1 2/5 inches long, half as wide. Rounded or cuneate at base, margin dentate, lustrous dark green above, pale below, bronzy in winter, glabrous but bearded near the midrib base, often persisting through the winter in the south, medium textured, semievergreen.

STEM: Fine textured, reddish brown, young growth pubescent, older exfoliating.

INFLORESCENCE: Pale pink to white, funnel-shaped flowers are arranged in loose terminal panicles that reach 3/4 inch long, becoming purple and persisting for months. Bloom is prolific on new growth from midsummer to early fall.

FRUIT: Single seeded, nonornamental, leathery achene.

OF SPECIAL INTEREST: The nectar in the flowers of glossy abelia is a favorite of hummingbirds.

SELECTED CULTIVARS AND VARIETIES: 'Francis Mason' Foliage variegated green and yellow.
'Prostrata' A prostrate growing, more compact cultivar, 1 1/2 to 2 feet tall by 4 feet wide.
'Sherwoodii' Dwarf selection 3 to 3 1/2 feet tall by

4 to 4 1/2 feet wide; more dense and compact in habit than the species; leaves smaller, glossy red becoming bright green when mature. Small, persistent bell-shaped flowers.

CULTURE: Soil: Best in well-drained, organically rich loam; a pH of 5.0 to 6.5 is optimal.
Moisture: Quite drought tolerant once established; occasional deep watering in summer may be to some benefit.
Light: Full sun to moderate shade (best growth in moderate shade).

PATHOLOGY: Diseases: Leaf spots, powdery mildew, root rots.
Pests: Root knot nematode.

MAINTENANCE: Annual pruning may be required to remove dead wood when grown in zones 6 to 7 where stems are frequently killed by winter cold.

PROPAGATION: Cuttings: Softwood cuttings can be rooted easily and benefit from 1,000 to 2,000 ppm IBA/talc treatment.

Figure 3–1 *Abelia × grandiflora* 'Prostrata' (life-size)

Seed: Sow when ripe for optimal germination or store at cool temperatures in airtight container up to one year prior to sowing. Seed propagation of glossy abelia is primarily of interest to the hybridizer.

A*cacia* (a-kā´shē-à) From Greek *akazo,* to sharpen. I'm not sure what the reference is to. This genus is huge, containing some 800 species of trees and shrubs, many of which are native to dry, tropical, and warm temperate regions.

FAMILY: FABACEAE (LEGUMINOSAE): Pea family.

A. redolens (red-a-lens) Having a pleasant odor, fragrance (Figure 3–2).

COMMON NAME: Trailing acacia.

NATIVE HABITAT: Australia.

HARDINESS: Zones 9–10.

SIZE: 1 to 4 feet tall, spreading 10 to 15 feet across.

HABIT: Low, trailing, shrubby ground cover.

RATE: Relatively fast; space plants from 1 to 2 gallon containers 4 to 6 feet apart.

LANDSCAPE VALUE: Excelling in desert landscapes for general cover and erosion control in moderate to large, level or sloping areas.

FOLIAGE: Evergreen, flat gray-green, simple, alternate, linear, to 3 1/2 inches long and 1/2 inch across, margin entire, longitudinal veination prominent, apex obtuse, base acute.

STEM: Trailing, greenish-brown.

INFLORESCENCE: Tiny, about 1/8 inch in diameter, dense, yellow globose heads, effective (but not very noticeable) in spring.

FRUIT: Tiny legume, not ornamentally significant.

CULTURE: Soil: Adaptable to well-drained sandy and rocky soils. Tolerant of high salinity and alkalinity.
Moisture: Water frequently during the first season. Thereafter, tolerance to drought is excellent and little or no irrigation is needed.
Light: full sun, tolerating light shade but becoming more open in habit.

PATHOLOGY: Diseases: Twig canker, leaf spots, powdery mildew, root rot.
Pests: Cottony cushion scale and other scales, caterpillars.

MAINTENANCE: Periodic shearing is needed to maintain a neat, compact appearance.

PROPAGATION: Cuttings: Softwood cuttings are probably the most common propagation method. Root cuttings are said to be effective also.
Seed: Soak ripe seed overnight in hot water (160°F) and sow.

A*caena* (ak-ē-nà or à-sēn´à) Greek, from *akanthos,* a thorn, many species with spiny calyxes. This genus is composed of 40 or more species of herbaceous or semiwoody, perennial trailing plants. Their usefulness in the landscape is generally limited to mild winter regions. Sometimes they perform well in protected sites as far north as New York City.

FAMILY: ROSACEAE: Rose family.

Ground covering types include too many species to describe in detail here. Therefore, the most commonly used species, *A. microphylla,* is discussed in detail, and a brief description of the others follows.

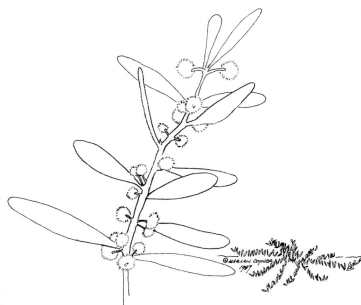

Figure 3–2 *Acacia redolens* (life-size)

Figure 3–3 *Acaena microphylla* (life-size)

A. microphylla (mi-krō-fil´à) Small leaved (Figure 3–3).

COMMON NAME: Redspine sheepburr.

NATIVE HABITAT: New Zealand.

HARDINESS: Zones 6 to 7.

HABIT: Prostrate, mat forming, herbaceous ground cover.

SIZE: 4 to 6 inches tall, spreading indefinitely.

RATE: Relatively fast; space plants from pint- or quart-sized containers 12 to 18 inches apart.

LANDSCAPE VALUE: General ground cover for small areas, especially useful between flagstones. Most common along the Pacific coast. No foot traffic.

FOLIAGE: Compound, 1 inch long, with 7 to 15 leaflets that are each about 1/8 inch long, rounded, crenate-dentate, purplish to olive or bronzy green, fine textured.

INFLORESCENCE: Insignificant, red stamened, globose, grayish heads on 2 inch long scapes.

FRUIT: Interesting and decorative crimson, globular burrs, effective from early to late summer.

OTHER SPECIES: *A. adscendens* (ad-scen´denz) Ascending; a taller species to 9 inches tall. This species has leaves that are blue-gray on reddish stems.

A. buchananii (bū-kan-a´ni-ī) After Buchanan; to 3 inches tall; foliage pale, jade green, compound with 11 to 13 small roundish leaflets.

A. caesiglauca (sē-zi-glâ-kà) Meaning blue-gray, the foliage color; silky; to 2 inches tall.

A. inermis (in-êr´mis) Unarmed, without thorns; to 3 inches tall; foliage like that of *A. microphylla*.

A. novae-zelandiae (nō-vē-zeel-and´é-ē) Of New Zealand; to 5 inches tall; foliage larger and richer green than *A. microphylla*.

A. sanguisorbae (san-gwi-sôr´bā) (Burnets) With healing properties. To 10 inches tall; foliage gray and hairy; stems also hairy.

Generally Applicable to All Species

CULTURE: Soil: Sandy, well-drained, moist or dry soil with a pH of 7.0 to 8.5 preferred.
Moisture: Will tolerate considerable drought. Additional watering is usually not necessary, although an occasional misting to remove dust is a good practice.
Light: Requiring full sun.

PATHOLOGY: No serious diseases or pests reported.

MAINTENANCE: Little or no special maintenance required.

PROPAGATION: Division: Divide and replant at any time of the year, keeping soil moistened until roots are well established.

*A*chillea (ak-i-lē´à or à-kil-ē´à) After Achilles, the Greek hero, who used the plant in medicine. This genus contains nearly 100 species, most of which are aromatic herbs. Many are considered to be weeds.

FAMILY: ASTERACEAE (COMPOSITAE): Sunflower family.

HABIT: Types for ground cover are often erectly oriented, spreading, herbaceous perennials. *A. millefolium* and *A. ptarmica* are rhizomatous.

SIZE: Ranging from 6 inches tall to about 4 1/2 feet tall, spreading indefinitely. Their habit of self-sowing often makes them invasive.

RATE: Relatively fast; set plants from pint- to quart-sized containers 8 to 16 inches apart.

LANDSCAPE VALUE: General ground covers for contained or casual areas where they can spread freely. Flowers are valued for the contrast that they provide against their own foliage and other greens in the landscape. Often used successfully as border or rock garden specimens. Occasionally, lower growing species are used as lawn substitutes along the West Coast, where they accept infrequent foot traffic. Species of *Achillea* listed here are considered relatively fire resistant.

FRUIT: Achene, without pappus, not of ornamental significance.

* * *

A. ageratifolia (a-jêr´ȧ-ti-fō´li-ȧ) Leaves like ageratum.

COMMON NAME: Greek yarrow.

NATIVE HABITAT: Balkan region.

HARDINESS: Zones 3 to 10.

SIZE: 4 to 10 inches tall.

FOLIAGE: Evergreen, alternate, fasicled or arranged in basal rosettes, entire, crenate, dentate, or pinnately lobed, oblanceolate to spatulate, pubescent silvery-gray on both sides, aromatic, to 1 1/2 inches long and 1/8 to 1/4 inch across. Texture is relatively fine.

INFLORESCENCE: Solitary heads to 1 inch in diameter, supported by a stalk that is 4 to 10 inches tall, corolla white, 4 to 5 lobed, effective in summer and early fall.

SELECTED CULTIVARS AND VARIETIES: 'Aizoon' A selection with leaves that are mostly entire.

* * *

A. filipendula (fil-i-pen´dū-lȧ) Foliage characteristics like the genus *Filipendula*.

COMMON NAME: Fernleaf yarrow.

NATIVE HABITAT: Europe and Asia.

HARDINESS: Zones 3 to 9.

SIZE: May reach 3 to 4 1/2 feet tall.

FOLIAGE: Alternately arranged, to 10 inches long, progressively reduced upward, linear to elliptic, 1 to 2 pinnatifid into linear-lanceolate toothed segments, pubescent with strong spicy odor. Fine fernlike texture, colored gray-green.

STEMS: Stiff, medium gray-green, erect.

INFLORESCENCE: Heads in dense, yellow, compound, convex corymbs that may reach 5 inches across, effective early to late summer and may be prolonged by removal of heads upon their senescence.

SELECTED CULTIVARS AND VARIETIES: Var. *alba* is like the species, but flowers are white.
× 'Coronation Gold' Growing lower than the species, good vigor and tolerance to heat; flowers abundant, mustard-yellow, in corymbs to 3 inches

across. Reportedly, this cultivar is the result of the cross *A. clypeolata* × *A. filipendula*.
'Golden Plate' Tall growing cultivar to 4 or 4 1/2 feet tall; large mustard-yellow flowers in corymbs to 6 inches in diameter.
'Parker's Variety' 3 1/2 feet tall, with flowers that are yellow in corymbs to 4 inches across.

* * *

A. millefolium (mil-le-fō´li-um) Thousand leaved (Figure 3–4).

COMMON NAME: Milfoil, common yarrow, sanguinary, thousand-seal, nose-bleed.

NATIVE HABITAT: Europe.

HARDINESS: Zones 2 to 9.

SIZE: To 3 feet tall.

Figure 3–4 *Achillea millefolium* 'Roseum' (life-size)

FOLIAGE: Finely 2 to 3 pinnate, lower leaves lanceolate to oblanceolate, to 8 inches long, petioles long; upper leaves lanceolate to linear, sessile; nearly evergreen, aromatic when crushed, very fine textured.

INFLORESCENCE: Corymbs round or flattish, 1/4 inch in diameter; flowers usually white, effective midsummer to early or midautumn.

OF SPECIAL INTEREST: Many different species of butterflies are attracted to the flowers of *A. millefolium.*

SELECTED CULTIVARS AND VARIETIES: All cultivars are generally less vigorous than the species. They are described as follows:
'Crimson Beauty' Flowers are bright red in 2 to 3 inch diameter heads.
'Fire King' Flowers rosy-red; in heads 2 to 3 inches across; stems to 18 inches; foliage silvery.
'Kelwayi' Flowers magenta-red.
'Red Beauty' Flowers red.
'Roseum' Flowers pink, foliage silvery-green.

* * *

A. ptarmica (tär'mi-kȧ) From Greek, *ptarmos*, sneezing, the dried flowers once used for snuff. The sneezewort.

COMMON NAME: Sneezewort, sneezeweed, white tansy.

NATIVE HABITAT: Europe, Asia, North America.

HARDINESS: Zones 2 to 8.

SIZE: To 2 feet tall.

FOLIAGE: Sessile, simple, linear to linear-lanceolate, 1 to 4 inches long, finely serrate to subentire, glabrous, relatively fine textured.

INFLORESCENCE: Heads to 3/4 inch across in a loose corymb that may reach from 3 to 6 inches across; disc florets greenish white, ray florets white, effective midsummer to early autumn.

STEMS: Glabrous to slightly pubescent.

SELECTED CULTIVARS AND VARIETIES: 'Angel's Breath' Abundant snow-white flowers.
'Perry's White' with double white flowers.
'Snowball' Also with double white flowers.
'The Pearl' A selection of 'Angel's Breath' that produces flowers more abundantly.

* * *

A. tomentosa (tō-men-tō'sȧ) With downy foliage.
COMMON NAME: Wooly yarrow.
NATIVE HABITAT: Throughout the temperate zone.
HARDINESS: Zones 2 to 9.

SIZE: 6 to 12 inches tall.

FOLIAGE: Evergreen, fernlike, alternate or in basal rosettes, twice-pinnately dissected, linear-lanceolate in outline, tomentose both sides, grayish, aromatic, to 2 inches long, leaflets to 1/4 inch long. Relatively fine textured.

STEMS: Angular, hairy.

INFLORESCENCE: Dense corymbs; heads to 3/16 inch across; disc and ray flowers yellow, corolla 4 to 5 lobed, effective late spring to early fall.

SELECTED CULTIVARS AND VARIETIES: 'Aurea' With darker yellow flowers.
'Moonlight' With pale yellow flowers and slower rate of spread.
'Nana' Dwarf selection with compact habit, and small white flowers.

OTHER CULTIVARS: Achillea 'Moonshine' has attractive yellow flowers against soft textured silvery-gray foliage. Often it is listed as a hybrid (*A.* × 'Moonshine'), while others list it as a cultivar of the species *A. taygetea.*

Generally Applicable to All Species

CULTURE: Soil: Well drained sandy loam of low fertility is best.
Moisture: Very tolerant of drought, best in hot, dry locations.
Light: Full sun.

PATHOLOGY: Diseases: Crown gall, powdery mildew, stem rot, rust.

MAINTENANCE: Divide clumps every 2 to 4 years to prevent overcrowding. Cutting flowers back in summer may produce extended bloom. Self-sowing may be prevented by removing flowers as soon as they begin to fade.

PROPAGATION: Cuttings: Cuttings taken in midsummer are easily rooted.
Division: Division of clumps is best exercised in spring or fall.
Seed: Progeny demonstrate marked variability, often being inferior where landscape merits are concerned. Seed propagation should not be practiced on a commercial scale unless the propagator is willing to explain this to the customer.

A diantum (ad-i-an'tum) Greek *a*, without, and *dianthos*, moistened, the fronds supposedly remaining dry even after being plunged under water. This genus of delicate ferns, commonly called the maidenhairs, is composed of 200 or more species. They are primarily native to tropical America, with a few species from temperate North America and East Asia.

FAMILY: POLYPODIACEAE: Polypody family.

A. pedatum (pe-dā´tum) Fronds like a bird's foot (Figure 3–5).

COMMON NAME: American maidenhair fern, five finger fern, northern maidenhair.

NATIVE HABITAT: North American woodlands from Alaska and Canada south to Georgia and Louisiana.

HARDINESS: Zones 3 to 8.

HABIT: Broad spreading, graceful fern.

SIZE: 18 to 24 inches, spreading 3 to 5 feet across.

RATE: Relatively fast spreading by creeping root stalks; space plants 8 to 14 inches apart from quart- to gallon-sized containers.

LANDSCAPE VALUE: Foreground plant for perennial or wildflower border. Combines well with large rocks and hostas or underneath trees and shrubs as an understory ground cover. Sometimes used to good effect overhanging a wall of a raised planter. Exceptionally graceful and rich appearing. No foot traffic.

FOLIAGE: Deciduous, twice pinnately compound, to 16 inches long by about 10 inches across, nearly orbicular in outline, forked into two spreading branches, each bears 4 to 12 or more horizontally held narrow 1-pinnate pinnea (leaflets), pinnules (subleaflets) are about 3/4 inch long, bluish green, fan-shaped, oblong with short smooth petioles.

STEM: About 20 inches long, shiny black or dark brown, smooth, base scaly, often conspicuous on ground after leaves have fallen.

SORI: Grouped 1 to 5 on upper margin of subleaflets; indusia, white to yellowish green, thin and rounded to linear in outline.

OTHER SPECIES: *A. capillis veneris* (southern maidenhair) is a soft, lacy, light green foliaged American native. Similar in habit to *A. pedatum*, this plant reaches only 18 inches tall and is hardy only in zones 7 to 10.

CULTURE: Soil: Rich, moist, high in humus, requiring a pH of 5.5 to 7.0.

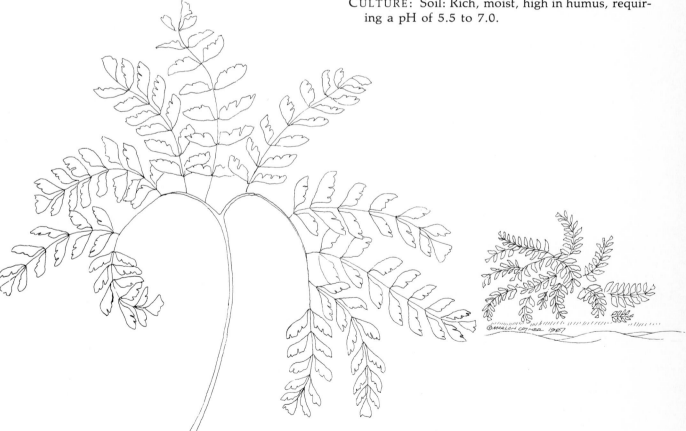

Figure 3–5 *Adiantum pedatum* (0.33 life-size)

Moisture: Best when supplied with liberal soil moisture and high humidity.

Dry conditions markedly stunt the growth.

Light: Light to moderately dense shade.

PATHOLOGY: See **Ferns: Pathology.**

MAINTENANCE: Little maintenance is required. In the north, remove dead material in spring prior to new growth. Annually, topdress with leaf mold to supply organic matter if grown in amended, naturally light textured soil.

PROPAGATION: Spores and division. See **Ferns: Propagation.**

*A**egopodium*** (ē-gō-pō´di-um) From Greek, *aego*, goat, and *podo*, foot, perhaps in reference to the shape of the leaves. This genus contains about five species of coarse, herbaceous perennials that spread by rhizomes.

FAMILY: UMBELLIFERAE: Parsely family.

A. podagraria (pō-da-grâr-i´á) Greek, *pod*, foot, and *agra*, a catching or seizure; hence goutweed as the common name (Figure 3–6).

COMMON NAME: Bishop's weed, goutweed, ashweed, ground ash, ground elder, herb gerard.

NATIVE HABITAT: Europe.

HARDINESS: Zones 3 to 9.

HABIT: Horizontal spreading, rhizomatous, herbaceous, ground cover.

SIZE: 6 to 14 inches tall, spreading indefinitely.

RATE: Extremely rapid spreading (invasive), slowed somewhat by drought or infertile soil. Space plants from 3 to 4 inch diameter pots 12 to 18 inches apart.

LANDSCAPE VALUE: Commonly used en masse as a general ground cover for large- or moderate-sized areas and is very effective for lending continuity and softening the overall texture of an area; but be sure that the area is surrounded by a border if its growth must be restrained. Excellent also for planting between barriers such as a building wall and sidewalk or in a perennial border when bounded by edging or walkway. The variegated cultivar contrasts sharply when planted beneath dark leaved shrubs such as purple leaved *Berberis* or *Prunus.* No foot traffic.

FOLIAGE: Biternately compound, basal and lower stem leaves with short, broadly expanding petioles, segments 1 1/2 to 3 inches long, medium to dark green, broad at the base, narrowing to a sharp point at the apex, margin coarsely serrate. Medium textured.

INFLORESCENCE: Small, white, nonshowy flowers are arranged in compound umbels in early summer on 1 1/2 to 2 feet high stalks. In general they are carrotlike in appearance and of little ornamental significance.

FRUIT: Ornamentally insignificant.

SELECTED CULTIVARS AND VARIETIES: 'Variegatum' A very popular silvery-white and green variegated cultivar that is somewhat less invasive than the species. Foliage is light green, irregularly edged with silvery-white, excellent for brightening areas of deep shade. This cultivar can also be grown in full sun, but often becomes unsightly as the leaf margins are prone to dessication and necrosis in the heat of midsummer.

Figure 3–6 *Aegopodium podagraria* 'Variegatum' (0.75 life-size)

CULTURE: Soil: Very adaptable, growing in anything from sandy loam to heavy clay. Tolerant of compacted soils and infertility, with most rampant growth in moist rich soils.

Moisture: Usually keeping soil on the dry side is desirable as it slows the rate of spread without sacrificing a luxuriant apearance.

Light: Full sun to dense shade.

PATHOLOGY: Generally trouble free.

MAINTENANCE: Mow periodically throughout the growing season whenever foliage begins to look untidy or, in the case of the variegated cultivar grown in full sun, the margins dessicate. Crisp new foliage will soon appear after mowing and plantings will become more dense and compact.

PROPAGATION: Division: Division can be successful at anytime. It is simplified by first mowing plantings to a uniform height of 3 to 4 inches to remove leaves. After transplanting, new shoots and foliage soon develop from the crown.

Seed: Seed is reported to be a viable means of propagation of the species.

Root cuttings: Actually rhizome cuttings, these can be used for propagating either the species or the variegated cultivar. Segments should contain a joint and be 1 to 2 inches in length. Cover lightly with soil and use no growth regulators; new shoots arise in 2 to 3 weeks.

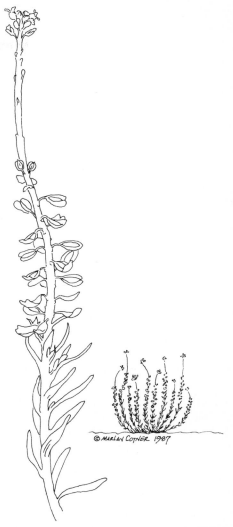

Figure 3–7 *Aethionema grandiflorum* 'Warley Rose' (life-size)

A*ethionema* (ē-thi-ō´nē-ma) Said to be derived from Greek, *aitho*, to burn, and *nema*, a filament, alluding to burnt appearance of the stamens; possibly also for the burning or acrid taste of some species. The genus is composed of 30 to 40 species of glabrous, annual or perennial, usually shrubby, evergreen herbs, many of which are low growing. *A. grandiflorum* and its cultivars are useful as ground covers.

FAMILY: CRUCIFERAE: Mustard family.

A. grandiflorum (gran-di-flō´rum) Large flowered (Figure 3–7).

COMMON NAME: Persian stone cress.

NATIVE HABITAT: Mediterranean region.

HARDINESS: Zone 6.

HABIT: Low, mounded, horizontally spreading, compact, miniature shrub.

SIZE: 6 to 8 inches tall, spreading 12 or more inches across.

RATE: Moderate; space plants from 1/2 gallon containers 12 to 14 inches apart.

LANDSCAPE VALUE: General matlike ground cover for smaller areas. Well suited to rock garden planting. No foot traffic.

FOLIAGE: Evergreen, evenly distributed on stems, 1 1/2 inches or somewhat longer, needlelike, linear-oblong, bluish-green.

STEM: Many stemmed, simple or highly branched, to 12 inches in length.

INFLORESCENCE: Terminal racemes; flowers pink, slightly fragrant, to 1/2 inch across, effective early to midsummer, overall similar to *Iberis sempervirens* in appearance.

FRUIT: Silicles, ovate or rounded, to 1/2 inch long, flat, 1 or 2 seeded, not of great ornamental merit.

SELECTED CULTIVARS AND VARIETIES: 'Warley Rose' Commonly grown hybrid of *A. grandiflorum* × *A. armenum.* The plant is dense, compact, and mounded, with steel-blue foliage. Southernmost hardiness limit is zone 8.
'Warley Ruber' Similar to 'Warley Rose,' but with darker flowers.

CULTURE: Soil: Best in light textured, excellent-draining loam. Seems adaptable to pH within a point or so of neutral.
Moisture: Relatively tolerant of drought, but intolerant of saturated soils.
Light: Full sun.

PATHOLOGY: Diseases: Root rot is devastating when grown in poorly draining soils.
Pests: Aphids are common but seldom serious.

PROPAGATION: Cuttings: Softwood cuttings are the standard means of propagation. Application of 1000 to 2000 ppm IBA/talc is beneficial but not essential.
Seed: Seed is occasionally used; however, seedlings reportedly display wide variation.

✦✦✦✦✦✦✦✦✦✦✦✦✦✦✦✦✦

A*gapanthus* (ag-à-pan′thus) From Greek, *agape*, love, and *anthos*, a flower; literally, love flower. The genus is composed of about nine species of herbs with thick rhizomes, tall spikes of many funnel-shaped flowers, and masses of sword-shaped leaves like those of daylilies.

FAMILY: LILIACEAE: Lily family.

SIZE: From about 20 inches to 3 feet tall.

HABIT: Horizontally spreading, thick rhizomed, basal leaved, herbaceous ground covers.

RATE: Relatively fast; space plants 18 to 24 inches apart from quart to gallon sized containers.

LANDSCAPE VALUE: Mass planted and used as general covers, agapanthuses are outstanding when in bloom and impart a graceful, softening effect with their arching foliage. Accent plantings of clumps in a border setting, rock garden, or among low growing ground covers such as *Hedera* or *Vinca* are also excellent uses of these plants. No foot traffic.

OF SPECIAL INTEREST: Flowers of agapanthuses are attractive to hummingbirds.

* * *

A. africanus (af-ri-kā′nus) From Africa.

COMMON NAME: African agapanthus, blue African lily, lily-of-the-Nile.

NATIVE HABITAT: Africa.

HARDINESS: Zones 9 to 10.

FOLIAGE: Basal, linear-lanceolate, to 20 inches long, to 1/2 inch wide, grouped 8 to 18, evergreen.

INFLORESCENCE: 12 to 40 terminally flowered umbels; flowers with corolla funnel shaped, deep violet-blue, to 1 1/2 inch long, flower stalks reaching 1 1/2 feet, effective late spring through fall.

FRUIT: Ornamentally insignificant.

SELECTED CULTIVARS AND VARIETIES: Var. *albidus* With white flowers.
Var. *variegatus* Leaves striped white.

* * *

A. inapertus (in-a-pêr′tis) Adjective meaning closed, although normally open. The application to this species is unclear.

NATIVE HABITAT: Africa.

HARDINESS: Zones 9 to 10.

FOLIAGE: Leaves grouped 5 to 8 per tuft, to 2 1/2 feet long and 2 inches wide, deciduous, stiff, glaucous.

INFLORESCENCE: Up to 100 flowered umbels; florets tubular, drooping, to 1 1/4 inches long, supported by 4 to 5 foot tall flower stalks, effective late spring through fall.

* * *

A. orientalis (ôr-i-en-tā′lis) Oriental, eastern (Figure 3–8).

COMMON NAME: Oriental agapanthus.

NATIVE HABITAT: Orient.

HARDINESS: Zones 9 to 10.

FOLIAGE: Originating basally, broadly linear, to 2 inches wide and 2 feet or more in length, entire, parallel veined with conspicuous midrib, evergreen, dark glossy green above, glabrous dotted white on both sides, grouped about ten per clump.

INFLORESCENCE: Size variable, 40 to 110 florets per umbel; flowers funnelform, blue, to 2 inches long, on stalks 4 to 5 feet tall, effective midspring to early fall.

SELECTED CULTIVARS AND VARIETIES: 'Albidus' Flowers white.
Var. *aurivittatus* Leaves striped yellow.
'Flore Pleno' Double flowered.
Var. *gigantus* With sturdy spikes that carry umbels of about 200 dark blue flowers.
Var. *leichtlinii* With deeper blue flowers.
'Mooreanus' Dark blue flowers and somewhat hardier than the others; possibly zone 7.

'Peter Pan' Dwarf selection that reaches a height of 8 to 12 inches; flowers deeper blue on 12 to 18 inch tall stalks.

'Nanus' Dwarf and compact cultivar.

'Queen Anne' Intermediate size with bright blue flowers.

'Rancho White' Similar to 'Peter Pan,' but with white flowers.

'Variegatus' With leaves almost entirely white.

OTHER SPECIES: *A. umbellatus* (um-bel-lā´tus) Flowers in umbels. Presumably this is identical to *A. africanus.*

Generally Applicable to All Species

CULTURE: Soil: Not particular to soil; sandy well aerated soil is ideal, but other heavier types are tolerated.

Moisture: Plants should receive enough water to maintain constant but relatively low soil moisture. Somewhat more water will be needed when plants are in bloom.

Light: Vegetative growth is fine in range of full sun to partial shade, but bloom is most impressive when in full sun.

PATHOLOGY: No serious diseases or pests reported.

MAINTENANCE: Divide plants every 5 to 6 years if they become overcrowded.

PROPAGATION: Division: Divide thick rhizomes and crowns in spring or fall.

Seed: Seed is reportedly a viable means of propagating the species; however, seed propagation is practiced to a much lesser extent than division.

A*juga* (a-jū´gȧ) or (aj´ū-gȧ) From Latin, *a*, no, and *zugon*, a yoke, in reference to the calyx lobes, which are equal, not bilabiate. This genus is composed of about 40 species of annual and perennial herbs. The breeding of ground cover types, unfortunately, is often poorly documented, and the tendency to hybridize and sport, with concomitant naming of very similar cultivars, has lead to great confusion in the nomenclature.

FAMILY: LAMIACEAE (LABIATAE): Mint family.

HABIT: Dense, mat-forming, horizontal spreaders. See species for type of spread.

RATE: Rate varies with species and cultivar. In general, plants from 2 to 3 inch diameter containers should be spaced 8 to 12 inches apart for *A. pyramidalis* and *A. reptans*, both of which spread fairly rapidly. Spacing for the moderate to relatively slow growing *A. genevensis* should be about 6 to 8 inches.

LANDSCAPE VALUE: Spreading types are most often used to cover small- to moderate-sized areas as a general cover. Often ajuga is grown in the shade as a substitute for grass. Planted surrounding and underneath trees and shrubs, ajuga adds interest and lends contrast. *A. genevensis* is most often used to cover small areas. None are tolerant of much foot traffic.

OF SPECIAL INTEREST: Honeybees and bumblebees have a great attraction to the flowers of all species of ajuga.

* * *

A. genevensis (jen-e-ven´sis) Of Geneva.

COMMON NAME: Geneva carpet bugle, bugleweed.

NATIVE HABITAT: Europe and Asia.

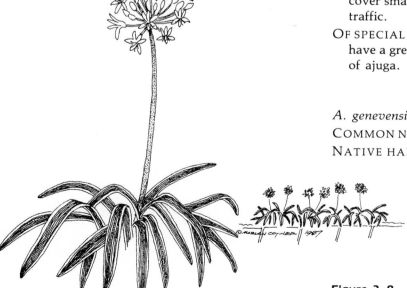

Figure 3–8 *Agapanthus orientalis* 'Variegatus' (0.125 life-size; flowers enlarged to show detail)

HARDINESS: Zones 3 to 9.

HABIT: Rhizomatous, herbaceous, basal leaved herb.

SIZE: Usually 2 to 4 inches tall, spreading indefinitely.

FOLIAGE: Nearly evergreen, hairy to almost glabrous, basal rosettes or opposite, lower leaves are oblong-spatulate, to 4 1/2 inches long and 2 inches wide. The texture is relatively coarse.

INFLORESCENCE: Six or more flowered verticillasters, bracts obovate, blue or violet (much like mints). Effective late spring to early summer.

FRUIT: Pairs of four net-veined nutlets, ornamentally insignificant.

SELECTED CULTIVARS AND VARIETIES: 'Alba' Flowers creamy white.
'Broekbankii' Reported to be similar to the species but dwarf in habit.
'Rosea' Flowers rose-pink, may sometimes be listed as 'Rosy Spires.'
'Variegata' Leaves mottled, creamy white and green.

*   *   *

A. pyramidalis (pi-ram-i-dā´lis) Inflorescences are pyramidal spikes.

COMMON NAME: Pyramidal carpet bugle (bugleweed).

NATIVE HABITAT: Central Europe

HARDINESS: Zones 5 to 9.

HABIT: Rhizomatous, basal leaved herb.

SIZE: Usually 3 to 4 inches tall, spreading about 12 inches across.

FOLIAGE: Opposite or basal rosettes, nearly glabrous or hairy, nearly evergreen, crenate or nearly entire, dark green, bronzy in cold season, spatulate, to 4 inches long by 2 inches wide, relatively coarse textured.

INFLORESCENCE: Crowded verticillasters, 4 to 8 flowered, calyx teeth as long as tube, corolla to 5/8 inch long, pale violet-blue, blooming in late spring to early summer.

FRUIT: Pairs of four net-veined nutlets, ornamentally insignificant.

SELECTED CULTIVARS AND VARIETIES: 'Alexander' compact, with deep bronze foliage.
'Metallica Crispa' Leaves purplish-brown, crinkled, margin crisped, flowers dark blue in late spring to early summer. Space closely; slow spreading.
'Nanus Compactus' Reported to be a dwarf sport of 'Metallica Crispa'; foliage deep-metallic steel-blue-green, crinkled with crisped margins.
'Pink Beauty' With pink flowers.

*   *   *

A. reptans (rep´tanz) Creeping (Figure 3–9).

COMMON NAME: Carpet bugleweed.

NATIVE HABITAT: Central Europe.

HARDINESS: Zones 4 to 9.

HABIT: Stoloniferous, rhizomatous, basal leaved herb.

SIZE: 3 to 4 inches tall, spreading indefinitely.

FOLIAGE: Semievergreen, basal rosettes or opposite, to 4 inches long and 1 1/2 inches wide, crenate or nearly entire, spatulate, ovate or obovate; mature leaves are glabrous and glandular dotted on both sides, and are usually nearly sessile.

INFLORESCENCE: Crowded verticillasters to 4 inches high, usually 6 flowered; bracts often tinged blue, calyx 1/4 inch long; teeth as long as the tube, corolla blue, blooming in late spring to early summer.

FRUIT: Four obovoid nutlets, not ornamentally significant.

SELECTED CULTIVARS AND VARIETIES: 'Alba' Flowers creamy white.
'Alba Variegata' White variegated foliage.
'Atropurpurea' Foliage bronzy purple; flowers dark purplish blue.
'Bronze Beauty' Similar to 'Atropurpurea,' but leaves wider and more intensely bronze.
'Burgundy Glow' Leaves burgundy with pink, creamy white, and green variegation; performs poorly in cold winter areas.
'Burgundy Lace' Likely the same as 'Burgundy Glow.'
'Compacta' Like the species, but denser in habit.
'Gaiety' ('Bronze Improved') Likely the same as 'Giant Bronze,' possibly with deeper foliage tones.
'Giant Bronze' Leaves larger, metallic-bronze and margins crisped.
'Giant Green' Like 'Giant Bronze,' but foliage bright green.
'Jungle Bronze' Taller, leaves bronze, margins crisped, flower spikes to 10 inches tall.
'Jungle Green' With leaves rounded, margins crisped, colors green.
'Multicoloris' Leaves mottled red, white, and yellow on green. Possibly the same as 'Rainbow' and 'Tortoise Shell.'
'Purpurea' Leaves purplish; flowers dark purple.
'Rosea' Flowers rose-pink.
'Rubra' Flowers purplish red.
'Silver Beauty' Foliage reported to be silvery metallic.
'Tottenhamii' Leaves turning bronze-purple in autumn; flowers purple.

'Variegata' or 'Albovariegata' Leaves mottled creamy-white, scorching when grown in full sun. Best color in partial shade. This cultivar self-sows, but does not come true; flowers should be removed when fading.

Generally Applicable to All Species

CULTURE: Soil: Best in well-drained sandy loam, but will tolerate heavier soils to some extent provided drainage is adequate. The pH is best at about 6.5. Moisture: Plants should be kept turgid to ward off insect and fungal diseases. Ajuga is not particular-ly drought tolerant, the roots being rather shallow and the foliage near the hot soil. Protect plantings from winter sun and wind as they easily dessicate in cold weather when snow cover is absent.
Light: Ajugas thrive in light conditions ranging from full sun to moderate shade.

PATHOLOGY: Crown rot: *Sclerotium delphinii* is a very serious disease affecting *ajuga*, especially those plants grown in poorly drained sites. The fungus enters the plants through the roots and crown, cutting off the transport of water. Hence, the first symptom appears as a sudden wilting, even when soil is moist. This disease seems most prevalent in spring and fall when ground is moist and temperatures reach about 50°F. Entire plantings have been destroyed, often attributed to "winter kill" in zones 5 and 6 with good snow cover, areas where all species are relatively hardy.

Mosaic: A virus of uncertain identity often infests ajuga, rendering it worthless as growth is discolored, contorted, and unhealthy.

Pests: Common pests throughout the growing season are aphids. Various aphids cause the leaves to roll under at the edges, hampering growth and making them unsightly. Because of the down turning of the leaves, control must include the use of a translocatable insecticide.

Root knot nematode: Common pest in field plantings. This nematode does not seem to greatly impair the host (causes nodulelike growths on the roots), but even so commercial growers must administer control as infected plants will not be salable.

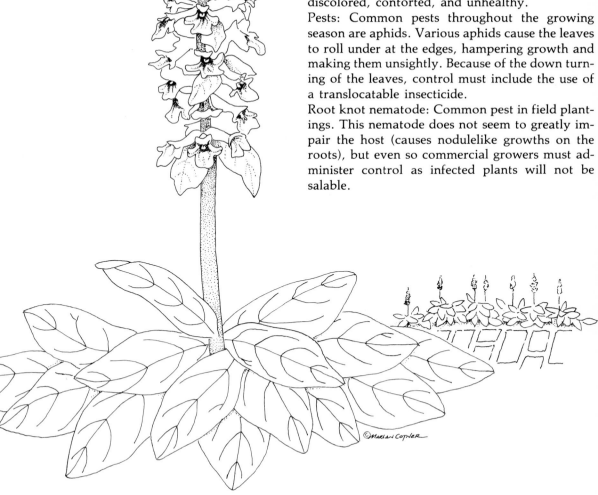

Figure 3-9 *Ajuga reptans* (life-size)

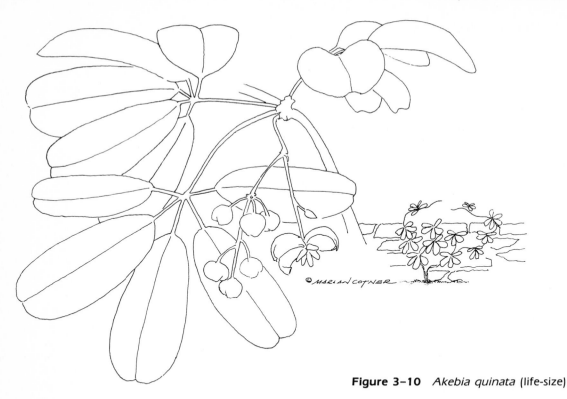

Figure 3–10 *Akebia quinata* (life-size)

PROPAGATION: Many cultivars of ajuga are self-sowing but often do not come true from seed. The situation can be remedied by encouraging dense growth through proper cultural practices.

Cuttings: Leaf and node cuttings are made from sections of stolons and are prepared in the same manner as unrooted plantlets; see *Stolons*.

Division: Clumps of mature crowns divide easily. The potting process (if desired) can be facilitated by removing roots with a knife and sticking the plantlets as though they were cuttings. In this case, transpiration should be kept low with intermittent mist or shading until rooted (approximately 1 week).

Stolons: Plantlets formed on creeping stems under long-day photoperiod in spring can be prepared as cuttings, dipped in a talc preparation of 1000 ppm IBA, and placed directly in the medium as cuttings. Kept under mist, plants root rapidly (often within 2 weeks), and the rapid growing varieties may produce mature, well-developed plants by fall.

*A**kebia** (a-kē´bi-à) Adaptation of the Japanese name, meaning unknown. The genus is composed of monoecious, twining vines.

FAMILY: LARDIZABALACEAE: Akebia family.

A. quinata (quin-ā´tà) Five lobed, the leaves (Figure 3–10).

COMMON NAME: Fiveleaf akebia, chocolate vine.

NATIVE HABITAT: China, Korea, Japan, introduced in 1844 by Robert Fortune.

HARDINESS: Zones 4 to 8.

HABIT: Twining vine, low spreading when unsupported; also spreading by rhizomes.

SIZE: 6 to 12 inches tall, spreading indefinitely when unsupported.

RATE: Relatively fast; space plants 3 to 4 feet apart from quart- to gallon-sized containers.

LANDSCAPE VALUE: General ground cover for large areas. Plant away from low shrubs as it can easily smother them. Tolerates infrequent foot traffic. Provides fair erosion control.

FOLIAGE: Alternate, palmately compound; 5 leaflets, each entire, 1 1/5 to 2 2/5 inches long, emarginate, rounded or broad cuneate at the base, semievergreen, fine textured, bluish-green in summer, very attractive.

STEM: Rounded, thin, green in youth becoming brown, heavily lenticeled, glabrous, leaf scars with 6 or more thraces in a broken ellipse.

INFLORESCENCE: Axillary racemes; flowers small, fragrant, generally inconspicuous, deep reddish-purple, with 3 sepals; male flowers toward apex of raceme, 6 stamened; female flowers below with 3 to 12 pistils, born in late spring.

FRUIT: Very interesting purple-violet colored, 2 1/4 to 4 inches long, sausage-shaped pods that ripen in early to midautumn to reveal black seeds.

OTHER SPECIES: *A. trifoliata* (threeleaf akebia) and *A. × pentaphylla* (*A. quinata × A. trifoliata*) are similar to *A. quinata* in habit, but differ in the number of leaflets, are generally less handsome, and are far less common in cultivation.

OF SPECIAL INTEREST: The fruit of *A. quinata* is reported to be edible, although hand pollination may be needed to ensure fruit formation.

CULTURE: Soil: Growth is best in sandy loam with moderate to high fertility, good drainage, and moderate moisture retaining capacity. A pH of neutral to more acidic is needed.
Moisture: Although soil is ideally kept moist, akebias can withstand moderate periods of drought. Planting in windy, exposed sites is not discouraged.
Light: Full sun or partial shade.

PATHOLOGY: Generally trouble free, with no serious disease or pest problems.

MAINTENANCE: Little maintenance is required. Clip overaggressive shoots as they outgrow their bounds.

PROPAGATION: Cuttings: Hardwood or softwood cuttings root readily. They should be made with the basal cut below a node for insertion into the medium so as to cover the node. Two sets of leaves above the medium is ideal. Treatment with 2000 to 3000 ppm IBA/talc is beneficial.
Layering: Natural and induced layering are an effective means of propagation.
Seed: Seed sown fresh (upon ripening) germinates quickly. Storing at a cool temperature is reported to induce double dormancy and necessitates 3 months of stratification at 50 °F to reverse. Others have reported one month cold stratification to be adequate to break dormancy.

✦✦✦✦✦✦✦✦✦✦✦✦✦✦✦✦✦✦

Alchemilla (al-kem-il´ă) From Arabic, *alkemelych,* alluding to the use of these plants in alchemy. The genus is composed of about 200 species of annual and perennial herbs.

FAMILY: ROSACEAE: Rose family.

A. vulgaris (vul-gā´ris) Common. (Figure 3–11).

COMMON NAME: Common Lady's mantle.

NATIVE HABITAT: Europe.

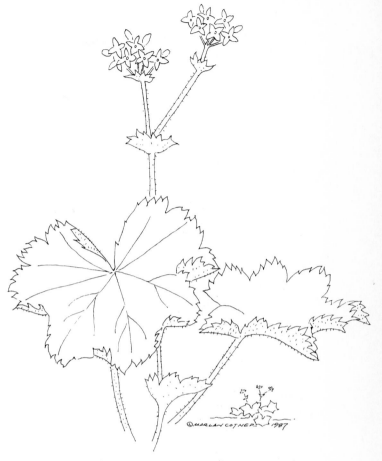

Figure 3–11 *Alchemilla vulgaris* (life-size)

HARDINESS: Zones 3 to 8.

HABIT: Clump forming, low, matlike ground cover with creeping rootstock.

SIZE: 6 to 12+ inches tall, spreading to 2 feet across.

RATE: Relatively slow; space plants from 3 to 4 inch diameter containers 8 to 10 inches apart.

LANDSCAPE VALUE: Mass-planted *A. vulgaris* makes a good general cover for large and small areas alike. Often it is used to edge borders and pathways. The pleated foliage adds an interesting feature and has a delicate texture. Combines well with pines, rhododendrons, azaleas, ferns, hostas, and many others. Tolerates no foot traffic.

FOLIAGE: Basal, evergreen, orbicular, 2 to 4 inches wide, individual leaves palmately veined, 7 to 11 shallowly toothed lobes per leaf, hairy to nearly glabrous, gray-green to green on both sides.

INFLORESCENCE: Compound cymes composed of small, yellow-green apetalous flowers; flowers bisexual, 4 to 5 sepaled, elevated above the foliage, regularly symmetrical, not greatly showy but providing some ornamental interest in mid to late spring.

FRUIT: Small achenelike, not ornamentally significant.

OF SPECIAL INTEREST: Flowers are commonly dried and used in arrangements. *A. vulgaris* is reported to have been used for adornment by the Virgin Mary, hence the common name Lady's mantle. Also, at one time it was used for medicinal purposes. In the garden the leaves tend to hold dew, which makes for a very pleasant sight on summer mornings.

OTHER SPECIES: *A. alpina,* mountain Lady's mantle; with silvery pubescence on the underside of leaves.

CULTURE: Soil: Best in moist, organically rich, well drained loam; neutral to slightly acidic pH.
Moisture: Not particularly drought resistant. Soil should be kept slightly moist.
Light: Performing equally well in full sun or partial shade.

PATHOLOGY: No serious diseases or pests reported.

MAINTENANCE: Remove old foliage in spring prior to new growth. A lawn mower set high is the best means if the planting is large. Plants reseed themselves and can become weedy, so it is wise to remove inflorescences shortly after flowers fade.

PROPAGATION: Division: Divide plants in spring or fall.
Seed: Seed usually germinates in 3 to 4 weeks at 60 to 70°F.

Ampelopsis (am-pel-op´sis) From Greek, *ampelos,* a vine, and *opsis,* resemblance; resembling a grape vine. The genus contains 20 species of deciduous, woody vines and shrubs that are native to North America and Asia.

FAMILY: VITACEAE: Grape family.

HABIT: Horizontal, low spreading, tendrilous vines that will climb if given support.

SIZE: 4 to 8 inches tall, spreading indefinitely when unsupported.

RATE: Relatively fast; space plants 3 to 4 feet apart from gallon-sized containers.

LANDSCAPE VALUE: When not given support, these rugged vines make a durable, dense ground cover

for use in large areas where they have room to spread. Very useful for soil retention on rocky slopes.

* * *

A. aconitifolia (ak-ō-nī´ti-fō´li-à) Aconitumlike, the leaves like those of aconitum.

COMMON NAME: Monkshood vine.

NATIVE HABITAT: Northern China, introduced in 1868.

HARDINESS: Zone 4.

FOLIAGE: Long petioled, deciduous, palmately 3 to 5 parted or compound, leaflets to 3 inches long, deep glossy green with mostly dentate or finely cut margins. Medium textured.

INFLORESCENCE: Perfect, greenish cymes with long peduncles; flowers small, 4 to 5 parted, blooming in late summer, but not greatly ornamental.

FRUIT: 1/2 inch diameter berry, dull orange-yellow, occasionally bluish prior to maturation, effective from early to mid autumn.

* * *

A. brevipedunculata (brev-i-ped-unk-ū-lā´tà) From Latin, *brevi,* short, and *pediculus,* little foot, meaning with a short flower stalk (Figure 3–12).

COMMON NAME: Porcelain ampelopsis.

NATIVE HABITAT: Northeastern Asia, introduced 1870.

HARDINESS: Zones 4 to 8.

FOLIAGE: Simple, medium textured, 2 1/2 to 5 inches long, alternate, broad-ovate in outline, acuminate pointed, 3 lobed, lateral lobes broadly triangular ovate, coarsely serrate, pilose below, petioles hairy and as long as blade or slightly shorter.

INFLORESCENCE: Cymes on long stalks, lying below foliage; flowers greenish, perfect, not of ornamental significance.

FRUIT: 1/4 to 1/3 inch diameter berry, lilac, green, turquoise, or sometimes whitish, suggestive of porcelain; colors frequently mixed within the same cluster; eventually they turn bright blue in maturity; effective in mid to late autumn.

SELECTED CULTIVARS AND VARIETIES: 'Elegans' Smaller leaves that are variegated white; less vigorous; reported to come true from seed. It is not a very attractive plant. The var. *maximowiczii* has leaves deeply lobed, 3 to 5 lobed; extremely vigorous; glabrescent.

Generally Applicable To Both Species

CULTURE: Soil: Tolerant of infertile, rocky soils, adaptable to most soils except those that do not drain well. A pH of 4.5 to 7.0 is suitable.
Moisture: Relatively tolerant to drought; do not hesitate to use these plants in a windy site.
Light: Growing well in a range from full sun to moderate shade.

PATHOLOGY: Generally the same diseases and pests that affect *Parthenocissus* also affect *Ampelopsis*.

MAINTENANCE: Generally, quite easy to maintain; mow plantings annually in spring to promote dense growth, and cut back stems as they outgrow their designated area.

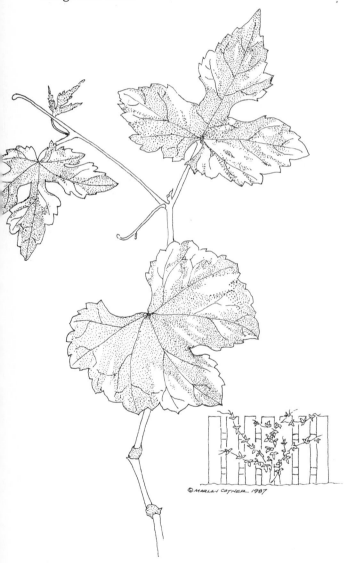

Figure 3–12 *Ampelopsis brevipedunculata* 'Elegans' (life-size)

PROPAGATION: Cuttings: Softwood and hardwood cuttings root readily; treatment with 3000 ppm IBA/talc is helpful.
Layering: Also acceptable for small-scale propagation. Dig rooted stem sections and transplant in spring or fall.
Seed: According to authorities, seed should not be allowed to dry out, but should be stratified immediately upon ripening at a temperature of 40°F for 3 months prior to sowing.

*A*ndromeda (an-drom´e-dá) Named after the Grecian princess who was bound to a rock and rescued by the hero Perseus. The genus contains only two species of low growing, evergreen shrubs.

FAMILY: ERICACEAE: Heath family.

A. polifolia (pol-i-fō´li-á) Smooth or polished, the leaves (Figure 3–13).

COMMON NAME: Bog rosemary.

NATIVE HABITAT: Northern Asia, Central Europe, northeastern North America; cultivated around 1786.

HARDINESS: Zone 2.

HABIT: Low, horizontal spreading, rhizomatous, shrub.

SIZE: 1 to 2 feet tall, spreading 3+ feet across.

RATE: Moderate; space plants from gallon-sized containers 2 to 3 feet apart.

LANDSCAPE VALUE: Excellent general covers in moderate- to larger-sized areas where soil is too moist for many other plants. No foot traffic.

FOLIAGE: Linear to oblong, to 1 1/2 inches long, 1/4 inch wide, evergreen, glaucous below, dark blue-green above, alternate, entire, glabrous both sides, leathery.

STEMS: Upright, branching is moderate to limited.

INFLORESCENCE: Pendulous umbels about 1 1/4 inch across; flowers 1/4 inch long, urn shaped, pinkish fading to white, corolla 5 lobed, stamens 10, effective early to midspring.

FRUIT: 5 valved capsule, not ornamentally significant.

SELECTED CULTIVARS AND VARIETIES: 'Compacta' With compact habit.
'Grandiflora Compacta' Large flowered with compact habit.
'Major' Taller and having broader leaves.

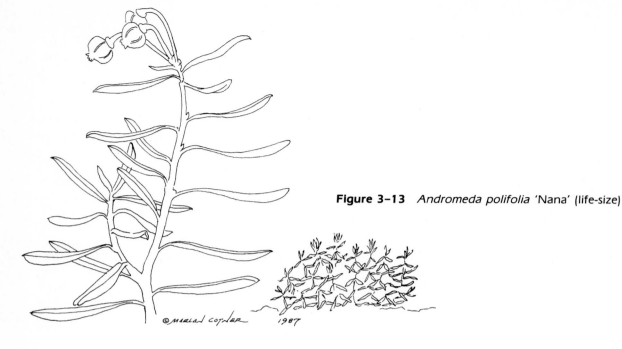

Figure 3–13 *Andromeda polifolia* 'Nana' (life-size)

'Minima' Decumbent or nearly prostrate growing.
'Montana' Lower and more compact.
'Nana' Dwarf, compact selection, to about 12 inches tall, very attractive.

CULTURE: Soil: Boglike conditions are best (i.e., high moisture, acidic, and rich in humus).
Moisture: Requires contantly moist soil.
Light: Best in full sun, but tolerant to partial shade.

PATHOLOGY: Pests: Aphids, thrips, mites. No serious diseases.

MAINTENANCE: Test soil pH periodically and amend as required.

PROPAGATION: Cuttings: softwood cuttings taken in late spring throughout summer root well and are aided with the use of a root inducing preparation of 8000 ppm IBA/talc. Firmer cuttings taken in late fall root as well if given bottom heat.
Division: Division of plantlets that are formed on creeping rootstalks in spring or fall is a simple means of propagation.

✛✛✛✛✛✛✛✛✛✛✛✛✛✛✛✛✛

A *nemone* (ȧ-nem´ō-nē) From Greek, *anemos*, wind, and *mone*, a habitation, some species enjoying windy places; hence windflower, the English name. The genus is composed of perennial herbs that are primarily native to the northern temperate zone.

FAMILY: RANUNCULACEAE: Buttercup family.

LANDSCAPE VALUE: Because of their height, anemones are well suited for use in back of lower growing ground covers or as border specimens. No foot traffic.

*   *   *

A. hupehensis (hū-pe-en´sis) From Hupeh, China.

COMMON NAME: Dwarf Japanese anemone, windflower.

NATIVE HABITAT: China.

HARDINESS: Zones 5 to 8.

HABIT: Relatively tall, multistemmed, herbaceous ground cover.

SIZE: 18 to 30 inches tall; spreading to 1 1/2 feet across.

RATE: Moderate; space plants from quart-sized containers 14 to 16 inches apart.

STEMS: 1 to 3 feet long, sparsely hairy.

FOLIAGE: Ternate, leaflets elliptic, 3-5 lobed (shallowly), large toothed, apex pointed, light green, deciduous.

INFLORESCENCE: 2 to 3 inches across, cymose; flowers with sepals numbering 5 to 6, spreading, rose colored, born late summer to early fall.

FRUIT: Achene, not ornamentally significant.

SELECTED CULTIVARS AND VARIETIES: *A. hup.* var. *japonica* Purplish, red, or pink flowered inflorescences to 3 inches across; reaching 2 1/2 feet tall.

A. *hup. jap.* 'Alba' Flowers single, white, 2 to 3 inches across.

A. *hup. jap.* 'Alice' Flowers rosy carmine.

A. *hup. jap.* 'Margarette' Flowers double, pink.

A. *hup. jap.* 'Queen Charlotte' Flowers semidouble, pink.

A. *hup. jap.* 'Whirlwind' Flowers semidouble, white.

OF SPECIAL INTEREST: There is considerable confusion and disagreement as to the classification of many of the above listed cultivars. Often they are listed under the hybrid group *Anemone × hybrida* as it is likely many of them are descendants of the cross *A. hupehensis × A. vitifolia.* Also, in the past *A. hupehensis japonica* was considered a separate species (i.e., *A. japonica*). Thus, it is still common to see cultivars listed as members of the questionable species *japonica.* There are many other cultivars than those listed above.

* * *

A. *pulsatilla* (pul-sả-til ́là) To shake, presumably in the wind (Figure 3–14).

COMMON NAME: European pasqueflower.

NATIVE HABITAT: Europe and Asia.

HARDINESS: Zones 5 to 8.

HABIT: Herbaceous, relatively dense ground cover.

SIZE: About 12 inches tall, spreading about 18 inches across.

RATE: Relatively slow to moderate; space plants from pint- to quart-sized containers 10 to 14 inches apart.

STEMS: Soft, hairy, 1/2 foot long when in bloom, reaching 1 1/4 feet tall by the time of fruiting.

LEAVES: Basal, developing following bloom, 4 to 6 inches long, thrice pinnate, leaflets 1 inch long, lobes linear.

FLOWERS: Bell shaped, solitary, erect, blue to reddish purple, about 2 inches wide, blooming early spring.

FRUIT: Fuzzy heads, quite interesting and showy in late spring and early summer.

OF SPECIAL INTEREST: The extract of the purple petals is green and has been used to dye Esater eggs. Several cultivars exist but are not listed here as they are rather uncommon.

Generally Applicable to All Species

CULTURE: Soil: Best in well-drained sandy loam that has enough organic matter to maintain moisture. A mildly acidic pH is preferred.

Figure 3–14 *Anemone pulsatilla* (life-size)

Moisture: Not known for drought tolerance; water frequently during prolonged hot, dry weather.

Light: Full sun to light shade.

PATHOLOGY: Diseases: Leaf spots, rhizome rot, flower spotting, downy mildew rust, smut, viral mosaic.

Pests: Aphids, black blister beetle, cutworms, flea beetle, fern nematode, root lesion nematodes.

MAINTENANCE: In zones 5 and 6, mulch plants in the fall to protect from winter cold.

PROPAGATION: Cuttings: Root cuttings should be made in fall. Make sections about 2 inches long, lay them in flats, and cover with about 1/2 inch of soil.

Division: Simply divide in spring or fall.

Seed: Germination takes from 5 to 6 weeks at 68 °F, and variability is great. Cultivars should be propagated by simple division or root cuttings. Plants hybridize freely and new cultivars are relatively easy to create.

Antennaria (an-te-nā´ri-à) From Latin, *antenna*, a sail yard, the hairs attached to the seed of the plant resembling the antennae (feelers) of insects. The genus is estimated to contain between 15 and 75 species of dioecious, small, white or gray perennial herbs. Male plants are few or absent in some species, and parthenogenesis is common.

FAMILY: ASTERACEAE (COMPOSITAE): Sunflower family.

LANDSCAPE VALUE: All species listed are excellent ground covers for covering small- to moderate-sized areas, and for use in a border or ornamental bed surrounding and below small shrubs. No foot traffic.

* * *

A. dioica (di-oy´kà) Meaning two houses, that is, male and female parts being on separate plants (dioecious) (Figure 3–15).

COMMON NAME: Common Pussytoes.

NATIVE HABITAT: Europe, Asia, and the Aleutian Islands.

HARDINESS: Zones 3 to 9.

HABIT: Prostrate, stoloniferous, matlike, spreading herbaceous ground cover.

SIZE: 8 to 12 inches tall, spreading indefinitely.

RATE: Moderate; space plants from 4 inch diameter containers 10 to 14 inches apart.

LEAVES: Semievergreen, in rosettes, spatulate, to 1 inch long, green and glabrate above, white-wooly beneath, arranged linearly on stems.

INFLORESCENCE: Heads, small, about 1/4 inch across and clustered 2 to 12; bracts white or pale pink, effective mid to late spring.

SELECTED CULTIVARS AND VARIETIES: 'Nywood' With bright red flowers.

Var. *rosea* With rose colored flowers.

'Tomentosa' Leaves white-tomentose, sometimes listed as *A. tomentosa*.

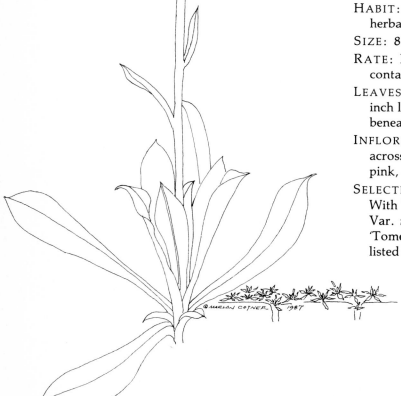

Figure 3–15 *Antennaria dioica* (life-size)

* * *

A. plantaginifolia (plan-tả-jin-i-fō ´li-ả) Plantago (plantain) leaved.

COMMON NAME: Plantainleaf pussytoes.

NATIVE HABITAT: United States from Maine to Georgia and west to Minnesota and Missouri.

HARDINESS: Zone 3.

HABIT: Low, herbaceous, stoloniferous ground cover.

SIZE: 8 to 10 inches tall, spreading indefinitely.

RATE: Moderate; space plants from 4 inch diameter containers 10 to 14 inches apart.

FOLIAGE: Basal rosette, elliptic to obovate, to 3 inches long, 3 to 5 nerved, tomentose becoming glabrous above, petioled, leaves on stems are lanceolate to linear, tomentose.

INFLORESCENCE: Heads are numerous, racemose to densely corymbose, involucral bracts are pale green or occasionally purplish with white apexes, effective mid to late spring.

* * *

A. rosea (rō ´zē-ả) Pinkish-red, the flowers. Previously considered a variety of *A. dioica*.

NATIVE HABITAT: Alaska to California and Colorado.

HARDINESS: Zone 4.

HABIT: Stoloniferous, herbaceous ground cover.

SIZE: 12 to 18 inches tall, spreading indefinitely.

RATE: Relatively slow; space plants from 4 inch diameter containers 8 to 12 inches apart.

LEAVES: Rosette, narrowly oblanceolate to spatulate, to 1 inch long. Leaves on stems are oblanceolate or linear.

INFLORESCENCE: Heads, several; involucre about 1/4 inch long; involucral bracts with white, pink, or rose apex, effective in spring.

Generally Applicable to All Species

CULTURE: Soil: Thriving in dry, rocky, infertile soil. They perform well in the pH range of 6.0 to 7.0 and likely will tolerate mild alkalinity.
Moisture: Very drought tolerant. Little or no supplemental irrigation is needed once plants are established.
Light: Full sun to light shade.

PATHOLOGY: No serious diseases or pests reported.

PROPAGATION: Cuttings: Cuttings are easily rooted in early summer.
Division: Simply divide plants in fall or spring prior to flowering.

***A**rabis* (ãr ´a-bis) From Greek, *arabis*, Arabia, that being the home of several species. This genus is comprised of more than 100 species of generally low, annual, biennial, or perennial herbs that are native to temperate North America and Eurasia.

FAMILY: CRUCIFERAE: Mustard family.

RATE: Varieties that are listed here spread at a rate that is moderate to rapid. Space them about 8 to 12 inches apart from containers of pint to quart size.

LANDSCAPE VALUE: Rock-cresses function well as soil stabilizers on rocky slopes. Planted between stepping stones or large stones in a rockery, they add interest through contrasting textures and colors. No foot traffic.

FRUIT: Long, narrow silique, not ornamentally significant.

* * *

A. alpina (al-pi ´nả) Growing on mountains above the limit of tree growth.

COMMON NAME: Alpine rock cress.

NATIVE HABITAT: Mountains of Europe.

HARDINESS: Zones 4 to 8.

HABIT: Moderate-sized, herbaceous, mat forming ground covers.

SIZE: To 16 inches tall, spreading to 10 inches across.

LEAVES: Basal, semievergreen, oblong or obovate, dentate, attenuate petioled, stem leaves cordate to auricled at the base, colored grayish green.

STEMS: Many branches from the base.

INFLORESCENCE: Loose racemes; outer sepals saccate at the base, petals white, to 3/8 inch long, effective in spring.

* * *

A. caucasica (kâ-kas ´i-kả) Native to the Caucasus Mountains (Figure 3–16).

COMMON NAME: Wall rock cress.

NATIVE HABITAT: Caucasus Mountains.

HABIT: Tufted, procumbent, herbaceous, ground cover.

SIZE: 6 to 12 inches tall, spreading 18+ inches across.

HARDINESS: Zone 3.

FOLIAGE: Grayish-green, semievergreen, basal, usually obovate tapering at base, leaves on stems auricled to sagittate at the base, 1 to 2 inches long, pubescent, arranged alternately; margins entire, slightly lobed or coarsely toothed toward the apex.

STEMS: Many branched, procumbent.

INFLORESCENCE: Loose racemes; flowers fragrant, white petaled, 3/8 to 5/8 inch long, petals numbering 4, held on 6 to 8 inch long stems, effective early to midspring.

SELECTED CULTIVARS AND VARIETIES: 'Floreplena' Striking, double, white flowered clone. Flowers born on 12 inch racemes.
'Snowcap' Large pure white flowers.
'Spring Charm' Large clear pink flowered selection.
'Variegata' Leaves with conspicuous creamy white, irregular patches that often revert to the normal color of the species. Another selection is said to have gold margins.

* * *

A. procurrens (prō-cur´ens) Latin, meaning to run forward, to project, or jut out; the growth habit.

NATIVE HABITAT: Southern Europe.

HARDINESS: Zone 4.

HABIT: Relatively low growing, stoloniferous, herbaceous ground cover.

SIZE: 12 inches tall, spreading indefinitely.

LEAVES: Basal, evergreen, oblong to lanceolate or obovate, acuminate, entire, leaves on stems are ovate and rounded at the base, colored glossy green.

INFLORESCENCE: Elongating racemes; flowers with 1/4 inch long white petals, effective early to midspring.

SELECTED CULTIVARS: 'Variegata' With striking white and green variegation.

Generally Applicable to All Species

CULTURE: Soil: Sandy, well-drained soil is best. Saturated soils often lead to fungal infections and poor floral display. A pH of 7.0 to 8.0 is preferred, but slightly acidic conditions are tolerated.
Moisture: Once established, these species are very tolerant to drought, although an occasional watering in summer may be beneficial.
Light: Full sun to light shade.

PATHOLOGY: Diseases: Club root, downy mildew, white rust, leaf spots.
Pests: Lily aphid.

MAINTENANCE: Shear or mow plantings after flowering to keep them looking neat and compact.

PROPAGATION: Cuttings: Softwood cuttings taken in early summer immediately after bloom, root easily under mist.
Division: Simply divide plants in spring or fall.
Seed: Seed is best sown in spring. Germination can be expected in about 3 to 4 weeks at 68°F.

✦✦✦✦✦✦✦✦✦✦✦✦✦✦✦✦✦✦

*A*rctostaphylos (ärk-tō-staf´i-los) From Greek, *arctos*, a bear, and *staphyle*, a bunch of grapes, the berries of some species being eaten by bears. This

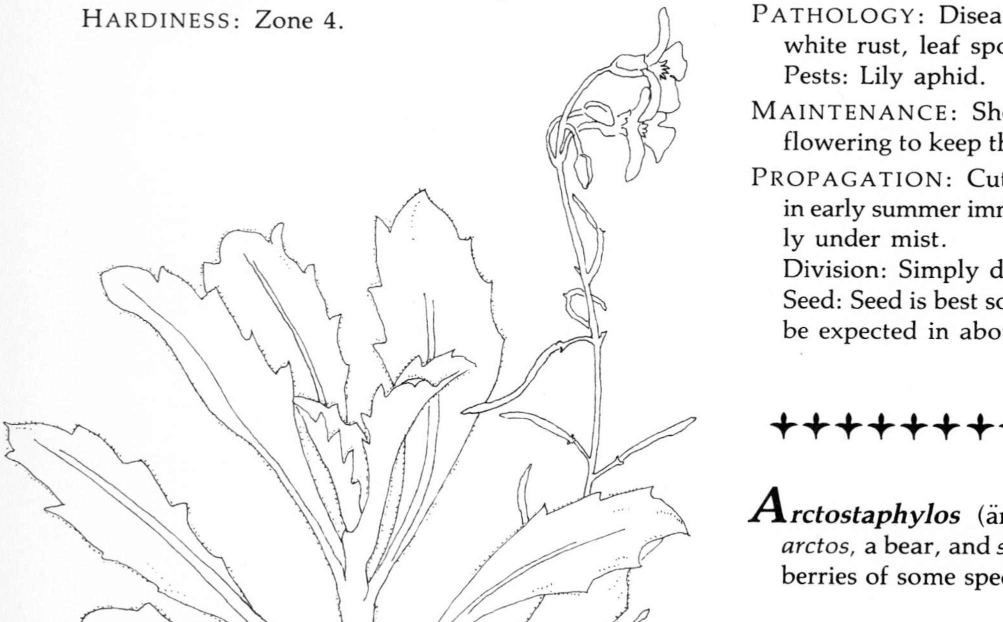

© MARLAN COTNER 1987

Figure 3–16 *Arabis caucasica* 'Snowcap' (life-size)

genus is composed of about 50 species of evergreen, woody, prostrate shrubs that are primarily native to western North America. Fruit of most species are a source of food for wildlife.

FAMILY: ERICACEAE: Heath family.

Ground cover types are 3 to 30 inches tall, spreading variably; see individual species.

*   *   *

A. edmundsii (ed-mundz´ i-ī) Meaning unknown.

NATIVE HABITAT: California.

HARDINESS: Zone 7.

COMMON NAME: Little Sur manzanita.

HABIT: Low, creeping, woody ground cover.

SIZE: Reaching about 12 inches tall and spreading to 6 feet across.

RATE: Relatively fast; space plants 2 to 2 1/2 feet apart from gallon-sized containers.

LANDSCAPE VALUE: Generally best for facing walls or evergreen shrubs.

MAINTENANCE: Shear plants annually to promote branching.

LEAVES: Alternate, simple, elliptic to broadly ovate, to 1 1/4 inches long, truncate to cordate base, grayish-green, evergreen.

STEMS: Creeping, many branched.

INFLORESCENCE: Terminal racemes or panicles; bracts leafy, pink; ovary glabrous, calyx 4 to 5 parted, not a heavy producer of flowers, born late winter to early spring.

FRUIT: Brown, berrylike drupe, effective in late autumn.

STEMS: Creeping and many branched.

SELECTED CULTIVARS AND VARIETIES: 'Danville' Early blooming; spreading to 12 feet across and ranging from 1/2 to 2 feet tall.
'Little Sur' With dense habit and slower rate of growth. Blooms in spring.

*   *   *

A. hookeri (hook-ēr´i) After Sir Joseph Hooker.

COMMON NAME: Hooker manzanita, Monterey manzanita.

NATIVE HABITAT: California.

HARDINESS: Zone 7.

HABIT: Procumbent, shrubby ground cover.

SIZE: 18 to 25 inches tall, spreading to 6 or more feet across.

RATE: Moderate; space plants about 3 to 4 feet apart from gallon-sized containers.

MAINTENANCE: Shear in early spring.

LANDSCAPE VALUE: Used on the West Coast as a general ground cover and soil retainer on slopes. No foot traffic.

LEAVES: Alternate, simple, elliptic to ovate or obovate, to 1 inch long, shiny, bright green, glabrous, evergreen.

STEMS: Procumbent, rooting as they touch, minutely pubescent.

INFLORESCENCE: White or pinkish, ovary glabrous, effective late winter and early spring.

FRUIT: Glossy, bright red, effective in late winter and early spring.

SELECTED CULTIVARS AND VARIETIES: Ssp. *franciscana*, with flowers and fruit larger than the species.
'Monterey Carpet' Reaching about 1 foot tall by 10 feet across; branches dark red; with white flowers.
'Wayside' Faster growing than the species, to 3 feet tall by 8 feet across.

*   *   *

A. pumila (pū´mi-là) Dwarf.

COMMON NAME: Dune manzanita.

NATIVE HABITAT: California, introduced into cultivation in 1933.

HARDINESS: Zone 7.

HABIT: Prostrate spreading, shrubby ground cover.

SIZE: Reaching about 1 to 2 1/2 feet tall.

RATE: Slow; space plants 2 to 4 feet apart from gallon-sized containers.

LANDSCAPE VALUE: Bank stabilizer, good when interplanted with taller shrubs in both moderate- and larger-sized areas.

MAINTENANCE: Mow or shear in early spring to keep plantings dense and compact.

LEAVES: Narrowly obovate to spatulate, to 1 inch long, dull green, evergreen.

INFLORESCENCE: White to pink in color, effective late winter to early spring.

FRUIT: Brown, berrylike drupe, effective in late winter and early spring.

*   *   *

A. uva-ursi (ū-và-ēr´si) Bear's grape, the bearberry (Figure 3–17).

COMMON NAME: Bearberry, billberry, bear's grape, barren myrtle, hog cranberry, creashak, mountain box, mealberry, sandberry, red bearberry, kinnikinick.

NATIVE HABITAT: North America, cultivated since 1800.

HARDINESS: Zones 2 to 7.

SIZE: 1 to 4 inches tall, spreading indefinitely.

HABIT: Prostrate, low, woody, creeping ground cover.

RATE: Relatively slow to moderate; space plants about 12 to 18 inches apart from 3 to 4 inch diameter containers.

MAINTENANCE: Mow in early spring to promote branching.

LANDSCAPE VALUE: Large or small scale, this ground cover is excellent around shrubs in informal and native settings. It also functions well in the pure sand of the lakeshore and controls erosion very well. Protect from winter sun and wind if snow cover is unreliable. Tolerates very limited foot traffic. This is the most frequently planted species of *Arctostaphylos*.

LEAVES: Simple, alternate, evergreen, obovate to spatulate, to 1 inch long, leathery, glabrous, sometimes fringed with hair, dark green above and shiny, lighter below, turning bronzy in winter and fall.

STEMS: Glabrous in youth or minutely hairy, older stems reddish to grayish, bark papery and exfoliating.

INFLORESCENCE: Terminal racemes; flowers white to pinkish, urn shaped, 1/4 inch long, 4 to 5 lobed, perfect, effective early through mid spring.

FRUIT: Berrylike, red, shiny, globose drupe, 1/4 to 1/2 inch long, 4 to 10 nutlets, dry and mealy, initially noticeable and green in late summer, ripening to red in fall and persisting through late winter.

SELECTED CULTIVARS AND VARIETIES: 'Big Bear' Leaves larger, more vigorous than the species.
'Massachusetts' Lower and more compact with more flowers and fruit.
'Point Reyes' Leaves darker green and closely spaced, supposedly more heat and drought tolerant.
'Radiant' Leaves light green and widely spaced, fruiting heavily.
'Vancouver Jade' Recent University of British Columbia Botanical Garden introduction, selected in Victoria, B.C., by E. H. Lohbrunner. This selection differs from the species in that it is more vigorous and flowers are born on branches that are semiupright.
'Vulcan's Peak,' Selected by Siskiyou Rare Plant Nursery; said to be more prolific in production of flowers and fruit. Leaves are rounded and overlapping.

OF SPECIAL INTEREST: Leaves have been reported to be used medicinally and for tanning in Sweden.

Generally Applicable to All Species

CULTURE: Soil: Well-drained sandy soil is best; a pH of 5.0 to 6.0 is acceptable, with the optimum about 5.5. In most cases, fertilizer should be acidic in reaction.

Moisture: Once established, these plants are highly drought resistant. Planting in exposed open sites in the wind is encouraged; however, leaves are likely to dessicate from the sun in cold winter areas if not covered with snow.

Light: Full sun to light shade is the acceptable range, with light shade preferable in hot summer areas.

PATHOLOGY: Generally quite pest free except for occasional problems with root rot, black mildew, leaf gall, leaf spots, and rust. I have personally not observed any serious pest problems.

PROPAGATION: Cuttings: Softwood and hardwood cuttings root slowly but in fairly high percentages. Sticking in small plugs and then transferring to

Figure 3–17 *Arctostaphylos* uva-ursi (0.75 life-size)

larger containers when rootbound is best, as plants do not transplant well when roots are disturbed. Treatment with 3000 ppm IBA/talc and bottom heat of 75 °F during rooting is beneficial. Early fall and early spring seem to be the best times to take cuttings. Research shows mycorrhizal fungal innoculation of the rooting medium has been effective with *A. uva-ursi*.

Seed: Seed may be stored for 1 year and then stratified at 40 °F for 3 months prior to sowing. Reportedly, germination is erratic, and sulfuric acid scarification increases germination percentages and uniformity. Hot water (160 °F) stratification has also been reported to be beneficial.

***A**renaria* (ãr-e-nā´ri-à) From Latin, *arena*, sand, that is, inhabiting sandy places. This genus contains approximately 150 species of low growing, annual or perennial, often matlike, herbaceous plants from the temperate northern hemisphere and arctic regions.

FAMILY: CARYOPHYLLACEAE: Chickweed or pink family.

HABIT: Low spreading, mat forming herbaceous ground covers.

RATE: Generally relatively fast; space plants from 2 1/4 to 4 inch diameter containers 12 to 16 inches apart.

FRUIT: Of no ornamental significance.

* * *

A. balearica (bal-ē-ãr´ik-à) Of the Balearic Islands.

COMMON NAME: Corsican sandwort.

NATIVE HABITAT: Balearic Islands, Corsica.

HARDINESS: Zone 6.

SIZE: 3 inches tall, spreading indefinitely.

LANDSCAPE VALUE: Excellent for use between flagstones. Tolerant of occasional foot traffic. Can be used to create wavelike appearance on unevenly graded terrain.

FOLIAGE: Opposite, broadly ovate, to 3/16 inch long, pubescent, thick, glossy green, petioles to 3/16 inch long, evergreen.

STEMS: Procumbent, to 2 1/2 inches long, branched, scabrid to glabrous.

INFLORESCENCE: Flowers solitary, sepals ovate and pubescent; petals to 1/4 inch long, white, numbering 5, effective in midspring.

SPECIAL CULTURE: Differing from the other species in that it grows as well in moderate shade as it does in full sun.

* * *

A. laricifolia (lãr-is-i-fō´li-à) *Larix* (larch) leaved. Syn. *Minuartia laricifolia.*

COMMON NAME: Larchleaf sandwort.

NATIVE HABITAT: Swiss Alps.

HARDINESS: Zone 5.

SIZE: 6 to 8 inches tall, spreading indefinitely.

LEAVES: Opposite, linear, rigid, curved, ciliate, semievergreen, awl shaped.

STEMS: Woody at the base, prostrate, rooting as they touch.

INFLORESCENCE: Cymes, 1 to 6 flowered, born on stems to 12 inches high; flowers showy, pedicels and sepals crisply hairy, sepals linear-oblong, 1/4 inch long, margin reddish, petals to 1/2 inch long, white; effective in early summer.

LANDSCAPE VALUE: Fine for use in the rock garden as a specimen.

* * *

A. montana (mon-tā´nà) Of the mountains.

COMMON NAME: Mountain sandwort.

NATIVE HABITAT: Portugal to central and northwest France.

HARDINESS: Zone 4.

SIZE: 2 to 4 inches tall, spreading 8 to 10 inches across.

FOLIAGE: Ovate, tiny, grayish-green, cascading, tapering to a point, evergreen.

INFLORESCENCE: Few flowered cyme; flowers white, borne in midspring.

LANDSCAPE VALUE: Use between flagstones or as a lawn substitute for small yards.

* * *

A. verna (vẽr´nà) Spring, time of flowering. Syn. *Minuartia verna* (Figure 3–18).

COMMON NAME: Moss sandwort.

NATIVE HABITAT: Arctic regions.

HARDINESS: Zones 3 to 9 or 10.

SIZE: 2 inches tall, spreading indefinitely.

LEAVES: Linear-lanceolate, very narrow, grasslike, about 3/4 inch long, opposite, produced on prostrate stems and to a lesser degree on flowering stems, light green, glabrous to glandular-pubescent, rarely scabrid, evergreen.

Figure 3–18 *Arenaria verna* (life-size)

STEMS: Prostrate, glandular-pubescent toward apex, reproductive stems elevated 2 to 3 inches.

INFLORESCENCE: Cymes, few to many flowered, on erect leafy stems; flowers starlike, fine petioled; petals white, longer than the ovate, glandular sepals. Born in midspring.

SELECTED CULTIVARS AND VARIETIES: 'Aurea' Foliage light yellow, often variable as it tends to be prone to reversion back to green.

LANDSCAPE VALUE: Exceptional for use in cracks between stepping stones, in a rock garden or as a lawn substitute for banks or uneven ground where turf maintenance is difficult.

Generally Applicable to All Species

CULTURE: Soil: Sandy, well-drained loamy soils are best. Clay or poor draining soils should be avoided. Seemingly adaptable to pH, likely best with moderate acidity.
Moisture: Not well suited to dry conditions; thus soil moisture should be maintained with regular watering. Root system is shallow.
Light: Generally best in full sun, but tolerant to light shade.

PATHOLOGY: Diseases: Damping off is common in northern regions during spring as plants do not tolerate slushy snow. Make sure to plant in well-draining soil. Others include leaf spot, powdery mildew, anther smut, rusts.
Pests: No serious pests reported.

MAINTENANCE: Mulch tender varieties in the north where snow cover is limited. Pine or spruce branches work well and should be removed in ear-

ly spring before new growth begins. *A. verna* may form mounds due to overcrowding. The situation is remedied by cutting patches out of the mounds, followed by tamping back to the soil surface where they will soon take hold again. All species are notorious for self-seeding and can become weedy. Thus it is wise to mow blossoms off before seed ripens.

PROPAGATION: Division: Simply dividing plants in spring or fall is the most common and easy way to propagate the sandworts.
Seed: Commercial supplies for some species are available. In general, seed should be sown in a greenhouse in spring and germination expected within 2 to 3 weeks at a temperature of about 60 °F.

✦✦✦✦✦✦✦✦✦✦✦✦✦✦✦✦

Armeria (är-mē´ri-à) Old Latin name for thrift. This genus is composed of approximately 35 species of low growing, evergreen, tufted herbs and shrubs.

FAMILY: PLUMBAGINACEAE: Plumbago family.

HABIT: Low, tuft forming herbaceous ground covers.

RATE: Slow to spread; space plants from 3 to 4 inch diameter containers 6 to 10 inches apart.

LANDSCAPE VALUE: Being rather small and tufted, thrifts are unsuitable to cover large areas efficiently; however, they provide excellent contrast and color accent when interplanted with other ground covers that will not overrun them. Also, they may be used alone in small areas as a general cover or, as commonly practiced, they may be arranged in geometric patterns, the shape of which they hold well. No foot traffic. Able to withstand salt and thus useful along the coast in seaside plantings.

FRUIT: Single seeded, not ornamentally significant.

*   *   *

A. juniperifolia (jö-ni-pĕr-i-fō´li-à) Leaves like the juniper.

COMMON NAME: Juniper thrift.

NATIVE HABITAT: Spain.

HARDINESS: Zone 3.

SIZE: To 2 inches tall, spreading to 4 inches across.

FOLIAGE: Rosette, to 3/4 inch long, awl shaped, triangular in cross section, sharply mucronate, evergreen.

INFLORESCENCE: Flowers arranged in 1/2 inch

diameter heads that are subtended by involucral bracts, bisexual, 5 parted, corolla white to pink, effective in midspring.

* * *

A. maritima (mȧ-rit´i-mȧ) Of the sea (Figure 3–19).
COMMON NAME: Common thrift, maritime thrift.
NATIVE HABITAT: Southern Greenland, Iceland, and northwestern Europe.
HARDINESS: Zones 3 to 8 or 9.
SIZE: 6 to 12 inches tall, spreading 8 to 12 inches across.
FOLIAGE: Evergreen, basal rosette, thin, grasslike, narrow, linear, single nerved, to 4 inches long, margin entire, glabrous or glandular dotted on both sides, grayish or green.
INFLORESCENCE: Heads to 1 inch across, solitary, on 2 to 12 inch scapes; flowers bisexual, interfloral bracts present, calyx funnelform and variously pubescent, 5 petaled, pink or whitish, effective midspring and continuing sporadically through summer.
SELECTED CULTIVARS AND VARIETIES: 'Alba' With white flowers.
'Bloodstone' Flowers intense red.
'Brilliant' With bright pink flowers.
'Californica' With larger flowers on short scapes.
'Dusseldorf Pride' Flowers reddish; habit compact.
'Lucheana' Deep crimson flowers in dense heads; habit is densely tufted, and plants reach 6 inches tall.
'Purpurea' With purple flowers.
'Royal Rose' Abundant pink flowers.
'Vindictive' Deep, rosy-reddish flowers.

Generally Applicable to Both Species

CULTURE: Soil: Best in light sandy loam; avoid heavy clay. Good drainage is important.
Moisture: Quite resistant to short periods of drought; usually watering is not necessary once established except in cases of prolonged drought.
Light: Full sun.
PATHOLOGY: Diseases: Stem rot is especially prevalent if soil does not drain well.
Pests: No serious pest problems reported.
MAINTENANCE: As tufts expand, they may become brown in the center. This is an indication that the plants need dividing and replanting.
PROPAGATION: Cuttings: Stem cuttings should be taken in early summer and treated with a fungicide before placing in rooting medium.
Division: Tufts are easily divided in spring or fall.

Figure 3-19 *Armeria maritima* (0.75 life-size)

Seed: Seed can be collected and used to propagate the species. It should be stored cool and sown the following spring.

$\boldsymbol{A}$*roniea* (ȧ-rō´nē-ȧ) Meaning unknown. A small genus of low, deciduous shrubs. *Hortus Third* lists three species.
FAMILY: ROSACEAE: Rose family.
A. melanocarpa (mel-an-ō-kȧr´pȧ) Black fruited (Figure 3–20).
COMMON NAME: Black chokeberry.
NATIVE HABITAT: Eastern United States.
HARDINESS: Zones 4 to 9.
HABIT: Low, broad spreading, suckering shrub.
SIZE: 1 1/2 to 3 feet high by 6 to 10 feet across.
RATE: Relatively fast; space plants from gallon-sized containers 3 to 4 feet apart.
LANDSCAPE VALUE: Mass planted, black chokeberry is an excellent general cover for large areas. It retains soil, and also makes a nice border plant for around the perimeter of a wooded location. No foot traffic.

Figure 3–20 *Aronia melanocarpa* (life-size)

LEAVES: Alternate, simple, short petioled, finely serrate, glabrous, elliptic to oblong, 1 1/2 to 3 1/2 inches long, 1/2 to 3/4 inch wide, lustrous dark green, turning red in autumn.

STEMS: Slender, glabrous, brownish.

INFLORESCENCE: Corymbs (hawthornlike), 1 to 1 1/2 inches across; flowers white, effective in late spring.

FRUIT: Black or purplish-black, berrylike pomes, 1/3 to 1/2 inch in diameter.

CULTURE: Soil: Grows well in most soils, including those that are dry and infertile. A pH range of 6.0 to 7.0 is optimal.
Water: Best when soil is maintained at a moderate degree of saturation, yet it is capable of withstanding short periods of drought.
Light: Full sun to moderate shade.

PATHOLOGY: Diseases: Bacterial blight, leaf spots, twig and fruit blights, rust.
Pests: Round-headed apple tree borer.

MAINTENANCE: Light shearing in spring is recommended to promote compactness.

PROPAGATION: Cuttings: Softwood cuttings root readily. A 3000 ppm IBA/talc treatment is beneficial.
Division: Division of rooted suckers is a simple means of propagating small quantities in spring or fall.
Seed: Stratify ripe seed for 3 months at about 40 °F prior to sowing.

A*rtemisia* (är-te-miz´i-à or -mish´i-à) Named for Artemis (Diana), one of the divinities of ancient Greece. This genus is composed of about 200 species of annual, biennial, and perennial herbs and shrubs, most of which are aromatic. The ground covering types listed here are generally herbaceous with semiwoody bases.

FAMILY: ASTERACEAE (COMPOSITAE): Sunflower family.

LANDSCAPE VALUE: Low growing species are most often used as ground covers for smaller areas, as specimens in rockeries, or for accent or edging in a border setting. Notable for their foliage; flowers in general are not very ornamental. Not tolerant of foot traffic, but quite tolerant of salt spray.

FRUIT: Achenes, of no ornamental significance.

*   *   *

A. abrotanum 'Nana' (ab-rōt´a-num) Latin name for southernwood.

COMMON NAME: Dwarf southernwood, silver spreader, old man.

NATIVE HABITAT: Eastern Europe.

HARDINESS: Zones 4 to 8.

HABIT: Low growing, wide spreading subshrub.

SIZE: 20 inches tall, spreading to 7 feet across.

RATE: Moderate; space plants from quart- or gallon-sized containers 2 to 3 feet apart.

FOLIAGE: Alternate, compound, 1 to 3 pinnately finely dissected; segments revolute, filiform or linear, finely textured, grayish, pungent when crushed.

STEMS: Highly branched, green, glabrous, aromatic.

INFLORESCENCES: Heads to 3/16 inch across, nearly globose, arranged in loose panicles, flowers yellowish-white, effective in late summer.

OF SPECIAL INTEREST: Aromatic foliage is used in toilet waters and has also been used as a moth repellant.

*   *   *

A. caucasica (kâ-kas´i-kå) Native to the Caucasus Mountains.

COMMON NAME: Caucasian wormwood, silver spreader.

NATIVE HABITAT: Caucasus region of the USSR.

HARDINESS: Zone 4.

HABIT: Low, mat forming ground cover.

SIZE: 3 to 6 inches tall, spreading to 2 feet across.

RATE: Relatively fast; space plants from 2 1/4 to 4 inch diameter containers 10 to 12 inches apart.

FOLIAGE: Alternate, to 1 1/4 inches long, palmately multifid into linear, acute segments, gray-green, silky-hairy, evergreen.

STEMS: Ascending, to 6 inches, tomentose.

INFLORESCENCE: Loose panicles of heads; to 1/4 inch in diameter; flowers yellow, effective in late summer.

*   *   *

A. frigida (frij´i-då) Cold, that is, frosty looking.

COMMON NAME: Fringed sagebrush, fringed wormwood.

NATIVE HABITAT: Midwestern United States.

HARDINESS: Zone 2.

HABIT: Upright growing, rhizomatous, horizontal spreading, matlike ground cover.

SIZE: To about 1 1/2 feet high, spreading 6 to 18 inches across.

RATE: Moderate; space plants from 2 1/4 to 4 inch diameter containers 8 to 12 inches apart.

FOLIAGE: Clustered, to 1/2 inch long, 2 to 3 ternately parted, linear or linear filiform segmented, silvery silky-pubescent, aromatic.

INFLORESCENCE: Panicles or racemes; heads to 1/8 inch across; flowers yellow, effective late summer.

*   *   *

A. schmidtiana (shmit-i-ā´nå) Meaning unknown, most likely a commerative (Figure 3–21).

COMMON NAME: Satiny wormwood, angel's hair.

NATIVE HABITAT: Japan.

HARDINESS: Zones 3 to 9.

HABIT: Erect, rhizomatous, horizontal spreading, matlike ground cover.

SIZE: 18 to 24 inches tall, spreading to 1 foot across.

RATE: Moderate; space plants from 3 to 4 inch diameter containers 8 to 12 inches apart.

FOLIAGE: Alternate, to 1 3/4 inch long, twice palmately divided into linear segments, covered with silky white pubescence, upper leaves becoming linear, aromatic and bitter tasting. Texture overall is quite fine.

INFLORESCENCE: Heads that reach 3/16 inch across are arranged in pyramidal panicles. Bloom occurs in later part of summer, but flowers are not considered of great ornamental worth.

SELECTED CULTIVARS AND VARIETIES: 'Nana' Mature height of 4 inches.
'Silver Mound' Seemingly no different than the cultivar 'Nana.' Very common in the trade. Good for use in smaller areas.

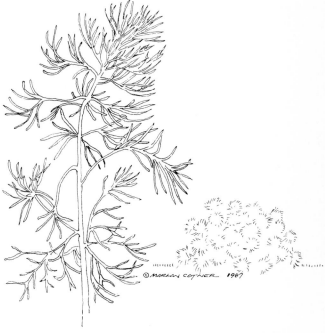

Figure 3–21 *Artemisia schmidtiana* 'Silver Mound' (life-size)

* * *

A. stellarana (stel-lâr-ā′nȧ) Starry; I am not sure what the reference is to.

COMMON NAME: Beach wormwood, dusty miller, old woman.

NATIVE HABITAT: Eastern North America.

HARDINESS: Zones 2 or 3.

HABIT: Rhizomatous, herbaceous, shrubby ground cover.

SIZE: 2 to 2 1/2 feet tall, spreading to 3 feet across.

RATE: Moderate; space plants from quart-sized containers 1 1/2 to 2 feet apart.

LEAVES: To 4 inches long, oblong to ovate, pinnately lobed, white wooly.

INFLORESCENCE: Compact racemes, to 1/4 inch in diameter, flowers yellow, born in summer.

OF SPECIAL INTEREST: The relatively high ability of this plant to tolerate salt makes it popular in seaside plantings.

Generally Applicable to All Species

CULTURE: Soil: Generally tolerant to poor, infertile, dry soils. In respect to pH, they usually perform well in the range from 4.5 to 7.5.
Moisture: Relatively tolerant to drought once established. Supplemental water is needed only during prolonged drought.
Light: Full sun is best, tolerant to light shade.

PATHOLOGY: Diseases: Rusts, damping off.
Pests: None significant.

MAINTENANCE: Following periods of heavy rainfall, plantings may tend to become ragged in appearance. At this time, it is wise to cut them back to stimulate fresh growth.

PROPAGATION: Cuttings: Softwood cuttings root fairly well if stuck in a very well drained medium and treated with a mild root inducing compound that contains fungicide to control damping off.
Division: Dividing plants in spring or fall is a common and effective means of propagation.

Arundinaria (ȧ-run-di-nā′ri-ȧ) From Latin, *arundo*, a reed. This genus is composed of about 30 species of semiwoody grasses from North America and south and eastern Asia.

FAMILY: BAMBUSACEAE: Bamboo family.

HABIT: Rhizomatous, spreading bamboos.

RATE: Moderate to relatively fast, often invasive; space plants from gallon-sized containers 1 to 2 feet apart.

LANDSCAPE VALUE: Exceptionally interesting plants for use as general covers in small- to moderate- (lower species) and moderate- to large-sized areas (taller species). Very functional near stream or pond and poolside for accent or specimen planting. Along walks of stone or brick where their growth can be contained, they make very nice edging in a more formal landscape, yet function well along paths and borders of informal settings as well. Attention should be given to checking spread with the use of edging that extends several inches into the soil should the confining of their growth be desired. They tolerate very little or no foot traffic. Extensive root systems bind soil.

OF SPECIAL INTEREST: Flowering occurs very infrequently, setting bamboos apart from the other grasses that flower annually. Often the flowering of a given species will occur over a broad geographic region simultaneously. Flowering often, but not exclusively, results in the death of the plant.

* * *

A. humilis (hū′mi-lis) Low growing, dwarf.

NATIVE HABITAT: Japan.

HARDINESS: Zones 6 or 7.

SIZE: 1 to 4 feet tall, spreading indefinitely.

FOLIAGE: Blades to 6 inches long, 3/4 inch wide, 6 to 10 nerved, long pointed, round at base, scarcely hairy, pale green, sheaths with 2 clusters of bristles at apex.

STEMS: To 3 feet long, very slender, narrowly fistulose, green, branching two or three times at each node; sheath purplish, turning green with age.

INFLORESCENCE: Spikelets, 6 to 12 flowered, large, compressed, rachilla joints thick, adpressed hirsute, glumes unequal, shorter than lemma; lemma papery, rather thin, born on stems, which usually die after seed is produced.

* * *

A. pumila (pū′mi-lȧ) Dwarf.

COMMON NAME: Dwarf bamboo.

NATIVE HABITAT: Japan.

HARDINESS: Zone 8.

SIZE: To 2 feet tall, spreading indefinitely.

STEMS: Slender, to 2 feet, sheaths strongly tinged purple in youth, becoming light green.

FOLIAGE: Blades to 6 inches long, 3/4 inch wide, 8

to 10 nerved, usually abruptly pointed, base rounded, serrulate, bright medium green, slightly pubescent on both sides.

* * *

A. pygmaea (pig-mē´á) Dwarf (Figure 3–22).
COMMON NAME: Dwarf bamboo.
NATIVE HABITAT: Japan.
HARDINESS: Zones 6 or 7.
SIZE: 6 to 10 inches tall in sun, to 18 inches in shade, spreading indefinitely.
FOLIAGE: To 5 inches long by 1/2 inch across, rounded at base, terminating in a sharp point, fringed with tiny bristles, coarsely pubescent on top, finer on bottom, bright green above, dull silvery-green below, petiole short but well defined; margins serrate, tending to wither and desiccate at the tip, margins, and midrib.
STEM: To 10 inches long, bright green from youth to maturity, purple and flattened at the top, very slender (about 1/16 inch across), cylindrical, internodes about 1 inch apart with prominent, bristled, purple-tinged nodes, having a waxy secretion around the base of the nodes, branching only once or twice.
SELECTED CULTIVARS AND VARIETIES: 'Akebeno' Longer and more slender branches with leaves that are crowded.
'Tanake' With longer leaves.

* * *

A. variegata (vãr-i-e-gā´tá) Varied leaf color.
COMMON NAME: Dwarf white stripe bamboo.
NATIVE HABITAT: Japan.
HARDINESS: Zones 6 or 7.
SIZE: Reaching about 3 feet tall, spreading indefinitely.
LEAVES: Blades to 8 inches long by 1 inch wide, 6 to 10 nerved, rounded at base, tapering to a fine point, somewhat folded at apex, one margin thickly endowed with bristles, the other only slightly bristled, dark-green and brilliant longitudinal variegation on top, less striking below. Variegation held throughout growing season.
STEMS: To 3 feet, internodes to 1 inch long, slender to 1/4 inch across, pale green, new shoots emerging white with tips green, sheaths thick and persistent, nodes not prominent, branches are nearly singular but occasionally paired.

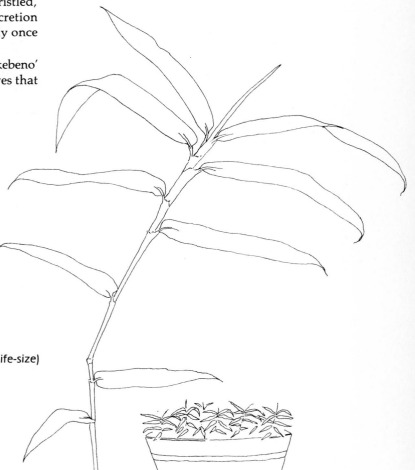

Figure 3–22 *Arundinaria pygmaea* (life-size)

* * *

A. viridistriata (vir-i-dis-stri-ā´tà) Striped green.

COMMON NAME: Dwarf yellow-stripe bamboo.

NATIVE HABITAT: Japan.

HARDINESS: Zone 7.

SIZE: 1 1/2 to 2 1/2 feet tall.

FOLIAGE: Blades to 8 inches long by 1 1/4 inches wide, rounded at base, tapering to a somewhat folded point, irregularly edged with minute bristles that are often more pronounced on one side than the other, light green striped longitudinally with rich yellow. Striping is most prominent in spring and early summer, reduced in shade.

STEM: To 2 1/4 feet tall, slender, tubelike, dark purplish-green, to 3/4 inch across, new shoots emerge pale creamy-yellow with purplish tips.

OF SPECIAL INTEREST: Tends to stay in a clump and therefore should be spaced somewhat closer than other species listed here.

SELECTED CULTIVARS AND VARIETIES: 'Nana' Two types exist, one reaches only about 1 foot high, the other grows from 1 1/2 to 2 feet tall. Both are slower spreading than the species.

Generally Applicable to All Species

CULTURE: Soil: Organically rich, fertile, loamy soils are the most acceptable. Good drainage is necessary. Standing water causes rotting.
Moisture: Thriving in, but generally not requireing, moist soil. Tolerant of considerable drought, but appearance is best when soil is kept moist.
Light: Full sun to moderate shade.

PATHOLOGY: Diseases: Smut.

Pests: Scale insects are the primary pest problem.

MAINTENANCE: Remove dead stems as they occur. Mow or use a sickle to reduce height and keep plantings looking neat, which can be done as needed.

PROPAGATION: Cuttings can be made from rhizomes in early spring. Cut rhizomes into 12 inch sections and place in flats; then cover with about 3 inches of medium. Keep moist and divide new shoots once they become well established.
Division: Simply divide well-rooted sections in early spring, being careful not to allow them to dry out. Division and transplanting are usually best conducted in spring.

*A*sarum (as´à-rum or a-sah´rum) An ancient Greek name, the meaning of which is unclear. The genus contains about 75 species of stemless, perennial, rhizomatous herbs.

FAMILY: ARISTOLOCHIACEAE: Wild ginger family.

HABIT: Low growing, dense ground covers that spread with creeping rootstalks or by crown expansion.

RATE: Generally the species listed here spread at relatively slow to moderate rates. Space plants from pint- or quart-sized containers 8 to 12 inches apart.

LANDSCAPE VALUE: Difficult to beat for woodland or naturalized gardens, the gingers are excellent, dense covers for about any sized area where the cultural conditions are appropriate. They look very nice when planted underneath trees, especially those that are native hardwoods. Wild flowers can be

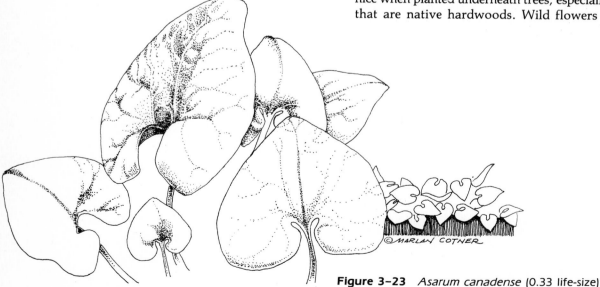

Figure 3–23 *Asarum canadense* (0.33 life-size)

combined for color. The gingers are certainly worthy of more attention than presently being given. No foot traffic.

*   *   *

A. canadense (kăn-a-den´sē) Of Canada (Figure 3–23).

COMMON NAME: Canadian wild ginger, wild ginger, snakeroot.

NATIVE HABITAT: Eastern North America from New Brunswick to South Carolina westward to Missouri.

HARDINESS: Zone 3.

HABIT: As above; rhizomatous.

SIZE: 4 to 6 inches tall, spreading indefinitely.

FOLIAGE: Deciduous, paired, to 6 inches across, kidney shaped, leathery dull gray-green, veins depressed, originating from rhizomes on petioles to 12 inches long, leaves and petioles both slightly pubescent.

FLOWERS: Single, arising from leaf axils; calyx bell shaped, purple or reddish-brown, usually obscured by foliage, effective early to midspring.

FRUIT: Fleshy, globose capsules.

OF SPECIAL INTEREST: Contact with this species commonly causes dermatitis to sensitive persons. The rhizomes smell much like the seasoning ginger and, reportedly, were once used as a substitute.

*   *   *

A. caudatum (kâ-dā´tum) Meaning unclear, likely in reference to the taillike projections of calyx lobes.

COMMON NAME: British Columbia wild ginger.

NATIVE HABITAT: From British Columbia to California.

HARDINESS: Zone 4.

HABIT: As above; rhizomatous.

SIZE: 7 inches tall, spreading to 2 feet across.

FOLIAGE: Paired, to 6 inches long, cordate, to 6 inches across, on petioloes to 7 inches long, dark green, evergreen, pungent when crushed.

FLOWERS: Single, brownish-purple, calyx lobes prolonged into tails to 2 inches long. Effective in spring.

FRUIT: Fleshy, globose capsules.

*   *   *

A. europeum (ū-rō-pē´um) Of Europe (Figure 3–24).

COMMON NAME: European wild ginger.

NATIVE HABITAT: Europe.

HARDINESS: Zones 4 to 8.

HABIT: As above; rhizomatous.

SIZE: 5 inches tall, spreading to 14 inches across.

FOLIAGE: Evergreen, leathery, dark glossy-green, paired, heart shaped, to 3 inches across, petioles to 5 inches long.

FLOWERS: Single, apetalous, calyx cup shaped, divided into three pointed lobes, greenish to purple or brown, 1/2 inch wide, often obscured by the foliage, in early to midspring.

FRUIT: Fleshy, globose capsules, usually hidden by the foliage.

*   *   *

A. shuttleworthii (shut´el-worth-i-ī) Memorial.

COMMON NAME: Mottled wild ginger.

NATIVE HABITAT: Virginia to Alabama.

HARDINESS: Zone 6.

HABIT: As above; rhizomatous.

SIZE: 8 inches tall, spreading to 14 inches across.

FOLIAGE: Thick, leathery, semievergreen (usually evergreen), single or paired, to 3 inches across, usually mottled with silvery-gray petioles to 8 inches, but often shorter.

FLOWERS: Sometimes this species is listed as *A. grandiflorum* as its flowers may reach 2 inches long. They are mottled outside and purple within. Effective in early summer.

OF SPECIAL INTEREST: Occasionally, species of *Asarum* (especially this species) are listed as members of the genus *Hexastylis*. Apparently, in the past, minor floral differences were recognized and the species were separated into two genera. Today, *Asarum* is recognized as the genus that properly encompasses species that used to be considered *Hexastylis*.

Figure 3–24 *Asarum europeum* (0.5 life-size)

SELECTED CULTIVARS AND VARIETIES: 'Callaway' Selected at Callaway Gardens, leaves are usually mottled green and dark green and are smaller than the species.

* * *

A. virginicum (ver-jin-ik´um) Of Virginia.

COMMON NAME: Virginian wild ginger.

NATIVE HABITAT: Virginia to North Carolina.

HARDINESS: Zone 5.

HABIT: As above; rhizomatous.

SIZE: 6 to 8 inches tall.

LEAVES: Elongated, heart shaped, evergreen, single, paired, or in groups of three, to 3 inches across, often mottled light and dark green, petioles to 7 inches long.

Generally Applicable to All Species

CULTURE: Soil: Soil rich in humus is required. Incorporation of peat moss or leaf mold is the best soil amendment. A pH range from 4.0 to 6.5 is acceptable.
Moisture: Plant gingers in sights that are sheltered from strong wind and keep soil moist with watering only as needed.
Light: In the north, gingers should be cultivated in light to dense shade. In the south, moderate to dense shade is preferred.

PATHOLOGY: No serious diseases or pests reported.

PROPAGATION: Cuttings: Cuttings can be made from root sections. Cover with 1/2 inch of soil, keep moist, and give mild bottom heat.
Division: Simple division of plantlets in spring or fall is the most common means of propagation.
Seed: Seed reportedly germinates readily in spring after sowing in summer followed by natural cold stratification of winter.

Asparagus (a-spar´i-gus) Ancient Greek name, said to be derived from Greek, *a*, intensive, and *sparasso*, to tear, alluding to the prickles of some species. The genus contains from 100 to 300 species of woody vines, perennial herbs, and woody shrubs.

FAMILY: LILIACEAE: Lily family.

A. densiflorus 'Sprengeri' (den-si-flō´rus) Meaning densely flowered. The cultivar name is for Sprenger, who was a professor of botany (Figure 3–25).

COMMON NAME: Sprenger asparagus.

HABIT: Tuberous, shrubby, horizontally spreading, vinelike ground cover.

SIZE: 2 to 3 feet tall when unsupported, spreading 2 to 3 feet across.

RATE: Slow to moderate; space plants from quart- to gallon-sized containers 3 to 4 feet apart.

NATIVE HABITAT: South Africa.

HARDINESS: Zones 8 to 10.

LANDSCAPE VALUE: Best used in small areas. Commonly used to excellent effect in raised beds where sprays of foliage can cascade over. Along a retaining wall or seawall, it is also exceptional. This plant should be used by itself as it has a tendency to climb upon other plants. No foot traffic. Seems to be quite tolerant of airborn salt.

FOLIAGE: Needlelike cladophylls (leaflike flattened branches) are solitary or grouped to six at a node (usually in threes), 3/8 to 1 1/4 inches long, slightly curved, one nerved, light green, evergreen.

STEMS: In feathery sprays, flexuous and pendulous, branched loosely, 3 to 6 feet in length, branches to 5 inches long.

INFLORESCENCE: Axillary racemes; flowers bisexual, 1/4 inch across, mildly scented but not of great ornamental worth, petals white to pale pink, blooming summer through fall.

FRUIT: Berry, 1/4 to 1/2 inch in diameter, green, turning bright red in winter.

OF SPECIAL INTEREST: The species *A. densiflorus* is rarely if ever seen in cultivation. The cultivar 'Sprengeri' is the most common, but others are available. 'Sprengeri Deflexus' is similar, but with broader cladophylls and a metallic appearance. 'Sprengeri Nanus' and Sprengeri Compactus' are shorter and less open. 'Sprengeri Robustus' is a more vigorous selection. Sprenger asparagus is quite valuable in the florist trade where it is used for greenery.

OTHER CULTIVARS: Although not as often looked on as a ground cover, the selection 'Myer's' (which grows more upright, is dark green, and appears to billow like green smoke) is exceptionally attractive when planted on about 3 foot centers.

CULTURE: Soil: Loamy soil with good drainage is prefered. Seems to be adaptable as to pH, but likely best in the range of neutral to slightly acidic.
Moisture: Will survive periods of drought; however, its appearance deteriorates. For best appearance, keep soil moderately moist. Root system has water storage nodules that help this plant cope with drought.

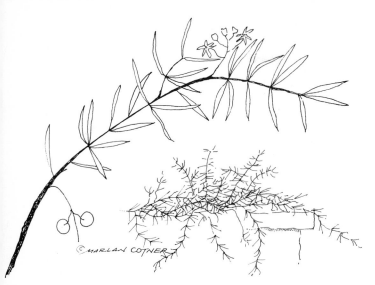

Figure 3–25 *Asaparagus densiflorus* 'Sprengeri' (0.5 life-size)

Light: Full sun to light shade. Better color is obtained when grown in full sun along its northern range.

PATHOLOGY: Diseases: Blight, root rot, crown gall. Pests: Asparagus fern caterpillar, aphids, thrips, and scales.
Physiological: Chlorosis often occurs as a result of overwatering in poorly draining soils.

MAINTENANCE: Clip plants back as they outgrow their bounds or become ragged looking.

PROPAGATION: Cuttings: Cuttings taken from side shoots in the spring root readily.
Division: Simply divide clumps and keep from drying out.
Seed: Seed is readily available commercially. It germinates in about 4 to 6 weeks at 70° to 85°F. Prior to sowing, seed coats should be cracked by mechanical scarification.

Asplenium (as-plē´ni-um) From Greek, *a*, not, and *spleen*, spleen, the black spleenwort, *(A. asiantum nigrum)*, once being regarded as a cure for diseases of the spleen. Commonly called the spleenworts, this is a very large genus of about 700 species of usually evergreen ferns of worldwide distribution. Several species of *Asplenium* may be of use as ground covers. Five of them are briefly described here.

FAMILY: POLYPODIACEAE: Polypody family.

HABIT: Low growing ferns.

RATE: Relatively slow to moderate rate of spread; space bareroot divisions or plants from pint- to quart-sized containers 6 to 8 inches apart.

LANDSCAPE VALUE: Excellent as graceful accent plants when planted randomly among other lower growing ground covers. Also functional alone as general covers for smaller areas. No foot traffic.

* * *

A. bulbiferum (bulb-if´ ẽr-um) Bulbil bearing.

COMMON NAME: Mother fern, hen and chickens fern, king and queen fern, parsley fern.

NATIVE HABITAT: New Zealand, Malaysia, Australia, and India.

HARDINESS: Zone 10.

SIZE: About 2 feet tall (ranging from 1 1/2 to 3 1/4 feet tall), spreading indefinitely.

GENERAL DESCRIPTION: Leaves medium green, to 4 feet long by 1 foot across, 2 to 3 pinnately compound; pinnae to 1 1/2 feet long, divided once again into obovate to oblong pinnules; especially interesting for the small bulbils that sprout and form new plantlets while still attached to the frond; evergreen.

* * *

A. platyneuron (plat-i-nẽr-on) Meaning broad nerved (Figure 3–26).

COMMON NAME: Ebony spleenwort.

NATIVE HABITAT: Eastern North America.

HARDINESS: Zone 3.

SIZE: 15 inches tall, spreading 5 to 6 inches across.

GENERAL DESCRIPTION: Leaves colored dark green, pinnately compound, semievergreen; sterile leaves to 15 inches long, erect, narrow, dark green, tapering from base to apex; fertile leaves are more numerous, lighter, with 28 or more pairs of leaflets; leaflets are narrow oblong, not quite opposite, but somewhat staggered, eared at their bases; petioles are dark, shiny, smooth, short, stiff, and brown; sori are short, straight, frequently confluent; indusia silvery in youth.

* * *

A. risiliens (ris´i-lens) Meaning unknown.

COMMON NAME: Blackstem spleenwort.

NATIVE HABITAT: From Pennsylvania to Mexico.

HARDINESS: Zone 6.

SIZE: 10 inches tall.

FOLIAGE: Evergreen, pinnately compound, tapering toward apex, lanceolate in outline, about 8 inches long, stiffly erect; leaflets opposite, eared at bases, more or less elliptic, margin toothed.

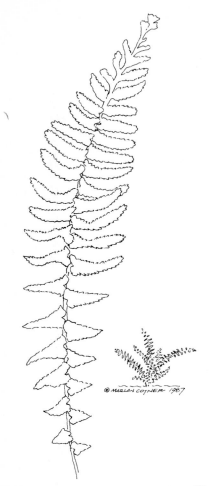

Figure 3–26 *Asplenium platyneuron* (life-size)

PETIOLE: Dark brown or black, shiny, erect, short, and stiff.

SORI: Oblong, located toward leaflet margins, prominent, confluent.

* * *

A. trichomanes (tri-kō´man-ēz or trik-om´e-nēz) A thin hair or bristle, in reference to many trichomes.

COMMON NAME: Maidenhair spleenwort.

NATIVE HABITAT: Eastern North America.

HARDINESS: Zone 3.

SIZE: 6 inches tall, spreading 4 to 5 inches across.

FOLIAGE: Evergreen, medium to dark green, clustered, pinnately compound, lanceolate, to about 8 inches long; fertile leaves upright, sterile leaves horizontal, both types dark green; leaflets are tiny, about 1/4 inch long, orbicular to oblong, margins slightly toothed, opposite, becoming crowded toward the apex.

PETIOLE: Short, wiry, brittle, dark purplish-brown.

SORI: Short, few, frequently confluent; indusia fragile.

Generally Applicable to All Species

CULTURE: Soil: Rich yet well drained soil is best, but these plants will tolerate dry, rocky soils (actually preferred by A. risilens). Tolerant to alkaline soils. Moisture: Constant moisture, but not saturation, is needed. Light: Moderate to heavy shade.

PATHOLOGY: See **Ferns: Pathology.**

MAINTENANCE: See **Ferns: Maintenance.**

PROPAGATION: Division: Simply divide plants in spring or fall when dormant. Spores: See **Ferns: Propagation.**

A*stilbe* (a-stil´bē) From Latin *a*, no, and *stilbe*, brightness, possibly in reference to many of the older species, which had relatively colorless flowers. This genus is made up of about 14 species of herbaceous perennials that often resemble *Spiraeas*. They are native to eastern Asia and North America.

FAMILY: SAXIFRAGACAE: Saxifrage family.

HABIT: Upright growing, low to moderately tall ground covers.

RATE: Generally quite rapid spreading; space plants from pint- to quart-sized containers 15 to 18 inches apart.

FRUIT: Dehiscent follicle, not ornamentally significant.

OF SPECIAL INTEREST: The flowers of many species are very showy and are valuable as cut flowers.

LANDSCAPE VALUE: Excellent as specimens around poolside or other areas (especially of public exposure) such as streamside or along a trail in naturalized sites. The flowers put on an outstanding show and can brighten the darkest corner of a border, fence, or wall. Astilbes are noteworthy for combining well with ferns, hosta, iris, and countless other herbaceous perennials. No foot traffic.

* * *

A. × *arendsii* (a-rends´i-ī) After George Arends (1862–1952) of Rosdorf, Germany, an avid hybridizer of astilbe.

COMMON NAME: Hybrid astilbe, Arend's astilbe, false spiraea, meadowsweet, goatsbeard.

HYBRID ORIGIN: *A. chinensis* var. *davidii* × *unknown species* (presumably hybrids of *A. japonica*).

HARDINESS: Zones 4 or 5.

SIZE: 2 to 3 1/2 feet tall, spreading 2 to 3 feet across.

FOLIAGE: 2 to 3 ternately compound; leaflets ovate-oblong, doubly serrate, color ranging from dark green to bronzy.

INFLORESCENCE: Erect, arching, dense panicles, 8 to 12 inches tall, supported by 2 1/2 to 3 1/2 feet high stems; flowers 4 to 5 petaled, perfect, small, calyx 4 to 5 parted, regularly symmetrical, effective early to midsummer.

SELECTED CULTIVARS AND VARIETIES: There are many cultivars in this hybrid species. Here are some of the more commercially popular.
'Avalanche' Flowers snowy white, stems slightly arching.
'Bremen' With deep carmine-rose flowers.
'Bridal Veil' Flowers white; height around 24 inches.
'Cattleya' Flowers light pink; height to 3 feet.
'Deutschland' Flowers pure white on 2 foot stems; compact habit and vigorous growth, to 24 inches tall.
'Diamond' Flowers white; height 24 to 30 inches.
'Etna' Flowers crimson; height 18 inches.
'Europa' Light pink flowers somewhat earlier than the species; foliage dark green.
'Fanal' Bright dark-red flowers in narrow panicles; flower stems 18 to 24 inches, effective mid to late summer.
'Ostrich Plume' Showy salmon pink nodding flowers; height to 30 inches.
'Red Sentinel' Carmine to deep purplish red, loose panicles; leaves reddish green, finely textured.
'Rheinland' Flowers carmine pink, height 24 inches.

* * *

A. astilboides (à-stil´boi-dēz) Typically astilbelike.

COMMON NAME: Goatsherd astilbe.

NATIVE HABITAT: Japan.

HARDINESS: Zones 4 or 5.

SIZE: 2 to 3 1/2 feet tall.

DISCUSSION: Although plants may be available under this name, the cultivation of this species is in question. It is generally believed to be a variant of *A. japonica*, having larger panicles of tightly arranged flowers.

* * *

A. chinensis 'Pumila' (chi-nen´sis) Of China.

COMMON NAME: Dwarf Chinese astilbe.

HARDINESS: Zones 4 or 5.

SIZE: 8 to 12 inches tall, spreading 8 to 12 inches across.

FOLIAGE: 2 to 3 ternately pinnate, leaflets ovate-oblong, margins doubly serrate, pubescent, 2 to 3 1/2 inches long.

INFLORESCENCE: Narrowly branched panicles; flowers with rosy colored petals that are linear-spatulate and longer than calyx; effective late summer.

* * *

A. × *crispa* (kris´pà) Crisped, the leaves.

COMMON NAME: Hybrid astilbe, crisped astilbe.

HYBRID ORIGIN: *A.* × *arendsii* × *A. simplicifolia*.

HARDINESS: Zones 4 or 5.

SIZE: 4 to 6 inches tall.

DISCUSSION: *A.* × *crispa* is reportedly a group of hybrids that is composed of generally small, low growing ground covers that have bronzy, deeply crisped foliage. They commonly bloom deep pink, white, or red. There are some authorities that include it among the hybrid group *A.* × *arendsii*.

* * *

A. japonica (jà-pon´i-kà) Of Japan.

COMMON NAME: Japanese astilbe.

NATIVE HABITAT: Japan.

HARDINESS: Zones 4 or 5.

SIZE: To 2 feet tall, spreading to 18 inches across.

FOLIAGE: 2 to 3 ternately compound, leaflets narrow to lanceolate-ovate, edged with sharp teeth, base cuneate.

INFLORESCENCE: Erect pyramidal-shaped panicles that are born terminally and in axils; flowers small, white, effective in early summer.

* * *

A. × *rosea* (rō´zē-à)

COMMON NAME: Rose astilbe.

HYBRID ORIGIN: *A. chinensis* × *A. japonica*.

HARDINESS: Zones 4 or 5.

FOLIAGE: Simple, ovate, deeply lobed or cut, to 3 inches long, colored glossy green.

INFLORESCENCE: Slender panicles; flowers white to pinkish, starlike, blooming in early fall.

* * *

A. × 'Sprite' (Figure 3–27) This exceptional cultivar, which is often listed as a cultivar of the species *A. simplicifolia*, is likely a hybrid, as foliage is compound rather than simple, as is typical of the species.

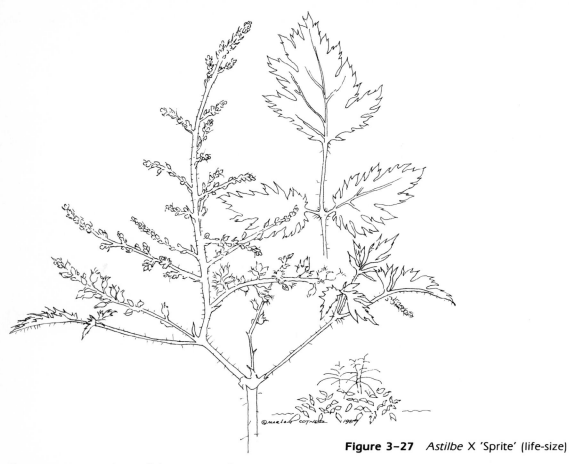

Figure 3–27   *Astilbe* X 'Sprite' (life-size)

SIZE: 6 to 10 inches tall by up to 2 feet across.

FOLIAGE: Very attractive bright, shiny green, 2 to 3 ternately compound, leaflets to 1 1/2 inch long by 1 inch across, margin doubly serrate and weakly ciliate, veins hairy on both sides, petiole bases red and hairy.

INFLORESCENCE: Stout panicles that ascend a few inches above the foliage mass; flowers light pink, born in early summer.

Generally Applicable to All Species:

CULTURE: Soil: Best in cool, well-drained soil that is rich in humus and acidic in reaction.
Moisture: Sensitive to drought, as they are relatively shallowly rooted. It is important to keep the soil moist, and if the soil does not drain well, plant them on elevated mounds.
Light: Full sun to moderate shade, although astilbes are usually at their best with light shade.

PATHOLOGY: Diseases: Powdery mildew, wilt, root rot.
Insects: Japanese beetle, red spider mite.

MAINTENANCE: Overall appearance and flower production are at their best if the plants are divided when they become crowded. Usually every 3 to 4 years is frequent enough. Shollow root system and vigorous growth demand attention to fertilization, which is best applied in two or three light applications annually, rather than a single heavy one. Providing a light organic mulch each fall is recommended if snow cover in the northern range is unreliable.

PROPAGATION: Division: Dividing plants in spring, summer, or fall is the most widely practiced means of propagation.
Seed: Plants can be grown from seed and usually will flower during their second year. Only species or hybrid groups (provided they are properly labeled) should be propagated by seed. Primarily, seed propagation is practiced by hybridizers.

✦✦✦✦✦✦✦✦✦✦✦✦✦✦✦✦✦

*A**thyrium** (a-thi´rē-um) Probably from Greek, *athyros* (doorless), referring to the late opening indusia of *A. filix-femina.* This genus is composed of about 25 species of deciduous ferns. They are native to various regions, but primarily to the tropics.

FAMILY: POLYPODIACEAE: Polypody family.

NATIVE HABITAT: Primarily to Europe and Asia.

HABIT: Graceful, broad, arching ferns, some creeping.

RATE: Relatively fast growing; space plants 10 to 14 inches apart from pint- to quart-sized containers, or 8 to 12 inches apart from bareroot divisions.

LANDSCAPE VALUE: Good ground covers for large or small places in a naturalized ornamental bed. Excellent when planted randomly among lower growing ground covers such as *Vinca minor* or *Ceratostigma plumbagingides.* No foot traffic.

A. filix-femina (fĭ´ liks-fem´ i-nȧ) Literally meaning lady fern, in reference to the delicate fronds, as compared to the male fern *(Dryopteris filix-mas).* (Figure 3–28).

COMMON NAME: Lady fern, female fern, backache brake.

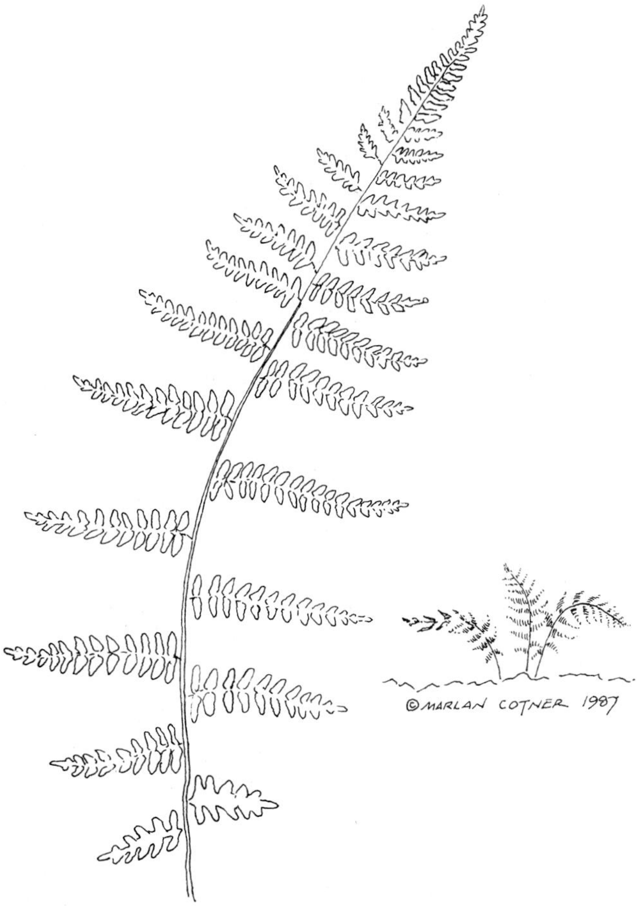

NATIVE HABITAT: Central and Northeastern North America.

HARDINESS: Zones 3 to 8.

SIZE: 18 to 36 inches, spreading 1 1/2 to 2 feet across.

FOLIAGE: Deciduous, large, to 30 inches long by 10 inches across, broadly lanceolate, pinnately twice compound; leaflets to 8 inches long by 1 1/2 inches across, bright green, short stemmed, dividied into about 12 pairs of subleaflets; subleaflets are variable, in opposite pairs, deeply cut and toothed, with once or twice forked veination.

PETIOLE: Smooth with some scales near base, vibrant green, relatively short, grooved (concave) on the front side.

SORI: Short, curved or horseshoe shaped, frequently located over veins; indusia are arching, hairy, and prominent.

SELECTED CULTIVARS AND VARIETIES: Many selections are available. A few include:
'Aristatum' With long points on pinnules.
'Crispum' With crisped fronds.
'Plumosum' Very finely divided.

OTHER SPECIES: *A. goeringianum* (gur-ring-gē-ā´ num) Leaves deciduous, dark green, drooping, pinnately twice compound, broadly lanceolate, to 1 1/2 feet long; leaflets toothed or cut. The cultivar 'Pictum' (Japanese painted fern) has leaves that are very interesting, with central gray stripe and purplish petioles.

SIZE: Growing to 1 1/2 feet, spreading indefinitely.

Generally Applicable to Both Species

CULTURE: Soil: Tolerates wide range of soil types, but best in moist, well-drained highly organic loam. Moisure: Keep soil moist during extended hot, dry weather. Plant out of windy locations.
Light: Light to moderately heavy shade.

PATHOLOGY: See **Ferns: Pathology.**

MAINTENANCE: Simply remove broken, dead fronds in the spring.

PROPAGATION: Division: Simply divide dormant clumps.
Spores: See **Ferns: Propagation.**

Figure 3–28 *Athyrium filix-femina* (0.20 life-size)

©MARLAN COTNER 1987

A*triplex* (at´ri-pleks) From Latin, *a,* no, and *traphein,* nourishment; several species native to arid, nonfertile desert soils. The genus consists of

Figure 3–29 *Atriplex semibaccata* (life-size)

about 100 species of widely distributed herbs and shrubs. Cultivation is limited to a few species.

FAMILY: CHENOPODIACEAE: Goosefoot family.

A. semibaccata (sem-ī-ba-kā′tȧ) Meaning unclear; literally, partially berried (Figure 3–29)

COMMON NAME: Creeping saltbush, Australian saltbush.

NATIVE HABITAT: Australia.

HARDINESS: Zone 8.

HABIT: Deeply rooted, mounding, horizontal spreading shrub that forms a low matlike gound cover.

SIZE: 1 to 1 1/2 feet tall, spreading to 6+ feet across.

RATE: Relatively fast, space plants from gallon-sized containers, about 3 1/2 to 4 1/2 feet apart.

LANDSCAPE VALUE: Planting as recommended usually results in a dense cover in one season. The brush Fire Safety Committee of Los Angeles California lists it among plants in the fire-resistant class, thus making it practical for planting along the foundations of houses and commercial buildings and for roadside planting in rest areas and on highway banks. No foot traffic.

FOLIAGE: Dense, gray, evergreen, fine textured, 1/2 to 1 1/2 inches long.

INFLORESCENCE: Clustered spikes; flowers small, ornamentally insignificant.

SELECTED CULTIVARS AND VARIETIES: 'Carto' More uniform in habit, reaching about 10 inches tall, and colored gray-green.

CULTURE: Soil: Soil must be well drained. Saltbush is very tolerant of high salt levels and infertility. Moisture: Extremely tolerant of drought. Deep watering in the heat of summer may be beneficial, but usually is not necessary. Light: Full sun to light shade.

PATHOLOGY: No serious diseases or insects reported.

MAINTENANCE: During the first season, water on a regular basis; afterward, little maintenance is required.

PROPAGATION: Reportedly, both cuttings and seed are viable means of propagation.

A*ubrieta* (au-brē′tȧ) or (a-brē′shi-ȧ) For M. Aubriet, a French botanical artist of the early 18th century. This genus consists of about 12 species of prostrate, mat forming, alpine, perennial herbs from Europe and Iran.

FAMILY: CRUCIFERAE: Mustard family.

A. deltoidea (del-toid′ē-ȧ) Three angled, from the Greek letter delta, and *oides*, meaning like, alluding to the triangular petals (Figure 3–30).

COMMON NAME: Purple rock cress, sometimes listed as *A. leichtinii*.

NATIVE HABITAT: Greece.

HARDINESS: Zones 4 to 9.

HABIT: Rhizomatous, horizontal spreading, herbaceous, matlike perennial that sometimes mounds off the ground.

SIZE: 3 to 6 inches tall by 1 to 1 1/2 feet across.

RATE: Moderate; space plants from 3 to 4 inch diameter containers 8 to 14 inches apart.

LANDSCAPE VALUE: Excellent between large stones in rock gardens, trailing over walls, or as a foreground plant in a perennial border or ornamental bed. Combines well with iberis, phlox, alyssum, and other herbs. Tolerates no foot traffic. Performs particularly well in the northwest coastal regions, reasonably well in other northern areas, and poorly in the south.

FOLIAGE: Evergreen, alternate, to 1 1/4 inch long by 1/4 inch across, rhombic to obovate-cuneate, hairy on both sides (hairs stellate or forked), glandular dotted both sides, and colored grayish.

INFLORESCENCE: Numerous, short terminal racemes that are elevated above the foliage; flowers are rose-lilac to purple, to 3/4 inch across, usually bisexual, regular, 4 petaled, 6 stamened, effective in midspring to early summer.

FRUIT: Broadly elliptic silique to 3/4 inch long. Not ornamentally significant.

SELECTED CULTIVARS AND VARIETIES: Many varieties are offered commercially: generally, they vary slightly in flower color. Listed here are some of the most common.

'Bengale' With semidouble, lilac-blue, purple, and red flowers.

'Graeca' Large light-blue flowers.

'Purple Cascade' With purple flowers.

'Moerheimii' Large mauve flowers appearing all summer.

'Red Cascade' Flowers red.

'Variegata' With cream variegated foliage.

'Leightinii' Flowers bright reddish-crimson, profusely born, dwarf habit.

'Bougainvillei' Dwarf and compact with violet flowers.

'Campbellii' With large purple flowers, somewhat taller than the species.

Figure 3–30 *Aubrieta deltoidea* (life-size)

CULTURE: Soil: Well-drained sandy loam is best; pH of 6.5 to 7.5 is acceptable.

Water: Relatively tolerant to drought; watering thoroughly prior to flowering is beneficial, especially if weather is hot and dry.

Light: Plants are best in full sun or light shade. Light shade is preferred in areas of hot summers.

PATHOLOGY: Diseases: Damping off may occur if planted in poorly drained soils.

Pests: Aphids and soil mealybugs.

MAINTENANCE: Remove dried flower stalks after bloom to enhance appearance.

PROPAGATION: Cuttings: Take cuttings immediately following bloom. They root quite well under mist with a 1000 ppm IBA/talc treatment in a very porous medium.

Division: simple division is usually most effective in spring.

Seed: Seed sown in spring germinates in 2 to 3 weeks and flowering occurs the following spring.

*A*urinia (aw-rin´i-á) Meaning unclear. A small genus of about seven species of perennial or biennial herbs that are native to central and southern Europe and Turkey.

FAMILY: CRUCIFERAE: Mustard family.

A. saxatilis (saks-a-til´is) Found among the rocks (Figure 3–31).

COMMON NAME: Golden tuft, basket of gold, golden alyssum, rock madwort, gold dust, madwort.

NATIVE HABITAT: Europe.

HARDINESS: Zone 3.

HABIT: Horizontally growing, low, mat forming, herbaceous ground cover.

SIZE: 6 to 10 inches tall, spreading about 12 to 18 inches wide.

RATE: Relatively slow; space plants from 3 to 4 inch diameter containers 10 to 14 inches apart.

LANDSCAPE VALUE: Fine for accent in border among lower covers such as *Sedum* or *Sempervivum* or as a colorful rock garden specimen. Often used to excellent effect as a general cover in terrace or atop retaining wall, where the cascading flowers provide a delightful display of springtime color. Plantings in the south frequently die due to excess heat. No foot traffic.

FOLIAGE: Evergreen, grayish, white tomentose on both sides, 2 to 5 inches long, about 1/2 inch across;

basal leaves spatulate, sinuate to repandidentate, arranged in tufted rosettes; stem leaves smaller, linear-oblanceolate, opposite, with long and deeply grooved petioles.

INFLORESCENCE: Dense panicles on 12 inch stems; flowers pale golden-yellow, regularly symmetrical, sepals 4, petals 4, about 1/4 inch across, born midspring.

FRUIT: Suborbicular, glabrous silique, not of ornamental value.

OF SPECIAL INTEREST: Reported to have been used as a cure for hydrophobia and according to myth, madwort possessed powers that calmed the troubled mind. I think it would be wise to find out if the plant is edible or not before giving it a try.

SELECTED CULTIVARS AND VARIETIES: 'Citrina' With many light yellow flowers.

'Compacta' Most frequently cultivated, this cultivar has light yellow flowers and grows more densely than the species.

'Golden Globe' Reaching only 6 inches tall, with bright yellow flowers.

'Plena' With double yellow flowers.

'Nana' More compact and lower growing than the species.

'Silver Queen' Flowers lemon-yellow; habit compact.

CULTURE: Soil: Nearly any soil is acceptable as long as it drains well. A pH from 5.5 to 7.0 is preferred.
Moisture: Quite tolerant of dry conditions, but best when soil is kept slightly moist.
Light: Full sun to light shade, flowering more profusely in full sun.

PATHOLOGY: Diseases are not a common problem. Pests: Aphids and soil mealybugs.

MAINTENANCE: Mow plantings following bloom to keep plantings looking neat and to promote compactness and stimulate new growth.

PROPAGATION: Cuttings: Softwood cuttings should be taken in late spring to early summer and stuck in a very porous medium under mist. Treating with 1000 ppm IBA/talc is to some benefit.
Division: Simple division is an effective means to propagate small quantities anytime.
Seed: Seeds germinate in 3 to 4 weeks at a soil temperature of 70° to 85°F. Sow seed in late summer, and mature flowering plants can be expected the following year. Cultivars, however, must be propagated vegetatively.

✛✛✛✛✛✛✛✛✛✛✛✛✛✛✛✛✛✛✛

*B*accharis (bak´a-ris) From Greek mythology, Bacchus, the god of wine, a spicy extract from some species having been used for mixing with wine. This genus is composed of about 350 species of dioecious, deciduous, evergreen shrubs that are native to North and South America.

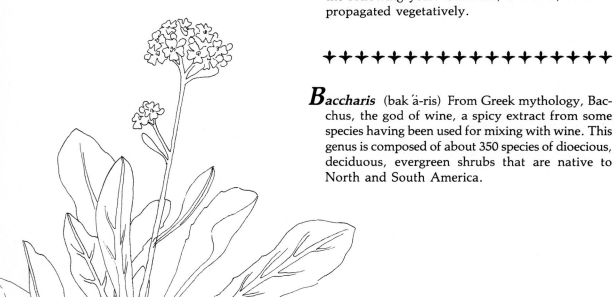

Figure 3–31 *Aurinia saxatilis* 'Compacta' (life-size)

FAMILY: ASTERACEAE (COMPOSITAE): Sunflower family.

B. pilularis (pil-ū-lā´ris) With globular fruit (Figure 3–32).

NATIVE HABITAT: Dry regions of Oregon and California.

HARDINESS: Zone 7.

HABIT: Low, prostrate shrub.

SIZE: 1 to 2 feet tall, spreading to 10 feet across.

RATE: Relatively fast growing; space plants 3 to 6 feet apart, from 1 to 2 gallon sized containers.

LANDSCAPE VALUE: Recommended as a fire-resistant plant for cover around building perimeters, as a bank stabilizer, and as general cover for large areas. Highly resistant to salt spray and thus quite useful in coastal regions. Considered by some to be the most dependable ground cover for desert regions. No foot traffic.

FOLIAGE: Evergreen, cuneate-obovate, to 3/4 inch long, dark green, alternate, slightly glutinous, sinuately toothed.

INFLORESCENCE: Heads, usually solitary, sometimes in panicles, originating from leaf axils at ends of twigs; flowers whitish-yellow in late summer, of little ornamental value.

FRUIT: Achenes 4 to 10 nerved, not of great ornamental significance. Female plants produce a messy cottony material; therefore, male clones are usually propagated.

SELECTED CULTIVARS AND VARIETIES: Var. *consanguinea* Reported to be more compact than the species.
'Pigeon Point' With larger, lighter green leaves and more aggressive nature. Reported to reach 9 feet wide in 4 years, with a mounding habit.
'Twin Peaks #2' With smaller dark green leaves and moderate growth rate; becoming 6 to 12 inches high by 6 or more feet across.

CULTURE: Soil: Best when planted in dry, well-drained, sandy soils with a pH of 4.5 to 7.5.
Moisture: Extremely drought tolerant, little or no supplemental water needed once established.
Light: Full sun required for best growth.

MAINTENANCE: Water regularly the first year until established, thereafter only occasionally to remove dust from the foliage. Shear or mow plants in early spring to keep them looking neat.

PATHOLOGY: Diseases: Black mold and rust.

PROPAGATION: Softwood and hardwood cuttings are the most common propagation methods. Generally, they are rooted in summer under mist and treated with a root inducing compound of 1000 to 3000 ppm IBA.

*B**ellis* (bel´is) From Latin, *bellus*, pretty. This genus is composed of about 15 species of annual and perennial herbs that are native to Europe, the Mediterranean region, and America.

FAMILY: ASTERACEAE (COMPOSITAE): Sunflower family.

B. perennis (per-en´is) Perennial (Figure 3–33).

COMMON NAME: English daisy, true daisy, marguerite (from Latin, meaning pearllike).

NATIVE HABITAT: East and west coasts of North America.

HARDINESS: Zones 6 to 10.

HABIT: Clump forming, horizontally spreading, low, tufted ground cover.

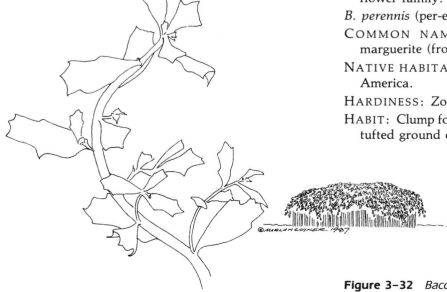

Figure 3–32 *Baccharis pilularis* (life-size)

SIZE: To 6 inches tall.

RATE: Relatively slow to moderate; often self-seeding; space plants from 2 1/2 to 3 inch diameter containers 8 to 10 inches apart.

LANDSCAPE VALUE: This species and its cultivars are best used as edging for paths and borders. They may also be used as facing to shrubs and can be effective as small-scale general covers where they combine well with spring flowering bulbous plants. They tolerate no foot traffic.

FOLIAGE: Arranged in a basal rosette, dark green, spatulate to obovate, 1 to 2 inches long, crenate-toothed margins, glabrous to pubescent.

INFLORESCENCE: Involucre, radiate, 1 to 2 inches wide, borne on 3 to 6 inch pubescent scapes, hemispherical or campanulate; involucral bracts are leaflike, equal or unequal, in 2 rows; disc flowers are yellow and bisexual; ray flowers are white or pink, female, and arranged in a single row. Blooming mid to late spring, then sparingly throughout summer and fall.

FRUIT: Achenes, not ornamentally significant.

OF SPECIAL INTEREST: *Bellis perennis* is the "true daisy" often referred to in literature, especially poetry.

SELECTED CULTIVARS AND VARIETIES: 'Dresden China' Light pink ray flowers.
'Montrosa' Dark red, double flowers.
'Prolifera' With secondary flower heads coming from axils of involucral bracts.
'Rob Roy' With crimson, double flowers.
'Rosea' With rose-pink ray flowers.
'Tuberosa' Rose colored, quilled ray flowers.

CULTURE: Soil: Best in organically rich, moist, fertile soils. Seems to be adaptable to pH.
Moisture: not very tolerant to drought; regular supplemental irrigation is needed.
Light: Full sun to light shade in cool climates; light to moderate shade in southern areas.

PATHOLOGY: Diseases: Blight, leaf spots, root rots, aster yellows, powdery mildew.
Pests: No serious pests reported.

MAINTENANCE: Divide plants every 2 to 3 years to stimulate growth. Often listed as hardy to zone 3 or 4, but will not survive unless transplanted to a cold frame or mulched heavily.

PROPAGATION: Division: Simple clump division is usually practiced in spring.
Seed: Seed should be sown in fall in a cold frame or unheated poly house. Germination occurs the following spring and seed may respond to light. Cultivars do not come true from seed.

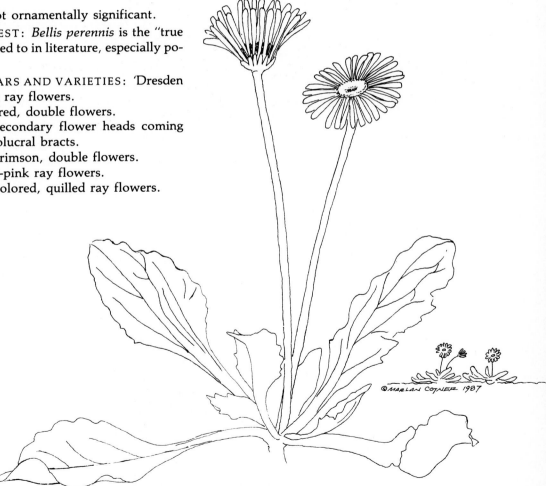

Figure 3–33 *Bellis perennis* (life-size)

✝✝✝✝✝✝✝✝✝✝✝✝✝✝✝✝✝

B*erberis* (bĕr´bĕr-is) From the Arabic name *ber-berys*. This genus is rather large, being composed of around 500 species of usually thorny shrubs. Most are native to South and North America, eastern Asia, Europe, and North Africa.

FAMILY: BERBERIDACEAE: Barberry Family.

LANDSCAPE VALUE: Low growing selections are excellent for dwarf hedging or specimen planting, especially in a rock garden setting. Mass planted, they can be used as general covers for larger areas and sloping banks. They do not tolerate foot traffic, nor can humans tolerate walking on them.

OF SPECIAL INTEREST: The wood, bark, and roots of many barberries are yellow and have been used in dyeing. While the fruit of barberries is often left untouched by birds, their thorny branches often serve as protective sites for nest construction.

*   *   *

B. thunbergii (thun-bĕr´ji-i) After Thunberg, a botanist. The species is a broad, often globose, thorny shrub with green foliage. It is suitable for specimen, border, or hedge planting and is widely cultivated. Recently, dwarf cultivars of this species have been accepted as ground covers (Figure 3–34). The species is described as follows:

NATIVE HABITAT: Japan.

COMMON NAME: Japanese barberry.

HARDINESS: Zones 4 to 8.

HABIT: Ground covering selections are generally upright branched, low spreading, dwarf shrubs.

RATE: Ground covering types generally grow very slowly, in many years only putting on about 1 to 2 inches of growth in the north. Space plants from 1 to 2 gallon sized containers on centers that roughly correspond to two-thirds of their expected width at maturity.

FOLIAGE: Alternate, simple, deciduous, rhombic-ovate, bright green in summer, changing to various shades of orange, red, and purple in autumn.

STEMS: Many stemmed, dark red in maturity, usually single spined, but occasionally multiple spined.

INFLORESCENCE: Umbellate fasicles, usually composed of 2 to 5 perfect, yellow flowers; flowers are 1/3 to 1/2 inch long, occasionally solitary, of limited ornamental worth, born midspring.

FRUIT: Ellipsoid berry, bright red, shiny, about 1/3 inch long, effective in fall and persisting into midwinter.

SELECTED CULTIVARS AND VARIETIES FOR GROUND COVER USE: 'Crimson Pygmy,' often listed as 'Atropurpurea Nana' and several other erroneous names, this selection reaches 1 1/2 to 2 feet high and 2 1/2 to 3 feet across. Foliage is deep purplish, dense, attaining its best color in full sun.

Figure 3–34 *Berberis thunbergii* 'Crimson Pygmy' (life-size)

'Globe' Green leaved, globose habit, reaching 2 to 3 feet high by 4 to 6 feet across.

'Kobold' Selected by VanKlavern of Boskoop, Holland, around 1960, this low growing barberry displays rich green foliage, flowers and fruits very little, and reaches 2 to 2 1/2 feet high.

'Minor' Originating at the Arnold Arboretum as a seedling in 1892, this selection is characterized by smaller foliage, fruits, flowers, and habit than the species. It is very compact, reaching about 3 1/2 feet high by 5 feet across.

* * *

B. × *stenophylla* 'Irwinii' (sten-ō-fil´á) Narrow-leaved.

COMMON NAME: Irwin's rosemary barberry.

HYBRID ORIGIN: *B. darwinii* × *B. empetrifolia.* Introduced 1969.

HARDINESS: Zone 5.

HABIT: Low, broad, shrubby ground cover.

SIZE: 18 inches tall by 1 1/2 to 2 1/2 feet across.

RATE: Moderate, space plants from gallon-sized containers 1 1/2 to 2 feet apart.

FOLIAGE: Evergreen, 1 inch long by 1/4 inch wide, alternate or whorled, mostly entire, apex 3 lobed, lobes cuspidate, oblanceolate in outline, base revolute, dark green above, grayish-green below, glandular dotted on both sides.

STEMS: Red-brown, somewhat hairy, with three parted spines.

INFLORESCENCE: Pendulous racemes composed of 7 to 14 yellow flowers; flowers are about 1/2 inch in diameter, regular, six petaled, six stamened, blooming in spring.

FRUIT: 1/2 inch diameter, globose, black to purplish berry.

Generally Applicable to Both Species

CULTURE: Soil: Tolerant to a wide variety of soils, with pH optimally from 6.0 to 6.5.
Moisture: Extremely tolerant to drought; little or no supplemental irrigation is needed once established.
Light: Full sun to light shade; foliage color is generally at its best in full sun.

PATHOLOGY: Generally, these selections are quite resistant to bacterial and fungal diseases. Pests include aphids, mealybugs, webworm, scales, and nematodes.

MAINTENANCE: Little or no maintenance is required.

PROPAGATION: Cuttings: Softwood cuttings in late spring or early summer are most easily rooted before they become stiff. A root inducing preparation of 2000 to 5000 ppm IBA/talc is helpful in speeding the process, but is not a necessity.

B*ergenia* (bẽr-gen´i-á) Named for Karl August von Bergen, an 18th century botanist and physician of Frankfort, Germany. The genus is composed of approximately 12 species of perennial herbs that have thick rhizomes and develop into clumps or broadly spreading colonies. They are, for the most part, native to the temperate regions of Asia. Many catalogs inaccurately list cultivars of Bergenia as cultivars of *Saxifraga.*

FAMILY: SAXIFRAGACEAE: Saxifrage family.

HABIT: Low, sometimes clump forming, laterally spreading herbaceous ground covers that produce new plants from rhizomes.

RATE: Moderate; space plants from pint- or quart-sized containers 12 to 15 inches apart.

LANDSCAPE VALUE: General ground covers for small- to moderate-sized areas, often used near the front of a border as facing or along walks as edging. Occasionally, they are attractively combined with hostas, ferns, daylilies, or liriope. No foot traffic. Very reliable and long lived.

* * *

B. cordifolia (kôr-di-fō´li-á) Heart shaped, the leaves (Figure 3–35).

COMMON NAME: Heartleaf bergenia, pig squeak.

NATIVE HABITAT: Siberia.

HARDINESS: Zones 2 or 3 to 8 or 9.

SIZE: To 12 inches tall, spreading 1 to 2 feet across.

FOLIAGE: Evergreen, shiny, medium green becoming purplish bronzy in fall, fleshy, thickened, conspicuously veined, 12 inches long, 6 to 8 inches wide, rounded or cordate at the base, crenulate-serrate margins, glabrous, frequently bullate (covered with small superficial undulations), with long and thick petioles that are sheathed at the base.

INFLORESCENCE: Densely nodding, scapose cyme that reaches up to 16 inches tall; flowers are clear rose colored, 1/4 to 1/2 inch across, bisexual, regularly symmetrical, with 5 short, broad sepals, petals 5 and rounded, 10 stamened, effective midspring for several weeks; in California, blooming again in winter, and in the north not blooming reliably; foliage is its main attribute, thus blooming or not this species is worth growing.

FRUIT: Capsule, not ornamentally significant.

SELECTED CULTIVARS AND VARIETIES: 'Alba' With white flowers.

'Morning Red' Foliage bronzy green; flowers dark purplish red.

'Perfect' With large purplish bronze leaves and purplish red flowers.

'Rothblum' With red flowers.

* * *

B. crassifolia (kras-si-fō ′ li-à) Thick leaved.

COMMON NAME: Leather bergenia, winter blooming bergenia.

NATIVE HABITAT: Siberia and Mongolia.

HARDINESS: Zones 2 or 3.

SIZE: To 12 inches tall, spreading 1 to 1 1/2 feet across.

FOLIAGE: Evergreen, basal, elliptic to oblong or obovate, to 8 inches long and 4 1/2 inches wide, shallowly serrate margins, glabrous and glandular dotted on both sides, medium green.

INFLORESCENCE: Scapose, upright arching panicles, to 18 inches high; flowers pink to purplish, to 3/4 inch in diameter, regular, petals 5, stamens 10, effective midwinter to early summer in warm climate, then blooming sparsely thereafter.

FRUIT: Capsule, of no ornamental significance.

HYBRIDS: There are many hybrid varieties of bergenia of which the origin is not always clear. Here are descriptions of some of the more common: 'Evening Glow' With dark green, large, rounded leaves and flowers that are rosy-red on 12 inch scapes; hardy to zone 3.

'Sunningdale' Flowers reddish pink on tall scapes; hardy to zone 3.

'Schmidtii' Large leaved and dark green with toothed margins; flowers clear pink and large.

Generally Applicable to Both Species

CULTURE: Soil: Bergenias perform well in any soil that is well drained with a pH ranging from slightly acid to slightly alkaline.

Moisture: Relatively tolerant to drought, yet in dry soils the growth is slowed considerably. Deep watering during drought periods to maintain soil in a slightly moistened condition is recommended. Similarly, because best performance is attained when soil is maintained relatively moist, planting in wind exposed sites is not encouraged.

Light: Full sun to light shade in the north, light to moderate shade in southern locations.

PATHOLOGY: Diseases: Leaf spot.

Pests: Aphids, thrips, mealybug, weevils, slugs, snails.

MAINTENANCE: Mulch plantings lightly in winter to shade the leaves from drying sun if snow cover is unreliable. Pine or spruce branches are ideal for this purpose.

PROPAGATION: Division: Most common method; usually carried out in spring or fall.

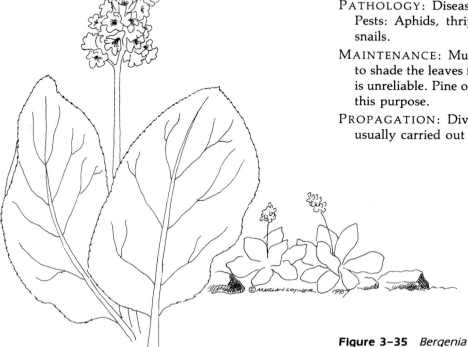

Figure 3–35 *Bergenia cordifolia* (0.33 life-size)

Seed: Collect seed as it ripens, store cool and dry, then sow the following spring. Chilling seed for a few weeks is helpful, and germination will occur over a period of 5 to 25 weeks at temperatures above 55°F.

✝✝✝✝✝✝✝✝✝✝✝✝✝✝✝✝

*B*runnera (brun-nē´ra) After Samuel Brunner, Swiss botanist. The genus is composed of only 3 species of perennial herbs from western Siberia and the eastern Mediterranean region.

FAMILY: BORAGINACEAE: Borage family.

B. macrophylla (mak-rō-fil´a) Large leaved (Figure 3–36).

COMMON NAME: Heartleaf brunnera, Siberian forget-me-not, Siberian bugloss, dwarf anchusa.

NATIVE HABITAT: Caucasus Mountains and Siberia.

HARDINESS: Zones 3 to 9.

HABIT: Dense, mounded, clump forming herbaceous ground cover.

SIZE: 18 to 24 inches tall, spreading 1 1/2 to 2 feet across.

RATE: Relatively slow; space plants from pint- to quart-sized containers 16 to 24 inches apart.

LANDSCAPE VALUE: Excellent when grown around the base of trees with coarse textured bark in naturalized settings. Also combines well with spring flowering bulbs, or grow alone as a general ground cover for moderate-sized areas. No foot traffic.

FOLIAGE: Deciduous, simple, alternate, basal leaves to 8 inches across, ovate, cordate, or reniform, long petioled, apex sharply pointed; leaf size and length of petiole decrease with distance from the stem until the upper leaves are about one-half the size of basal leaves and become sessile; all are dark blackish green and coarse textured.

STEMS: Slender, to 18 inches long, arising from central crown.

INFLORESCENCE: Coiled, one-sided cymes that are elevated above the foliage mass, uncoiling as flowers open; flowers are 1/8 to 1/4 inch wide, starlike, symmetrical, blue, effective mid to late spring.

FRUIT: Nutlets in groups of 4, not ornamentally significant, dispersed by ants.

CULTURE: Soil: Well-drained rich loam is best; tolerates relatively acidic soils.
Moisture: Needs constant soil moisture; irrigate as needed.
Light: Full sun to moderate shade; leaves may desiccate at the margins when grown in full sun.

PATHOLOGY: Crown rot can be counted upon should soil not drain adequately. No serious pests are reported.

MAINTENANCE: Every 2 to 3 years, it is wise to divide plantings to prevent overcrowding, especially if grown in confined areas. Yearly mulching in the spring with organic compost is usually beneficial.

PROPAGATION: Cuttings: Root cuttings are usually taken spring or fall.
Division: Commonly practiced in spring or fall.

Figure 3–36 *Brunnera macrophylla* (0.33 life-size); flowers enlarged to show detail.

Seed: Seed is commonly sown in fall in a cold frame or in seed trays in an unheated greenhouse or seed bed.

*C*alluna (kȧ-lū́ná) From Greek, *kalluno*, to cleanse, alluding to the use of branches of this plant as a broom. This genus is composed of a single species with many cultivars and varieties. They are usually low, shrublike, and quite colorful when in bloom.

FAMILY: ERICACEAE; Heath family.

C. vulgaris (vul-gā́ris) Common (Figure 3-37).

COMMON NAME: Heather, common heather, scotch heather, ling.

NATIVE HABITAT: Asia Minor and Europe.

HARDINESS: Zones 4 to 6 or 7.

HABIT: Laterally spreading alpine shrubs, often mound shaped.

SIZE: The species ranges from 18 to 36 inches tall, spreading 3 to 4 feet across. Cultivars vary somewhat in height and spread.

RATE: Slow to moderate growing; space plants 12 to 16 inches apart from quart- to gallon-sized containers.

LANDSCAPE VALUE: Exceptionally interesting general ground covers, especially when variously colored varieties are combined to cover large areas on open slopes. Often useful in borders and as edging plants. Will not tolerate foot traffic.

FOLIAGE: Scalelike, overlapping, closely packed, opposite, simple, four ranked, sessile, oblong-ovate, 1/25 to 1/8 inch long, base sagittate, puberulous to mostly glabrous, evergreen, medium green. Winter color varies from green to bronzy.

STEMS: Upright, branching, thin, terete; pith continuous, narrow, rounded in cross section.

INFLORESCENCE: One to 10 inches long, terminal, spikelike, one-sided racemes; flowers are perfect, urn shaped, calyx four lobed, corolla campanulate, about 1/8 inch across, rosy or purplish-pink in mid to late summer. Bees are attracted to the flowers in great numbers.

FRUIT: Four valved capsule in fall. Of no ornamental significance.

SELECTED CULTIVARS: There are probably in excess of 600 cultivars of *Calluna*, far more than can be mentioned here. Some of the most popular cultivars are listed here. Bloom times are given by

Figure 3-37 *Calluna vulgaris* (life-size)

the months of blooming period should they differ from the species. 'Alba' Flowers are white.

'Aurea' Flowers purple, blooming in August to October; height 18 inches; foliage is yellow in summer and russet in winter, this being its chief attribute.

'Country Wicklow' Flowers double and pink from August to October; inflorescence is 3 to 6 inches high.

'Cuprea' Flowers purple in August to October; height about 12 inches; foliage is golden, turning bronzy.

'Foxii Nana' Flowers purple from August to October; mature height of only about 4 inches; foliage bright green.

'H. E. Beale' Flowers pale pink from August to October; vigorous grower.

'J. H. Hamilton' Flowers double, pink in August to October; height about 10 inches; foliage dark green.

'Mair's Variety' Flowers white; mature height about 2 feet tall.

'Mrs. Ronald Gray' Very low, prostrate creeper, to about 4 inches tall; pale flowers; foliage emerald green.

'Nana Compacta' Foliage bright green; flowers pink; about 6 inches tall at maturity.

'Plena Multiplex' Flowers double, pink, effective from August to October; height to about 18 inches.
'Searlei Aurea' Flowers white from August to October; height about 12 inches.
'Robert Chapman' Flowers rose-purple in August to September; foliage is greenish-yellow and reddish in winter; height to about 10 inches.
'Tomentosa' Flowers lavender; height to about 10 inches.

CULTURE: Soil: Well-drained, infertile, sandy loam with good water holding capacity and a pH of 4.5 to 6.0 is desirable.
Moisture: Moderately tolerant of drought, but some shelter from strong winds is desirable. Water thoroughly during hot, dry weather.
Light: Full sun to light shade; flowering best in full sun.

PATHOLOGY: Diseases: Phytophthora wilt, leaf blight.
Pests: Aphids, Japanese beetle, two spotted mite, oystershell scale, chafer beetle larvae. Also, rabbits like to chew on the ends of stems.

MAINTENANCE: Do not overfertilize as leggy, open growth will be the result. Plants are best when grown in infertile soil. Mow or shear after flowering to increase density and compactness.

PROPAGATION: Cuttings: Take cuttings in late summer or fall from nonflowering shoots and treat with 1000 to 2000 ppm IBA/talc and 60°F bottom heat.
Seed: Expect variability. Cultivars must be propagated asexually.

✚✚✚✚✚✚✚✚✚✚✚✚✚✚✚✚✚

*C*ampanula (kam-pan ′ū-là) From Latin, *campanula*, meaning a little bell. This large genus is composed of about 300 species of annual, biennial, and perennial herbs. They are generally native to the northern hemisphere in the Caucasus, Balkan, and Mediterranean regions. Some are small and rosette forming, others are erect; some are slow spreading, and others invasive. Listed here are four of the most common species that make good ground covers. There are many more that are worth considering for this purpose. *Wyman's Gardening Encyclopedia* is a good starting point should one be interested in learning about other species.

FAMILY: CAMPANULACEAE: Bluebell family.

HABIT: Generally low spreading, carpetlike, trailing or tufted, herbaceous ground covers.

RATE: All bellflowers listed here have the potential to become invasive and should be planted where their spread can be controlled. Plants from pint- or quart-sized containers generally should be spaced 8 to 16 inches apart.

LANDSCAPE VALUE: These species and their cultivars make good general covers in smaller areas where their spread can be controlled. They are also useful as edging to walking paths and borders and work out quite well for facing dark-leaved shrubs, as their bright summer flowers are effective in lending color at a time when most shrubs are finished blooming. No foot traffic.

*   *   *

C. carpatica (kär-pat ′ik-à) Carpathian (Figure 3–38).

COMMON NAME: Carpathian bellflower, tussock bellflower, Carpathian harebell.

NATIVE HABITAT: Carpathians.

HARDINESS: Zones 3 to 9.

HABIT: Low, tufted, herbaceous ground cover.

SIZE: 12 to 18 inches tall, but generally 8 to 10 inches tall; spreading to about 18 inches across.

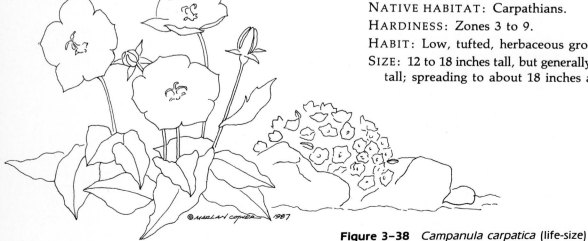

Figure 3–38 *Campanula carpatica* (life-size)

FOLIAGE: Ovate-triangular to broadly lanceolate, about 2 inches long, long petioled, deeply serrate, light green.

FLOWERS: Held on slender, elongated axillary pedicles, flowers bell-like, hence the common name; corolla blue-lilac, 1 to 2 inches across, held above foliage, broadly campanulate, effective midsummer.

SELECTED CULTIVARS AND VARIETIES: 'Alba' with white flowers.
'Blue Carpet' Flowers deep blue; lower growing and more compact than the species.
'Blue Clips' With large violet-blue cupped flowers; blooming midsummer to early fall.
'China Doll' With large mauve flowers; lower growing, to about 8 inches tall.
'Wedgewood Blue' With pale blue-violet flowers; reaching only about 6 inches tall.
'Wedgewood White' With white flowers; reaching only about 8 inches tall.
'White Carpet' Low, compact, with white flowers.
'White Clips' Similar to 'Blue Clips' but flowers white.

* * *

C. elatines var. *garganica* (ē-lā´tinz) Meaning unknown.

COMMON NAME: Adriatic bellflower, Dalmatian bellflower.

NATIVE HABITAT: Region around the Adriatic Sea.

HARDINESS: Zone 6, often successful to Zone 4 if given good drainage.

HABIT: Trailing herbaceous ground cover.

SIZE: 6 to 8 inches tall, spreading to 12 inches across.

FOLIAGE: Long petioled, cordate-ovate to orbicular, to 1 inch long by about 1 inch across, margin sharply toothed, grayish green.

INFLORESCENCE: Racemes or narrow panicles; flowers starlike, violet-blue with white center, blooming early summer to early fall.

RATE: Slower growing; space somewhat closer than other species.

* * *

C. portenschlagiana (pôr-ten-shlā-gē-ā´na) Commemorative.

COMMON NAME: Dalmatian bellflower.

NATIVE HABITAT: Yugoslavia.

HARDINESS: Zone 4.

HABIT: Tufted, herbaceous ground cover that spreads by rhizomes.

SIZE: 6 to 9 inches tall.

FOLIAGE: Semievergreen, alternate, orbicular to cordate, margins crenate to serrate, glabrous to glandular dotted both sides, 1 to 1 1/2 inches long by about the same across, medium to dark green.

INFLORESCENCE: Racemes; flowers regular, about 3/4 inch across, purple, from midspring to late summer and sometimes extending to fall.

* * *

C. poscharskyana (po-shär-she-ā´na) Commemorative.

COMMON NAME: Serbian bellflower.

NATIVE HABITAT: Yugoslavia.

HARDINESS: Zone 3.

SIZE: Reaching 12 inches tall, but usually 4 to 6 inches tall.

HABIT: Trailing herbaceous ground cover.

FOLIAGE: Evergreen, to 1 1/2 inches long by the same width, alternate, irregularly toothed, cordate to ovate or orbicular, glabrous and glandular dotted both sides, medium green.

STEMS: Slender, trailing, to 2 feet in length.

INFLORESCENCE: Racemes or panicles; flowers with corolla of light lavender-blue, broadly funnelform, 1 to 1 1/4 inches long, effective early to midsummer.

Generally Applicable to All Species

CULTURE: Soil: Well-drained sandy loam is ideal; adaptable to pH ranging from slightly acid to slightly alkaline.
Moisture: Not especially tolerant to drought; watering on a weekly basis during summer is adequate to maintain optimal appearance in most areas.
Light: In northern and coastal areas, they are best in full sun, while light shade is preferred in warmer areas.

PATHOLOGY: Diseases: Crown rot, leaf spots, rusts, powdery mildew, root rot.
Pests: Foxglove aphid, onion thrips, slugs.

MAINTENANCE: Generally maintenance free, except for having to occasionally cut back as they outgrow their bounds.

PROPAGATION: Division: Simple division in spring or fall is the standard means of propagation.
Seed: The species and even some cultivars (although incorrectly identified as such when propagated sexually) come true from seed. Generally, seed from commercial suppliers is sown in spring and germinates in 2 to 3 weeks at temperatures ranging from 70° to 85°F.

++++++++++++++++++

Carissa (ká-ris´á) Meaning unknown. The genus is made up of about 34 species of evergreen, highly branched, often spiny, small trees and shrubs. They are native to the Old World tropics.

FAMILY: APOCYNACEAE: Dogbane family.

C. grandiflora (gran-di-flō´rá) Large flowered (Figure 3–39).

COMMON NAME: Natal plum, amatungulu.

NATIVE HABITAT: South Africa.

GROUND COVERING TYPES: The species *C. grandiflora* is a tall (to 18 feet), dense, many branched shrub. It is not used for ground cover but, rather, low growing selections of the species are useful for this purpose. They are generally described as follows:

HARDINESS: Zones 7 to 10.

HABIT: Low, spreading shrubs.

SIZE: 18 to 30 inches tall, spreading 3 to 5 feet across.

RATE: Slow to establish, then moderate thereafter; space plants from gallon-sized containers 2 to 3 feet apart.

LANDSCAPE VALUE: Often used next to a wall or in a terrace where the branches can cascade over, or along a walkway, patio border, or building entrance for edging. Tolerant to salt spray, thus of value to the coastal garden. No foot traffic.

FOLIAGE: Evergreen, opposite, entire, ovate, to 3 inches long by 2 inches wide, leathery, dark green.

STEMS: Many branched, thorny.

INFLORESCENCE: Terminal or pseudoaxillary clusters; flowers fragrant, to 2 inches across, corolla tube to 1/2 inch long, salverform, bisexual, waxy white, effective in spring.

FRUIT: Ovoid-ellipsoid berry to 2 inches long, scarlet, 6 to 16 seeded.

SELECTED CULTIVARS AND VARIETIES: 'Boxwood Beauty' Mounding habit to 18 inches tall, compact, with dark green leaves.
'Green Carpet' Small dark green leaves; height 12 to 18 inches; spreading 3 to 4 feet across.
'Horizontalis' Stems trailing, leaves arranged very densely, reaching 18 to 24 inches tall and spreading to 4 feet across.
'Minima' Dwarf, to about 12 inches tall; leaves and flowers reduced in size.
'Prostrata' Low and spreading; to 2 feet tall by 5 feet across.
'Tuttlei' Very compact, mounding to about 3 feet and spreading to 5 feet across; heavier flowering and fruiting.

CULTURE: Adaptable to most soils but benefited by incorporation of humus. Tolerating saline soils.
Moisture: Tolerance to prolonged drought is fair, although plants perform and look better when occasionally watered thoroughly.
Light: Full sun to moderate shade.

Figure 3–39   *Carissa grandiflora* 'Tuttle' (life-size)

PATHOLOGY: No serious diseases or pests reported.

MAINTENANCE: As they arise, prune out the occasional branches that grow upward.

PROPAGATION: Cuttings: Cuttings are ordinarily rooted under mist in early summer with the aid of 3000 ppm IBA/talc.

*C*arpobrotus (kär-pō-brō′tus) From Greek *karpos*, a fruit, and *brotus*, edible (the fruit is edible). This genus is composed of about 29 species of succulent subshrubs, usually with prostrate growth habit. Along with the plants in the genera *Cephalophyllum, Delosperma, Drosanthemum, Lampranthus, Malephora, Mesembryantemum, Oscularia, Ruschis,* and others, species of *Carpobrotus* are often commonly called ice plants.

FAMILY: AIZOACEAE: Carpetweed family.

HABIT: Low, fleshy, trailing ground covers.

RATE: Relatively fast growing; space plants 16 to 24 inches apart from pint- to quart-sized containers.

LANDSCAPE VALUE: Often used as general covers for larger areas. Excellent along coast, where they are able to withstand shifting sand. They function well as soil binders for moderate slopes, but on very steep slopes, the weight of the plants may cause them to pull away from their crowns.

*   *   *

C. *chilensis* (chil-en′sis) Meaning unclear.

COMMON NAME: Trailing sea fig, California beech plant.

NATIVE HABITAT: Australia.

HARDINESS: Zones 9 and 10.

SIZE: 12 to 18 inches tall, by 8 + feet across.

FOLIAGE: 1 1/4 to 2 inches long, fleshy, triangular, straight-keeled without teeth, green, sometimes with reddish tones.

STEMS: Sturdy, prostrate, rooting as they creep to 3 feet or more in length.

FLOWERS: Solitary, to 3 1/2 inches across, nearly sessile or sessile, sepals unequal and leafy, corolla magenta, fairly sparse, borne in summer, opening in the sun.

FRUIT: 8 to 10 celled, fleshy, indehiscent capsule, edible, not of ornamental significance.

*   *   *

C. *edulis* (ed′ū-lis or e-du′lis) Edible, the fruit (Figure 3-40).

COMMON NAME: Trailing Hottentot fig.

NATIVE HABITAT: South Africa.

HARDINESS: Zones 9 and 10.

SIZE: 12 to 18 inches tall, by 8 + feet across.

FOLIAGE: Evergreen, succulent, 3 to 5 inches long by 1/2 inch wide, spreading, opposite, entire, somewhat curled, keeled beneath which is finely serrate and reddish, three angled, linear, glabrous and glandular dotted all sides, gray-green, reddish under stressed conditions.

STEMS: To 3 or more feet long, prostrate, branching.

FLOWERS: Solitary, to 4 inches across, regular, many petaled; corolla light yellow, rose-pink, or purple; borne in summer, opening in the sun.

Figure 3-40 *Carpobrotus edulis* (life-size)

FRUIT: 10 to 15 celled, fleshy, indehiscent capsule, edible but not of great ornamental value.

OTHER SPECIES: Other species that are less common, but useful as ground covers include: *C. acinaciformis* (gound, Hottentot fig) With rich magenta flowers.
C. deliciosus (four fig, goukum) Flowers magenta to purple.
C. muirii With smaller leaves and pink-purple flowers.

Generally Applicable to All Species

CULTURE: Soil: Adaptable to most soils, excellent drainage a necessity.
Moisture: Able to withstand moderate periods of drought once established. Benefits from occasional thorough watering.
Light: Full sun.

PATHOLOGY: No serious diseases or pests reported.

MAINTENANCE: Little or no special maintenance required.

PROPAGATION: Cuttings: Easily rooted throughout the year.
Divisions: Simply divide plants at any time.

✛ ✛ ✛ ✛ ✛ ✛ ✛ ✛ ✛ ✛ ✛ ✛ ✛ ✛ ✛ ✛ ✛ ✛

*C*eanothus (sē-ȧ-nō´thus or kē-an-ō´thus) Ancient Greek name, supposedly applied to a now unknown plant by Theophrastus, the Greek philosopher. This genus is composed of around 55 species of evergreen and deciduous shrubs or small trees. Sometimes they are thorny. Native to North America.

FAMILY: RHAMNALES: Buckthorn family.

HABIT: Prostrate, spreading shrubs.

LANDSCAPE VALUE: Excellent general covers for larger areas on flat or sloping surfaces. Often used for accent in formal landscapes. Best suited for coastal landscapes. No foot traffic.

*   *   *

C. gloriosus (glō-ri-ō´sus) Glorious, showy (Figure 3–41).

COMMON NAME: Point Reyes ceanothus, Point Reyes creeper.

NATIVE HABITAT: Southern California.

HARDINESS: Zone 7.

SIZE: To 24 inches tall, spreading to 5 feet across, rooting as it creeps.

Figure 3–41   *Ceanothus gloriosus* (life-size)

RATE: Relatively fast; space plants from gallon-sized containers 3 to 4 feet apart.

FOLIAGE: Evergreen, opposite, broadly elliptic to round, to 1 1/2 inches long, glossy dark green, leathery, usually with spiny-toothed margins.

INFLORESCENCE: Fluffy heads; flowers bisexual, lavender-blue, effective in spring.

FRUIT: 3 lobed capsule, not ornamentally significant.

SELECTED CULTIVARS AND VARIETIES: 'Anchor Bay' Somewhat lower and more compact than the species.
exaltatus 'Emily Brown' Height 18 to 36 inches by 6 to 10 feet across; leaves dark green; flowers blue-violet.
Var. *porrectus* Low sprawling, 12 to 30 inches tall by 7 or more feet across; fine textured, with deep blue flowers.

*   *   *

C. griseus horizontalis (gri-sē´us or gris-ē´us) Meaning unknown.

COMMON NAME: Carmel creeper, Yankee Point ceanothus.

NATIVE HABITAT: California.

HARDINESS: Zone 8.

SIZE: 18 to 30 inches tall, spreading 8 to 15 feet across.

RATE: Relatively fast, space plants from 1 to 2 gallon sized containers 4 to 6 feet apart; space different cultivars according to width of mature spread.

FOLIAGE: Evergreen, alternate, broadly ovate, to 1 3/4 inches long, undulate, revolute base, pubescent below, dark glossy green.

INFLORESCENCE: Panicles, to 2 inches long; flowers blue, effective in spring.

FRUIT: 3 lobed capsule, not ornamentally significant.

SELECTED CULTIVARS AND VARIETIES: 'Compacta' Foliage small and densely arranged; reaching only 1 foot tall and spreading to only 3 feet across. 'Hurricane Point' More vigorous, height to 4 feet tall and spreading to 30 feet across with open habit. 'Yankee Point' To 3 feet high and spreading to as much as 15 feet across.

* * *

C. prostratus (pros-trā´tus) Lying flat.

COMMON NAME: Squaw-carpet ceanothus, mahala mat.

NATIVE HABITAT: West coast of United States.

HARDINESS: Zone 7.

SIZE: 3 to 6 inches tall by up to 10 feet across.

RATE: Relatively fast; space plants from 1 to 2 gallon sized containers 3 to 4 feet apart.

FOLIAGE: Evergreen, opposite, obovate, coarsely spiny-toothed, leathery textured, dark green, glossy.

INFLORESCENCE: Terminal clusters; flowers small, blue, effective in spring.

FRUIT: Similar to other species.

* * *

C. thyrsiflorus var. *repens* (thr̄-si-flō´rus) Inflorescence thyrsoid.

COMMON NAME: Creeping blue blossom ceanothus.

NATIVE HABITAT: Pacific Coast from Oregon to California.

HARDINESS: Zone 8.

SIZE: 18 to 24 inches tall, spreading to 5 feet across.

RATE: Moderate; space plants from gallon-sized containers 3 to 3 1/2 feet apart.

FOLIAGE: Evergreen, alternate, oblong, to 2 inches long, finely toothed, glossy green.

INFLORESCENCE: Panicles; flowers small, blue, effective in spring.

FRUIT: Similar to other species.

OTHER SPECIES: Other species that have ground covering potential include: *C. divergens confusus* Commonly called Rincon ceanothus; habit is low, dense, mat forming; reaching 3 inches tall and spreading to 3 feet across.

C. maritimus Generally described as reaching 2 1/2 feet tall and spreading to 6 feet across. Foliage is green above and whitish or grayish below; flowers white to pale lavender in color.

Generally Applicable to All Species

CULTURE: Soil: Light, sandy, well-drained soil with a pH of 6.5 to 7.0 is preferred.
Moisture: Quite tolerant of short periods of drought; two deep waterings per month in summer is usually adequate, and overwatering should be avoided as root rot will occur.
Light: Full sun (best) to light shade.

PATHOLOGY: Leaf spots, powdery mildew, rust, root rot, crown gall.
Pests: Scale insects.

MAINTENANCE: Relatively free of special maintenance. An annual shearing in spring will help keep them neat and compact.

PROPAGATION: Cuttings: Cuttings root readily when taken in midsummer to early fall, and treatment with a root inducing preparation has proved beneficial.
Seed: Seed should be soaked 12 hours in hot water (190°F), then stratified for 3 months at 40°F prior to sowing.

*C*ephalophyllum (sef-à-lō´phil-um) Meaning unclear. This genus is composed of more than 70 species of dwarf succulents with low, spreading or decumbent habit. They are native to South Africa.

FAMILY: AIZOACEAE: Carpetweed family.

C. alstonii (all´stun-i-ī) Memorial (Figure 3–42).

COMMON NAME: Mesembryanthemum, ice plant, vygie.

NATIVE HABITAT: South Africa.

HARDINESS: Zones 9 to 10.

HABIT: Low, clump forming, succulent ground cover.

SIZE: 3 to 5 inches tall, spreading 15 to 18 inches across.

RATE: Moderate; set plants from 3 to 4 inch diameter containers 6 to 10 inches apart.

LANDSCAPE VALUE: General cover for medium-sized areas. Level or gently sloping dry ground or rock gardens are natural sites for this species. No foot traffic.

Figure 3–42 *Cephalophyllum alstonii* (life-size)

FOLIAGE: Erect, inward curving, 2 3/4 inches long by 3/8 inch across, nearly cylindrical, apex briefly acuminate, gray-green, dark dotted, evergreen.

STEMS: Prostrate, succulent, cylindrical, rooting as they creep, gray-green, becoming gray with age.

FLOWERS: Solitary, to 3 1/4 inches in diameter, corolla rich wine-red, effective in late winter to midspring, opening in the sun.

FRUIT: Many-celled 12 to 20 valved capsule, valves winged.

SELECTED CULTIVARS AND VARIETIES: 'Red Spike' Clumping with a height of 3 to 5 inches, spreading 15 to 18 inches across; foliage bronzy-red, clawlike, erect; flowers bright cerise red.

OTHER SPECIES: Many other species of *Cephalophyllum* are worth using as ground covers. Among them are the following:
C. anemoniflorum Procumbent, with salmon-pink flowers.
C. aureorubrum Decumbent; glaucous-green foliage; yellow to rose colored flowers.
C. cupreum Low growing; apricot to coppery colored flowers.
C. procumbens Low, trailing; foliage glaucous-rose; flowers golden-yellow.
C. subulatoides Trailing; light gray-green foliage; flowers purple-red with yellow center.

Generally Applicable to All Species

CULTURE: Soil: Adaptable to a wide range of well-drained soils.
Moisture: Withstanding short periods of drought, periodic thorough watering is beneficial.

C*erastium* (ser-ras'ti-um) From Greek, *keras*, a horn, the seed capsules of some appearing like horns as they emerge from the calyx. This genus is composed of about 60 species of annual and perennial herbs. Sometimes they are woody at the base and are usually low and mat forming.

FAMILY: CARYOPHYLLACEAE: Chickweed or pink family.

C. tomentosum (tō-men-tō'sum) Downy (Figure 3–43).

COMMON NAME: Snow-in-summer.

NATIVE HABITAT: Europe.

HARDINESS: Zone 3.

HABIT: Low, creeping, matlike ground cover.

SIZE: 3 to 6 inches tall, spreading to 12 inches across.

RATE: Relatively fast; space plants from 3 to 4 inch diameter containers 10 to 12 inches apart.

LANDSCAPE VALUE: General cover for both large and small areas, excellent on sunny banks, and low enough to be of use between stepping stones. No foot traffic.

FOLIAGE: Evergreen, simple, opposite, entire, oblong or oblanceolate to lanceolate, sometimes spatulate, to 1 inch long and 1/4 inch wide, densely covered with white hairs, soft textured; color is frosted looking gray-green.

STEMS: Pubescent, squarish, 6 to 10 inches long.

INFLORESCENCE: Terminal cymes, 3 to 15 flowered, each 1/2 to 3/4 inch wide, 5 petaled, regular, white, borne profusely in mid-late spring.

FRUIT: Cylindrical capsule, dehiscent, often curled, not ornamentally significant.

CULTURE: Soil: Sandy loam, well drained with low fertility is best.
Moisture: Very tolerant to drought once established; however, appearance is enhanced greatly if watered during dry periods.
Light: Requires full sun.

PATHOLOGY: Diseases: Damping off is a common fungal problem if soil is poorly drained.
Insects: Aphids, mealybug, soil mealybug.

MAINTENANCE: Mow in spring before growth starts. This will rejuvenate growth, promote compactness, and make plantings look neat. Mow again following bloom to get rid of dried flowers.

PROPAGATION: Cuttings: Softwood cuttings root readily in a very porous medium; watch for signs of damping off. The use of 1000 ppm IBA/talc may speed the process.
Division: Fall is generally most successful.
Seed: Commercial supplies are available. Germination takes 2 to 4 weeks.

Ceratostigma (ser-at-ō-stig´mȧ) From Greek, *keras*, a horn, and *stigma*, alluding to the hornlike branches of the stigmas. This genus contains about eight species of perennial herbs and shrubs. They are native to China, the Himalayas, Ethiopia, and Somalia.

FAMILY: PLUMBAGINACEAE: Leadwort family.

C. plumbaginoides (plum-ba-ji-noi´dez) Plumbagolike (Figure 3–44).

COMMON NAME: Blue ceratostigma, blue leadwort, Chinese leadwort, dwarf plumbago, plumbago.

NATIVE HABITAT: China.

HARDINESS: Zone 5 or 6 to 9. Mulching or good snow cover usually necessary in Zone 5.

HABIT: Low, erectly branched, rhizomatous ground cover.

SIZE: Reaching from 6 to 10 inches tall, spreading indefinitely.

RATE: Moderate to relatively fast; space plants from 2 to 4 inch diameter containers 12 to 18 inches apart.

LANDSCAPE VALUE: Excellent contrast is derived when plumbago is planted underneath shrubs and small trees of different color or texture. Often it is used for edging borders and walks. Combines well with other ground covers such as ajuga, vinca, and sempervivum. No foot traffic. Excellent late-season floral display.

FOLIAGE: Nearly evergreen, alternate, entire, simple, oblanceolate to obovate or spatulate, glabrous and glandular dotted both sides, ciliate with reddish hairs, to 1 1/2 inches long by about 3/4 inch across, medium green, becoming bronzy and tinged purplish after frost, then turning brown soon after.

STEMS: Branching, herbaceous becoming woody at the base, glabrous or sparsely hairy, somewhat

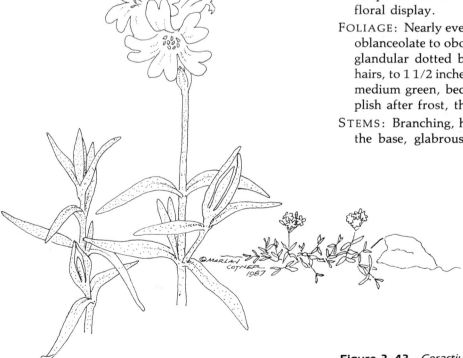

Figure 3–43 *Cerastium tomentosum* (life-size)

zigzag striped longitudinally green and red; new
shoots are late to emerge in spring.

INFLORESCENCE: Dense terminal axillary clusters;
flowers with persistent reddish bracts; corolla about
1/2 inch across, stellate and salverform, regular,
5-lobed, brilliant gentian blue, prolonged blooming period from midsummer to frost, opening with
mid-morning sun.

FRUIT: 5 valved capsule, not of ornamental significance.

CULTURE: Soil: Rich, well-drained, acidic loam is
best, but this plant is quite adaptable provided
drainage is good.
Water: Periodic watering in hot, dry weather is
beneficial, but usually not necessary for the plant's
survival.
Light: Full sun to moderate shade.

PATHOLOGY: Diseases: Root rot is prevalent in
poorly draining soils.
Pests: Aphids often feed on new growth, but the
damage that they do is insignificant.

MAINTENANCE: Mow as low to the ground as
possible in early spring before new growth begins.

PROPAGATION: Cuttings: Cuttings should be taken
in midsummer and rooted in a fairly porous
spagnum-based medium. They root rather easily,
but benefit from the use of a rooting compound of
3000 ppm IBA/talc. Discontinue mist as soon as
rooted or root rot is sure to occur.
Division: Simply divide plants at any time.

Chamaemelum (kam-e-mel´um) From Greek,
chama, ground, likely in reference to the habit of
its species (i.e., growing close to the ground). The
genus is composed of three species of perennial
herbs that are native to western Europe and the
Mediterranean region.

FAMILY: ASTERACEAE (COMPOSITAE): Sunflower family.

C. nobile (nō´bil) Noble, referring to the fact that the
flowers are very large for a plant so small (Figure
3–45).

COMMON NAME: Roman camomile, Russian
camomile, garden camomile.

NATIVE HABITAT: Europe, Azores, and North
Africa.

HARDINESS: Zone 4.

HABIT: Low, horizontally spreading, dense, herbaceous ground cover.

SIZE: 3 to 6 inches tall, rarely to 1 foot; spreading
to 12 inches across.

RATE: Moderate to relatively fast; space plants from
pint- or quart-sized containers 10 to 14 inches apart.

LANDSCAPE VALUE: Can be grown in large or small
areas as a general use ground cover. Occasionally,
it is treated as a lawn substitute where it is mowed
regularly, yet it tolerates only limited foot traffic.
It is also useful between stones in rock gardens.

FOLIAGE: Evergreen, alternate, entire, finely
dissected, to 2 inches long, 2 to 3 pinnately divided;
segments each about 1/8 inch long, linear-bristly
or filiform, bright green, aromatic.

STEMS: Decumbent or ascending, downy.

INFLORESCENCE: Heads to 1 inch across that are
elevated on stems to 1 foot, some buttonlike in appearance, others daisylike, receptacle conical; disc
flowers yellow, ray flowers white, effective in
summer.

Figure 3–44 *Ceratostigma plumbaginoides* (life-size)

FRUIT: 3 angled, smooth achene, not ornamentally significant.

OF SPECIAL INTEREST: Flower heads are used in medicinal herb preparations and, when dried, to make tea.

SELECTED CULTIVARS AND VARIETIES: 'Flore Pleno' With double flowers.

CULTURE: Soil: Best in light textured, sandy soils that have excellent drainage; tolerates slight alkalinity.

Moisture: Roots are deeply set, which helps to make this plant very drought tolerant; additional water is needed only during extended dry periods.

Light: Full sun to light shade.

PATHOLOGY: No serious diseases reported. Pests include slugs and snails.

MAINTENANCE: Mow plantings in spring to rejuvenate growth and in summer after bloom to remove dried flowers.

PROPAGATION: Division: Simple division is usually carried out in spring or fall.

Seed: The species is easily propagated by seed, which is naturally self-sowing.

Cistus (sis´tus) Ancient Greek name for the rock rose. The genus is composed of about 17 species of low and medium-sized woody shrubs. Most of them are evergreen or semievergreen and are native to the Mediterranean region.

FAMILY: CISTACEAE: Rock rose family.

HABIT: Low, spreading shrubs.

RATE: Relatively fast; space plants from gallon-sized containers 2 1/2 to 3 1/2 feet apart.

FRUIT: 5 to 10 valved capsules, not ornamentally valuable.

LANDSCAPE VALUE: Considered to be fire resistant and thus well suited to use along walks in public places such as parks and entrances to buildings. Similarly, they are well suited to face building foundations, and the fact that roots penetrate deeply makes them useful to bind soil on sloping ground. No foot traffic. Tolerant of high salt levels.

* * *

C. crispus (kris´pus) Curly, the leaves being wavy.

COMMON NAME: Wrinkleleaf rock rose.

NATIVE HABITAT: Western Mediterranean region.

HARDINESS: Zone 7.

SIZE: 18 to 24 inches tall, spreading 4 to 5 feet across.

FOLIAGE: Evergreen, simple, sessile, oblong to elliptic, to 1 1/2 inches long, undulate, 3 nerved, villous-tomentose, deep green, slightly aromatic.

INFLORESCENCE: Cymes, to 7 flowered; flowers to 2 inches across, sepals 5 and densely villous, corolla red-pink with yellow center, effective late spring to midsummer.

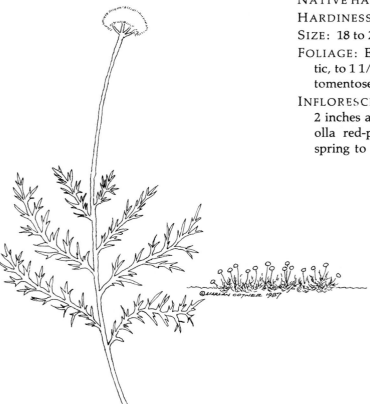

Figure 3–45 *Chamaemelum nobile* (life-size)

Figure 3–46 *Cistus salvifolius* (life-size)

*　　*　　*

C. salvifolius (sal-vi-fō´lē-us) Sagelike, the leaves (Figure 3–46).

COMMON NAME: Sageleaf rock rose.

NATIVE HABITAT: Western Mediterranean region.

HARDINESS: Zone 7.

SIZE: 2 feet tall, spreading to 6 feet across.

FOLIAGE: Elliptic to oblong, with petioles to 1 3/4 inches long, stellate-tomentose, scabrous and rugose above, gray-green, evergreen.

INFLORESCENCE: Terminal cymes or solitary, to 3 in a group; flowers to 2 inches across, sepals 5, corolla white with yellow spots at the base of petals, borne profusely from late spring to midsummer.

OTHER SPECIES: *C. × hybridus* (hīb´ri-dus) Hybrid group. Commonly called hybrid rock rose, this hybrid originated from the cross *C. populifolius × C. salviifolius*; it is hardy northward to zone 7, reaches 18 to 30 inches tall, and is characterized by evergreen, ovate, to 2 inch long leaves and 1 1/2 inch diameter flowers that are white with yellow centers.

Generally Applicable to All Species

CULTURE: Soil: Best in well-drained, coarse, slightly alkaline soil.

Moisture: Extremely drought tolerant once established. In areas where temperatures are moderated by the ocean, they will seldom, if ever, require irrigation. In hot summer areas, occasional deep watering is helpful.

Light: Full sun.

PATHOLOGY: Not hindered by serious diseases or pests.

MAINTENANCE: Occasionally, prune out dead branches, and shear to encourage growth and compact habit.

PROPAGATION: Cuttings: Softwood cuttings may be rooted in early summer.

Layering: Rooted plantlets are normally dug in spring.

Seed: Sow seed in containers as bare root seedlings do not transplant well.

*C**omptonia** (komp-tō´ni-à) Named after Henry Compton, Bishop of London. The genus is composed of only one species.

FAMILY: MYRICACEAE: Myrtle family.

C. peregrina (pâr-e-gri´nà) Exotic, foreign (Figure 3–47).

COMMON NAME: Sweetfern.

NATIVE HABITAT: North America from Nova Scotia to North Carolina, westward to Indiana and Michigan.

HARDINESS: Zone 2.

HABIT: Fernlike, low growing shrub.

SIZE: 2 to 4 feet tall, spreading from 4 to 8 feet across.

RATE: Moderate; space plants from gallon-sized containers 2 1/2 to 3 1/2 feet apart.

LANDSCAPE VALUE: Extensive matlike root structure makes this plant an excellent soil binder for use along highway banks or on naturalized slopes. Relatively tolerant to high levels of salinity. No foot traffic.

FOLIAGE: Deciduous, alternate, simple, pinnatifid, linear-oblong in outline, deeply notched, to 4 1/2 inches long by 1/2 inch across, pubescent, dark green, aromatic.

STEMS: Yellowish to green or reddish brown in youth, later yellowish or brown, pubescent with resinous dots, aromatic.

INFLORESCENCE: Male flowers are arranged in termial catkins up to 3/5 inch long; female flowers 1/2 to 1 inch in diameter, arranged in rounded clusters, yellow green to brown, effective in midsummer.

FRUIT: Glabrous, brown, 3/16 inch long, olive-brown nutlets, effective late summer to midautumn.

SELECTED CULTIVARS AND VARIETIES: Var. *asplenifolia*, having smaller leaves, less pubescent and with smaller catkins than the species.

CULTURE: Soil: Tolerating a range of soil types from moist, highly organic soils to infertile, dry, and sandy soils. Slightly acidic reaction is preferred (pH 5.5 to 6.5 is optimal).

Moisture: Once established, this plant is quite tolerant to drought. An occasional deep watering in summer may be beneficial.

Light: Full sun.

PATHOLOGY: Diseases: Rust.

No serious pests reported.

MAINTENANCE: No special maintenance needs; an annual light shearing may be desired to keep plants compact.

PROPAGATION: Cuttings: Root cuttings are said to be an effective method. Reportedly, it is best to dig them in late winter or early spring. Sections should be about 4 inches long and at least 1/16 inch wide. Stem cuttings root poorly except when they are quite soft. Treatment with 3000 ppm IBA/talc is helpful.

Division: Remove large divisions from main clump and transplant immediately.

Seed: Seed germinates poorly unless acid scarified and then soaked in gibberellic acid at 5000 ppm for 24 hours.

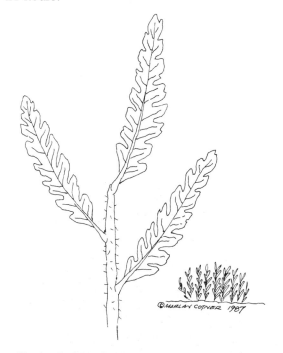

Figure 3–47 *Comptonia peregrina* (life-size)

+++++++++++++++++++

*C*onvallaria (con-va-lā´ri-à) From Latin, *convallium*, a valley, the natural habitat of the lily of the valley; a genus of three species of perennial, rhizomatous herbs native to temperate regions.

FAMILY: LILIACEAE: Lily family.

C. majalis (mà-jā´lis) Of May, in reference to the flowering period (Figure 3–48).

COMMON NAME: Lily of the valley, conval lily, mayflower, muguet (Canada), Lilia-o-ke-awawa (Hawaii).

NATIVE HABITAT: Europe, Asia, and Eastern N. America.

HARDINESS: Zones 3 to 7.

SIZE: 6 to 8 inches tall, spreading indefinitely.

HABIT: Low, lilylike, horizontally spreading, upright, stemless, rhizomatous ground cover.

RATE: Moderate to relatively fast; space divisions with 2 to 3 pips (large underground buds) 6 to 8 inches apart or 8 to 12 inches apart for plants from 2 to 3 inch diamteter pots.

LANDSCAPE VALUE: Commonly used as a general cover for large or small areas. Performs well underneath trees (supplement with a little more fertilizer than otherwise) and combines with low shrubs, ferns, and ivies. Its quality of binding soil makes it particularly useful on shaded slopes. No foot traffic.

FOLIAGE: Deciduous, grouped 2 to 3, arising from pips, 4 to 8 inches long, 1 to 3 inches wide, lanceolate-ovate to elliptic, entire, parallel veined, glabrous, medium to dark green, beginning to deteriorate in late summer and becoming progressively less attractive until necrosis in fall. Often they will die back before the first frost.

INFLORESCENCE: Terminal, erect, one-sided racemes; flowers bell shaped, nodding, 1/4 inch across, bracts lanceolate, bisexual, regularly symmetrical, corolla waxy, white, slightly recurved, fragrant, blooming mid to late spring.

FRUIT: Seldom fruiting as cross pollination is necessary and most plantings arise from a single clone. When they do fruit, an orange-red, 1/4 inch diameter, many seeded, round berry is formed, and is effective in fall. It reportedly is poisonous, and for this reason small children should not be allowed to play near plants that have fruit.

OF SPECIAL INTEREST: In the past the rhizomes and roots of this plant were used to treat heart maladies. In larger doses, these same plant structures are

Figure 3–48 *Convallaria majalis* (0.66 life-size)

poisonous, as are leaves and flowers and water in which plants have been stored. Apparently, larger doses cause irregular heartbeat and digestive and mental disorder. The toxic substances are reported to be digitalislike glycosides.

SELECTED CULTIVARS AND VARIETIES: 'Aurea Variegata' Rare, foliage longitudinally variegated green and creamy-yellow.
'Fortin's Giant' Leaves larger (to 10 inches) and flowers larger as well.
'Flore pleno' With double flowers.
'Fortunei' Also with larger leaves and flowers.
'Prolificans' With double flowers.
'Rosea' Flowers light purplish-pink, less vigorous.

CULTURE: Soil: Tolerant to most soils, but best in moderately acidic, organically rich, moist, well-drained loam.
Moisture: Will survive extended periods of drought without supplemental water; however, best appearance is attained when leaves are kept turgid with adequate supplemental watering as needed.
Light: Light to dense shade is best; in full sun the foliage usually turns yellow or brown in mid-summer.

PATHOLOGY: Diseases: Stem rot (pansies are also a host, so do not plant the two near each other), anthracnose, leaf spot, leaf blotch, crown rot; blossom failure occurs occasionally when soil does not drain well or when too much nitrogen is applied.
Pests: Lily of the valley weevil, nematode, Japanese weevil, slugs.

MAINTENANCE: In fall, remove dead leaves and topdress lightly with compost to supplement the soil. If flower production should decrease over time, dig up pips in spring and divide. Replant on 6 to 8 inch centers.

PROPAGATION: Division: Growth is rhizomatous with the pips on the distal end. Division is accomplished by separating the pips (with roots attached).

*C*oprosma (kop-ros´mȧ) From Greek *kopos*, dung, and *osme*, a smell, in reference to the unpleasant odor of some species. In this genus are about 90 species of evergreen shrubs and small trees that are native to Australia, New Zealand, and various islands of the Pacific.

FAMILY: RUBIACEAE: Madder family.

HABIT: Types for ground cover are low, prostrate shrubs.

LANDSCAPE VALUE: Valuable as bank covers near the shore, as they withstand wind and sea spray. Sometimes they are used along the base of a fence or low hedge for facing. No foot traffic.

*   *   *

C. × *kirkii* (kirk-i-ī) Honoring Sir John Kirk (1832–1922), English botanist (Figure 3–49).

COMMON NAME: Creeping coprosma, Kirk's coprosma.

HYBRID ORIGIN: *C. acerosa* × *C. repens* occurring naturally in areas where the two parents grow in close proximity.

HARDINESS: Zones 8 to 9.

SIZE: 2 to 4 feet tall, spreading to 10 feet across.

RATE: Moderate; space plants from 1 or 2 gallon sized containers 3 to 4 feet apart.

FOLIAGE: Evergreen, opposite, entire, quite variable but mostly oblong to oblanceolate, to 1 1/2 inches long and 1/4 inch across, midrib conspicuous, glabrous both sides, medium to yellow green with margins reddish.

STEMS: Stiff, pubescent, densely branched.

INFLORESCENCE: Cymes or flowers that are solitary; flowers are dioecious, funnelform, about 1/2 inch across, regular, with 4 to 5 lobed white corolla, 4 to 5 stamens, not of much ornamental worth.

FRUIT: Oblong, to 1 1/4 inch long, bluish, red-speckled, translucent, 2 celled drupe.

SELECTED CULTIVARS AND VARIETIES: 'Kirkii' A many branched prostrate to semierect selection that reaches 4 feet tall and 6 feet across. Foliage is linear-oblong to linear-lanceolate or narrowly linear-obovate and reaches 1/2 to 1 1/2 inches long by 3/16 to 1/3 inch across.

* * *

C. petriei (pē-trē´i) After Petrie.

COMMON NAME: Petrie's coprosma.

NATIVE HABITAT: New Zealand.

HARDINESS: Zone 7.

SIZE: To 3 feet tall, spreading to 6 feet across.

RATE: Moderate; space plants from gallon-sized containers 2 to 3 feet apart.

FOLIAGE: Minute, evergreen, linear-elliptic or oblong, to 1/4 inch long, hairy in youth, leathery, olive green.

STEMS: Slender, many branches, pubescent.

FLOWERS: Solitary, dioecious, not ornamentally significant.

FRUIT: Globose drupe, to 3/8 inch across, purplish-red or pale blue translucent color.

Generally Applicable to All Species

CULTURE: Soil: Tolerant of most soil types. Moisture: Once established, highly tolerant of drought, needing only occasional summer watering. One should not hesitate to use these plants in windswept locations. Light: Full sun to light shade.

PATHOLOGY: No serious diseases or pests.

MAINTENANCE: Shear branches lightly in spring to promote new growth and branching.

PROPAGATION: Cuttings: Softwood cuttings taken in early summer are most often used to propagate these species.

Coreopsis (kō-rē-op´sis) From Greek, *koris*, a bug or tick, and *opsis*, a resemblance, in reference to the appearance of the seed. This genus is composed of more than 100 species of annual and perennial herbs that are native to North and South America and Africa.

FAMILY: ASTERACEAE (COMPOSITAE): Sunflower family.

C. auriculata (ä-rik-ū-lā´ta) Ear shaped, the leaves (Figure 3–50).

COMMON NAME: Dwarf-eared coreopsis, dwarf-eared tickseed.

NATIVE HABITAT: Eastern United States from Virginia to Florida.

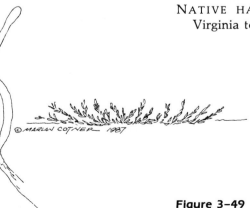

Figure 3–49 *Coprosma* X *kirkii* (life-size)

HARDINESS: Zone 4 or 5 to 9.

HABIT: Low, spreading, stoloniferous, matlike, herbaceous ground cover.

SIZE: 6 to 12 inches tall, spreading 12 + inches across.

RATE: Relatively slow; space plants from pint-sized containers 10 to 14 inches apart.

LANDSCAPE VALUE: Excellent general cover for small- to moderate-sized areas. No foot traffic.

FOLIAGE: Evergreen or semievergreen, simple or with 1 to 2 basal lobes, alternate, ovate to lanceolate, 2 to 5 inches long, about 1 inch wide, commonly pubescent below, glabrous above, margins ciliate, petioles reddish at base, medium green.

INFLORESCENCE: Solitary heads that range from 1 1/2 to 2 inches wide, on long stems; disc flowers

yellow, fertile; ray flowers yellow and sterile, both types with 4 to 5 lobed petals, borne in profusion spring to fall.

FRUIT: Winged achenes, not ornamentally significant.

SELECTED CULTIVARS AND VARIETIES: 'Nana' To 8 inches tall.

CULTURE: Soil: Tolerant of various soils, ranging from richly organic loams to very sandy loam, as long as drainage is good.
Moisture: Relatively tolerant to drought, requiring only infrequent summer watering.
Light: Full sun to light shade.

PATHOLOGY: Diseases: Leaf spot, rust, blight, root and stem rots, powdery mildew, wilt, beet curley top virus.
Pests: Aphids, leafhopper that is believed to inject a toxin in the pedicels that causes their necrosis, four-lined plant bug, spotted cucumber beetle, soil mealybug, mites.

MAINTENANCE: Mow plantings after they finish blooming to rid them of dried flower heads.

PROPAGATION: Division: Simply divide clumps in spring or fall.
Seed: Germination of seed takes from 2 to 3 weeks at 65° to 70°F.

*C*ornus (kôr´nus) From Latin, *cornu*, horn, in reference to the hard nature of the seed. This genus is composed of about 45 species of primarily shrubs, small trees, and rarely herbs. They are native to North America, Europe, Asia, South America, and Africa.

FAMILY: CORNACEAE: Dogwood Family.

* * *

Figure 3–50 *Coreopsis auriculata* (life-size)

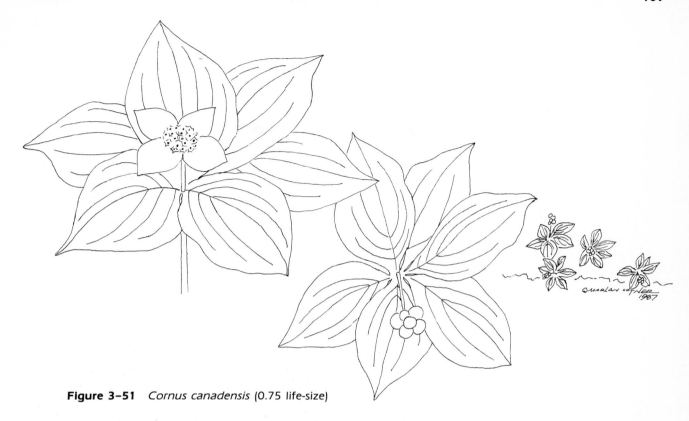

Figure 3–51 *Cornus canadensis* (0.75 life-size)

C. canadensis (kan-a-den´sis) Of Canada. (Figure 3–51).

COMMON NAME: Dwarf cornel, bunchberry.

NATIVE HABITAT: Mountainous woodlands of eastern North America and eastern Asia.

HARDINESS: Zones 2 to 7.

HABIT: Low, rhizomatous, carpeting, herbaceous ground cover.

SIZE: 4 to 9 inches tall, spreading indefinitely.

RATE: Relatively slow; space plants from 3 to 4 inch diameter containers or from sods 10 to 12 inches apart.

LANDSCAPE VALUE: Exceptionally attractive general cover or facing for small deciduous shrubs and trees, especially in a naturalized setting. Its spring and summer light requirements make it quite compatible with a deciduous canopy (see light requirements under culture). No foot traffic.

FOLIAGE: Deciduous, whorled in groups of 4 to 6, simple, ovate to elliptic or obovate, 1 to 3 inches long, 1/2 to 3/4 inch inch wide, petioled, glabrous or slightly adpressed pubescent, 5 to 7 parallel veined, margin entire, from dull to glossy dark green, yellow to red fall color.

STEMS: Herbaceous, upright, spreading by woody rhizomes.

INFLORESCENCE: Dense heads, subtended by involucre of 4 to 6 large, petallike, showy white bracts; bracts are oval shaped, 1/3 to 1 inch long, borne in fours; flowers are greenish-yellow, effective midspring to early summer.

FRUIT: Bright red, berrylike drupes, 1/4 inch across, clustered, effective late summer to midautumn, highly ornamental.

OF SPECIAL INTEREST: The fruit is reported to be a favorite food of ruffed grouse.

CULTURE: Soil: Needing moisture holding, highly organic, acidic soil.
Moisture: not tolerant to hot or dry conditions; plant out of wind and keep the soil moist with irrigation as needed; syringing with water midday may extend the southern range.
Light: Light to moderate shade is needed in summer, while in springtime more sun is desirable.

PATHOLOGY: No serious diseases or pests reported.

MAINTENANCE: Occasional topdressing with acidified compost is a good practice to follow if growing this plant in amended soil.

PROPAGATION: Division: Relatively large divisions are needed to insure success. Spring seems to be the most favorable time.
Seed: Seed should be depulped, sown immediately before drying out, and covered with 3/4 inch of soil.

Stratified naturally outdoors, seed germinates the following spring, and plants take 3 years to mature and bloom. It is important to keep the seed moist during the entire procedure.

* * *

C. sericea 'Kelseyi' (sẽr-ē-sē´a) Meaning silky.

COMMON NAME: Kelsey's red-osier dogwood.

NATIVE HABITAT: The species is native to swampy areas of the eastern United States.

HARDINESS: Zones 2 to 7.

HABIT: Compact, low, spreading, stoloniferous, multistemmed, shrub.

SIZE: 18 to 30 inches tall by 2 feet across.

RATE: Moderate; space plants from gallon-sized containers about 1 1/2 feet apart.

LANDSCAPE VALUE: Fine for use as a dwarf shrub or massed in front of taller growing shrubs as a facing. No foot traffic.

FOLIAGE: Deciduous, opposite, simple, ovate to oblong-lanceolate, base rounded, petioled, margin entire, bright green above, pale green below, fall color brown to purple, or withering without any notable color.

STEMS: Slender, bright red in youth, becoming brownish red with age; either way, the contrast with snow is excellent in northern areas.

INFLORESCENCE: Flat topped cymose panicles, 1 1/2 to 2 inches across; flowers off-white, small, effective but not very showy, and not always occurring. When present they are born from midspring to early summer.

FRUIT: Two seeded, globose, white drupes that are seldom observed.

OTHER SPECIES: *C. mas* 'Nana' is a dwarf form of cornelian cherry dogwood. Habit is similar to *C. serecia* 'Kelseyi', slower growing and generally vertically branched.

CULTURE: Soil: Very adaptable to a wide range of soil types; at its best in moist, acidic to neutral soil and tolerant to swampy, saturated soils.
Moisture: By no means a plant that tolerates drought well; leaves quickly lose turgor in hot, dry weather. Water deep and often during summer.
Light: Full sun to light shade.

PATHOLOGY: Diseases: Leaf spot, crown canker, powdery mildew, twig dieback, blight of flower and stem, root rot, collar rot.
Pests: Borers, leaf miners, scales, leafhoppers, sawfly, whitefly.

MAINTENANCE: Shear annually to keep dense and compact.

PROPAGATION: Propagation is easy by either softwood or hardwood cuttings. Treatment with 3000 ppm IBA/talc is useful in either case. If softwood cuttings are rooted under mist, be sure to use a porous medium to help prevent disease.

*C*oronilla (kôr-ō-nil´la) Meaning a little crown, from Latin, *corona*, a crown or garland, in reference to the arrangement of the flowers. The genus contains about 20 species of shrubs and herbs that are native to the Old World.

FAMILY: LEGUMINOSAE: Pea Family.

C. varia (vãr-ē´a) Varying in color, the flowers (Figure 3–52).

COMMON NAME: Crown vetch; vetch refers to a group of legumes.

NATIVE HABITAT: Europe.

HARDINESS: Zones 3 to 10.

HABIT: Rhizomatous, sprawling, rather loose appearing ground cover.

SIZE: 12 to 24 inches tall, spreading indefinitely.

RATE: Relatively fast; may self-seed into surrounding areas. Space plants from 2 to 3 1/2 inch diameter containers 10 to 14 inches apart.

LANDSCAPE VALUE: Use for fast cover on large open areas. Its extensive root system makes this plant a valuable bank stabilizer, often used along highways. Accepts very limited foot traffic.

FOLIAGE: Deciduous, alternate, pinnately compound, fine textured, with 5 to 12 pairs of oblong, entire, 1/2 inch long emerald-green leaflets.

STEMS: Sprawling, to 2 feet.

INFLORESCENCE: Dense crownlike umbels, supported on lengthy axillary peduncles; flowers about 1/2 inch long, calyx 5 toothed, corolla light pink and white, effective from early summer to early fall.

FRUIT: Erect, long beaked, slender legume, not of ornamental significance.

SELECTED CULTIVARS AND VARIETIES: 'Penngift' A more vigorous selection.

CULTURE: Soil: Tolerant of most well-drained soils, growth is slowed somewhat in infertile soils; however, this is often desirable. A pH range of 6.0 to 7.5 is preferred.
Moisture: Tolerant to drought once established. Crown vetch is usually not damaged (just slowed) by hot dry periods of moderate duration.
Light: Full sun to light shade.

PATHOLOGY: No serious diseases or pests reported.

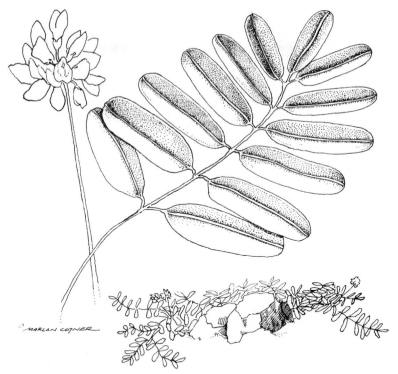

Figure 3–52 *Coronilla varia* (2X life-size)

MAINTENANCE: Mow plantings to the ground in late fall and bag the clippings. This prevents accumulation of dried stems, which can become a fire hazard and shelter for rodents.

PROPAGATION: Division: Simple division of individual or multiple crowns is effective anytime as long as moisture is adequate.

Seed: Seed is commercially available and inexpensive. Germination takes 2 to 3 weeks at 40°F and only about 1 week at 70°F. It is best to inoculate seed with the appropriate nitrogen fixing bacteria (most often supplied with the seed) at the time of sowing.

✝✝✝✝✝✝✝✝✝✝✝✝✝✝✝✝✝

Correa (kôr-rē´ȧ or kôr´ē-ȧ) Named after Jose Correa de Serra, a Portuguese botanist. This small genus is composed of about 11 species of trees and shrubs from Australia.

FAMILY: RUTACEAE: Rue or citrus family.

C. pulchella (pul-kel´ȧ) Beautiful, from Latin (Figure 3–53).

COMMON NAME: Australian fuchsia; flowers similar to fuchsia.

NATIVE HABITAT: Australia.

HARDINESS: Zones 9 to 10.

HABIT: Low, spreading shrub.

SIZE: 2 to 3 feet tall, spreading to 8 feet across.

RATE: Moderate; space plants from 1 to 2 gallon sized containers 3 to 4 feet apart.

LANDSCAPE VALUE: Useful as a general ground cover for large areas, especially on slopes. Avoid planting next to walls where heat and sun may be reflected. No foot traffic.

FOLIAGE: Evergreen, opposite, oblong, elliptic to ovate, to 1 1/2 inches long by 1/2 inch across, margin entire, mostly glabrous and dark green above, hairy and gray-green below, glandular dotted on both sides.

STEMS: Pinkish and hairy.

INFLORESCENCE: Cymose clusters; flowers regular, pendulous, about 1/4 inch across; corolla tubular, pinkish-red, 4 petaled, effective late fall to midspring.

FRUIT: Dehiscent capsule, 4 parted with 1 or 2 seeds per section. Not of ornamental worth.

CULTURE: Soil: Adaptable to most soils as long as drainage is good; tolerant of infertile and rocky soils.

Moisture: Relatively tolerant to drought; water

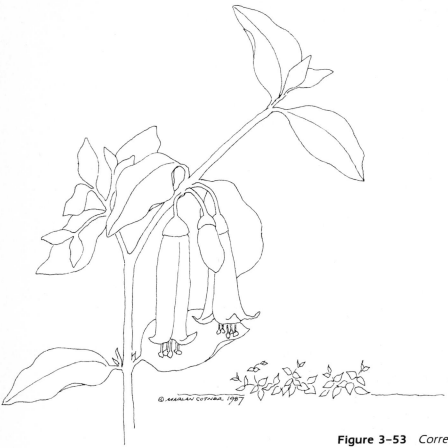

Figure 3–53   *Correa pulchella* (life-size)

thoroughly but only occasionally in hot dry periods. Avoid overwatering as tolerance to saturated soils is poor.

Light: Full sun to light shade.

PATHOLOGY: No serious diseases or pests reported.

MAINTENANCE: Occasional shearing keeps it looking neat and compact. Fertilizer should be applied sparingly, if at all.

PROPAGATION: Cuttings: Cuttings should be rooted in a very porous medium.

*C*otoneaster (kō-tō-nē-as´tẽr) From *cotoneus*, old Latin name for quince, and *aster*, probably a corruption of *ad instar*, meaning a likeness, that is, quincelike. The genus is composed of about 50 species of woody shrubs native to temperate Old World regions.

FAMILY: ROSACEAE: Rose family.

*   *   *

C. adpressa (ad-pres´sà) Close, pressed-down growth; or fruits closely pressed against the branch.

COMMON NAME: Creeping cotoneaster.

NATIVE HABITAT: Western China.

HARDINESS: Zones 4 to 7.

HABIT: Prostrate, creeping, rather coarse textured shrubs.

SIZE: 4 to 6 inches, sometimes to 10 inches high by 4 to 6 feet wide.

RATE: Relatively slow growing; space plants about 3 feet apart from gallon-sized containers.

LANDSCAPE VALUE: Excellent as a rock garden specimen or foundation facer. No foot traffic.

FOLIAGE: Deciduous, alternate, simple, 3/16 to 5/8 inch long, short petioled, broadly ovate, margins wavy, apex blunt or nearly pointed, tip bristled, dull dark green, glaucous, somewhat hairy below in youth. Fall color is sometimes reddish.

STEMS: Prostrate, stiff, slightly pubescent, rooting as they contact the soil.

FLOWERS: Solitary or in pairs, nearly sessile, corolla is 1/4 to 3/8 inch in diameter, pinkish, not of great ornamental significance, blooming late spring to early summer.

FRUIT: Subglobose, berrylike pome, 1/4 inch in diameter, bright red, usually containing two nutlets, effective late summer to early fall.

SELECTED CULTIVARS AND VARIETIES: 'Little Gem' A dwarf selection, reaching only about 6 inches high and spreading only about 1 1/2 feet wide, seldom flowering or fruiting. Space about 12 to 16 inches apart.
'Praecox' (early cotoneaster) More vigorous, reaching 3 feet high by 6 feet wide, leaves and fruit somewhat larger. Flowers are pink with a hint of purple.

*   *   *

C. apiculatus (ā-pik-ū-la´tus) From Latin *apicul*, tipped with a short abrupt point, in reference to the apex of the leaves (Figure 3–54).

COMMON NAME: Cranberry cotoneaster.

NATIVE HABITAT: Western China.

HARDINESS: Zones 4 to 7, farther south results in severe insect problems.

HABIT: Wide spreading shrub with stiff branching pattern. Tends to mound up in time.

SIZE: 2 to 3 feet high by about 5 or 6 feet across.

RATE: Growth rate moderate; space plants about 4 feet apart from gallon-sized containers.

LANDSCAPE VALUE: Often planted along wall where branches can gracefully cascade over. Fine

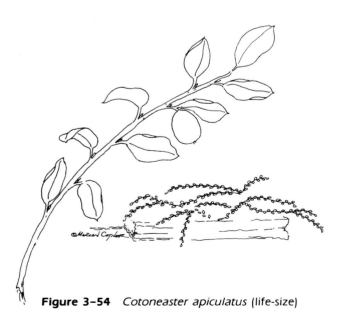

Figure 3–54 *Cotoneaster apiculatus* (life-size)

general cover for banks or large flat areas and around building foundations. Avoid planting along walks in public settings as debris accumulates and is difficult to remove. No foot traffic.

LEAVES: Deciduous, simple, alternate, suborbicular to orbicular-ovate, 1/4 to 3/4 inch long and about as wide, end pointed, margin wavy, shiny dark green and glabrous above, slightly gray pubescent below, bronzy-red or purplish in fall, often until late November.

STEMS: Reddish-purple with herringbone branching pattern, but less regular than *C. horizontalis*; adpressed pubescence.

FLOWERS: Usually solitary, perfect, corolla pinkish, small, effective late spring to early summer, not of great ornamental significance.

FRUIT: 1/4 to 1/3 inch diameter, cranberry red, globose pome, highly ornamental in late summer to early fall.

SELECTED CULTIVARS AND VARIETIES: 'Blackburn' A more compact form.
'Nana' Dwarf and compact, reaching 12 inches tall by 4 feet wide.
'Tom Thumb' A miniature selection that reaches 4 inches tall and 10 inches across.

*   *   *

C. congestus (kon-jes´tus) Crowded, the growth habit dense.

COMMON NAME: Pyrenees cotoneaster.

HABIT: Small, mound forming, ground hugging shrub.

SIZE: To 3 feet high by 6 feet across.

RATE: Slow growing; space plants 1 1/2 to 2 feet apart from gallon-sized containers.

LANDSCAPE VALUE: Use as a rockery specimen or as a general cover for small- to medium-sized areas; very little maintenance once established, as it grows very slowly.

COMMON NAME: Pyrenees cotoneaster.

NATIVE HABITAT: Himalayas.

HARDINESS: Zones 5 to 9.

LEAVES: Evergreen, ovate, very small, less than 1/2 inch long, bluish-green above, whitish below, glabrous.

FLOWERS: Usually solitary, pinkish, about 1/4 inch across, born in spring.

FRUIT: Brilliant red, 1/4 inch diameter, globose pomes that are effective in fall.

SELECTED CULTIVARS AND VARIETIES: 'Likiang' Lower and more compact.

* * *

C. conspicuus decorus (kon-spik´ū-us) From Latin meaning visible, remarkable, striking, conspicuous.

COMMON NAME: Necklace cotoneaster.

NATIVE HABITAT: Western China.

HARDINESS: Zones 6 to 8.

HABIT: Usually prostrate or low spreading.

SIZE: 1 1/2 to 2 feet tall by 2 to 3 feet across.

RATE: Moderate rate of growth; space plants 1 1/2 feet apart, from gallon-sized containers.

LANDSCAPE VALUE: Best used in rock garden as specimen as this species is somewhat loose and does not prevent weed penetration as well as others. No foot traffic.

LEAVES: Evergreen, simple, alternate, entire, elongate-elliptic, to 3/8 inch long, bristled below, dark green above.

FLOWERS: Solitary, white, borne in spring, but not of great ornamental importance.

FRUIT: Round to obovoid, 3/8 inch diameter red pome.

* * *

C. dammeri (dam´ĕr-i) Meaning unclear (Figure 3–55).

COMMON NAME: Bearberry cotoneaster.

NATIVE HABITAT: Central China.

HARDINESS: Zones 5 to 8.

HABIT: Very low, dense, prostrate shrub.

SIZE: 12 to 18 inches tall by 6 feet (or more) across.

RATE: Grows relatively fast; space plants 2 to 3 feet apart from gallon-sized containers.

LANDSCAPE VALUE: Among the finest evergreen groundcovers for mass planting to cover slopes and flat areas. Excellent for surrounding low shrubs and trees in an ornamental bed. No foot traffic.

LEAVES: Semievergreen, alternate, simple, ovate to rounded ovate, about 1 inch to 1 1/4 inches long by 1/4 to 5/8 inch wide, leathery, dark shiny green and glabrous above, slightly reticulate below, color changing to deep purplish in fall and winter.

FLOWERS: Solitary or in pairs, corolla white, 1/3 to 1/2 inch in diameter, quite pleasing, borne in midspring.

FRUIT: 1/4 inch wide, bright red pome, not usually produced in great quantity, but very attractive.

SELECTED CULTIVARS AND VARIETIES: 'Coral Beauty' (syn. 'Pink Beauty,' 'Royal Beauty') Low growing to about 16 inches tall, white flowers are borne profusely in early spring and result in many coral-red fruit in fall. Supposedly slightly more cold hardy than the species.

'Joergle' Selected for copious flower production and orangish-red fruit.

'Lowfast' Very fast spreading, with slightly smaller leaves.

'Major' More vigorous, with leaves 1 1/2 inches long.

'Moon Creeper' Very low, dense covering selection with dark, shiny green leaves.

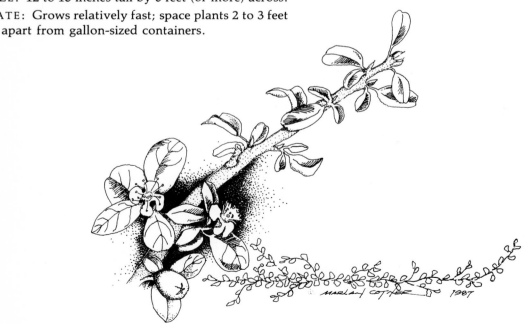

Figure 3–55 *Contoneaster dammeri* (0.75 life -size)

'Skogholmen' A low-growing vigorous selection from Sweden. Spring floral display is excellent, but fruit production is minimal.

* * *

C. horizontalis (hor-i-zon-tā´lis) Horizontal, habit of growth is horizontal.

COMMON NAME: Rock spray cotoneaster.

NATIVE HABITAT: Western China.

HARDINESS: Zones 4 or 5 to 7; growing farther south usually results in severe insect infestations.

HABIT: Low, dense, horizontally spreading shrub.

SIZE: 2 to 3 feet tall by 8 to 10 feet wide.

RATE: Slow to moderate growing; space 4 to 5 feet apart from gallon-sized containers.

LANDSCAPE VALUE: Plant next to wall in planter or terrace to take full advantage of stiff but graceful, arching branches. Suitable for covering large areas and banks and sometimes espaliered against a fence or wall.

LEAVES: Deciduous to semievergreen, alternate, simple, entire, short petioled, suborbicular to broadly elliptic, end pointed with bristle tip, 1/4 to 1/2 inch long, base wedge shaped, dark green, shiny and glabrous above, slightly hairy below, fall color is reddish-purple.

STEMS: Hairy, stiff, horizontal, often arching, secondary branches forming a herringbone pattern.

FLOWERS: Single or paired, perfect, inconspicuous alone but numerous and rather showy, en masse, borne late spring to early summer.

FRUIT: Glossy, bright red, 1/5 to 1/4 inch diameter globose pome, becoming effective in early fall and persisting well into winter.

SELECTED CULTIVARS AND VARIETIES: 'Purpusilla' Extremely prostrate with 1/4 inch long leaves; more compact in habit; with an ellipsoid fruit.
'Robusta' Heavy fruiting; more vigorous with more upright growth habit than the species.
'Saxatilis' Sparse fruiting form with compact and more prostrate habit and small leaves.
'Variegata' Slow growing; leaves edged with white, turning pink in autumn.

* * *

C. microphyllus (mi-krō-fil´us) Small leaved.

COMMON NAME: Little-leaved cotoneaster, small-leaved cotoneaster.

NATIVE HABITAT: Himalayas.

HARDINESS: Zones 5 to 9.

HABIT: Low, spreading or prostrate shrub.

SIZE: 3 feet high by 6 feet across.

RATE: Moderate growing; space plants 4 to 5 feet apart from 1 to 2 gallon-sized containers.

LANDSCAPE VALUE: Good covers for medium to large areas and banks, looking very pleasant when grown in an elevated planter or terrace. No foot traffic.

LEAVES: Evergreen, simple, alternate, ovate, to 5/16 inch long, about 1/4 inch wide, base obtuse, shiny dark green and glabrous above, gray pubescent below.

STEMS: Gray-brown, glabrous.

FLOWERS: Solitary or paired, sometimes in threes, perfect, corolla five petaled, white, effective in late spring and early summer.

FRUIT: Bright shiny red, 1/4 inch diameter pome.

SELECTED CULTIVARS AND VARIETIES: 'Cochleatus' Low growing, dwarf selection.
'Thymifolia' Very compact habit with pink flowers.

* * *

C. salicifolius cultivars (sal-is-i-fō´li-us) Willow (salix) leaved. The species is a large evergreen of spreading habit and height to 15 feet. Many cultivars, however, are prostrate growing and make excellent ground cover plants.

COMMON NAME: Willow-leaved cotoneaster.

NATIVE HABITAT: Western China.

HARDINESS: Zones 5 to 9.

SIZE: 12 to 15 inches tall, generally, and 5 or more feet across.

RATE: Moderate growth rate; space plants 3 to 4 feet apart from 1 to 2 gallon sized containers.

LANDSCAPE VALUE: Good general covers for along walks, on banks, or in large open areas. Often used in ornamental plantings surrounding low conifers and bordered by timbers or railroad ties, which they gracefully trail over. No foot traffic.

HABIT: Low spreading or prostrate shrubs.

LEAVES: Semievergreen or evergreen, narrowly lanceolate, 1 1/2 to 3 inches long, glabrous and lustrous green above, tomentose below.

FRUIT: Bright red, 1/4 inch diameter pomes, effective in fall and winter.

INFLORESCENCE: Wooly corymbs to 2 inches in diameter; flowers are small, white, effective in late spring or early summer.

SELECTED CULTIVARS AND VARIETIES: 'Autumn Fire' (syn. 'Herbstfeuer') A fine evergreen ground cover of open habit; foliage is glossy green and fruit scarlet; reaches 2 to 3 feet high (usually less), spreading to 8 feet, with foliage becoming purplish-red in late fall.

'Gnom' A low selection with mounding habit.

'Parkteppich' Similar, but taller growing than 'Autumn Fire.'

'Repens' Horizontal, trailing selection with shiny, dark green foliage and small red fruit (syn. 'Avondrood,' 'Dortmund,' and 'Repandens').

'Saldam' much like 'Autumn Fire,' but leaves stay green throughout the winter.

'Scarlet Leader' Low, trailing selection that reaches 6 to 12 inches high. Foliage is dark, shiny green, and resistance to insects and disease is very good.

Generally Applicable to All Species

CULTURE: Soil: Generally, best in fertile, medium, or light loam with good drainage. The pH is best between 5.5 and 7.0, but they will tolerate somewhat more acid or alkaline conditions.
Moisture: Generally, quite tolerant to dry conditions. Occasional deep watering is beneficial, but avoid overwatering as cotoneasters do not tolerate saturated soils.
Light: Full sun to light shade.
PATHOLOGY: Diseases: Leaf spots, canker, fire blight; *C. adpressa* and *C. microphylla* are said to be resistant to fire blight.
Pests: Scales (greedy, olive parlatoria, oystershell,

San Jose); cotoneaster webworm, sinuate pear tree borer, pear leaf blister mite, aphids (very common on new growth), root knot nematodes.
MAINTENANCE: Remove stray, broken, or dead branches as they occur; very low growing selections may be mowed to thicken their growth. Avoid stepping on their crowns. Others should be sheared annually.
PROPAGATION: Cuttings: Softwood cuttings in summer root readily under mist. A talc preparation of 3000 to 5000 ppm. IBA is helpful.
Layering: Simple layering is effective but time consuming.
Seed: Scarify for 90 minutes in concentrated sulfuric acid, then stratify for 3 to 4 months at 40°F, or stratify moist at 60° to 75°F for 3 to 4 months and follow with cold stratification at 40°F for 3 to 4 months prior to sowing.

*C*otula (kō´tew-là) From Greek, *kotyle*, a cup, possibly in reference to the cuplike heads of some of the species. The genus is composed of about 60 species of annual, biennial, and perennial herbs. They are aromatic when crushed and are native to various locations in the southern hemisphere.
FAMILY: ASTERACEAE (COMPOSITAE): Sunflower family.
C. squalida (skwol´i-dà) Squalid or lowly, the flower heads dingy (Figure 3–56).
COMMON NAME: New Zealand brass buttons.
NATIVE HABITAT: New Zealand.
HARDINESS: Zone 8.
HABIT: Dense, mat forming, rhizomatous, herbaceous ground cover.
SIZE: 2 to 4 inches tall, spreading to 2 feet across.
RATE: Relatively fast to invasive; space plants from 3 to 4 inch diameter containers 8 to 12 inches apart.
LANDSCAPE VALUE: Suitable for growing between stepping stones or as an alternative to turf or *Dichondra* in small scale lawns, as it tolerates moderate foot traffic.
FOLIAGE: Evergreen, fernlike, arranged in clusters at the nodes, soft to the touch, hairy, narrowly obovate in outline, deeply pinnatifid to pinnately dissected, to 2 inches long, bronzy green.
STEMS: Wiry, hairy, prostrate, and rooting as they trail.
INFLORESCENCES: Solitary heads, unisexual, to

Figure 3–56 *Cotula squalida* (life-size)

5/16 inch in diameter, buttonlike in appearance; flowers yellow, effective in summer.

FRUIT: Achene of little ornamental value.

OTHER SPECIES: *C. atrata* is a creeping, 6 inch tall, fleshy leaved, herbaceous perennial ground cover.

CULTURE: Soil: Best in moist, rich loam.
Moisture: Not markedly resistant to drought; regular summer watering is recommended.
Light: Full sun to light shade.

PATHOLOGY: No serious diseases or pests reported.

MAINTENANCE: Mow off flowers as they fade to neaten appearance.

PROPAGATION: Layering: dig up self-layering stems that have developed roots in spring or fall.

*C*rassula (cras´ū-lȧ or cras-ul-ȧ) From Latin, *crassus*, thick, referring to the thick or fleshy leaves. This genus is rather large, being composed of around 300 species of succulent, perennial herbs and shrubs. Most are native to Africa.

FAMILY: CRASSULACEAE: Stonecrop family.

C. multicava (mul-ti-cā´vȧ) Meaning much hollowed or with many cavities. The exact reference is uncertain (Figure 3–57).

NATIVE HABITAT: Southern Africa.

HARDINESS: Zones 9 to 10.

HABIT: Low, spreading, succulent ground cover.

SIZE: 8 to 10 inches tall, spreading indefinitely.

RATE: Rapid spreading; space plants from pint- or quart-sized containers 18 to 24 inches apart.

LANDSCAPE VALUE: General cover for small- to medium-sized areas. No foot traffic.

FOLIAGE: Opposite, succulent, evergreen, obovate, 1 to 3 inches long, 3/4 to 1 1/2 inches across, without mucro, base obtuse to retuse, with petiole, dark green.

STEMS: Fleshy, erect or somewhat decumbent, to 12 inches long, rooting as they touch the ground.

INFLORESCENCE: Thyrsoid to nearly umbellate, with bulbils, flowers 4 parted, corolla light pink, blooming late winter and spring.

FRUIT: Not ornamentally significant.

CULTURE: Soil: Tolerant to almost any type of soil, good drainage being essential.
Moisture: tolerant of considerable drought, as leaves and stems store great amounts of water.
Light: Full sun to light shade.

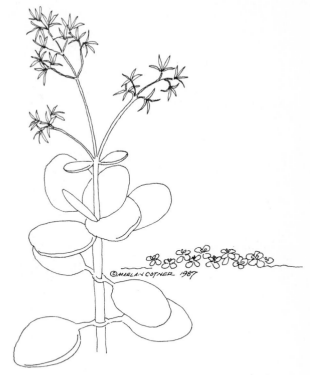

Figure 3–57 *Crassula multicava* (life-size)

PATHOLOGY: Diseases: Anthracnose, leaf spot, root rot.
Pests: Mealybugs and cyclamen mite.

MAINTENANCE: Trim shoots back as they outgrow their bounds.

PROPAGATION: Cuttings: Stem or leaf-bud cuttings can be taken anytime.
Division: Simple division is effective year round.
Seed: Reported to be a viable means of propagation. Ordinarily, germination takes 2 to 4 weeks at 75 °F.

*C*yclamen (si´klȧ-men or sik´lȧ-men) The name is a contraction of Greek, *kyklaminos*, from *kylos*, circle, in reference to the coiled stem of the seed vessel. The genus is composed of around 15 species of tuberous herbs that are native to central Europe and the Mediterranean region to Iran. Some species are common florists' plants.

FAMILY: PRIMULACEAE: Primrose family.

C. hederifolium (hed-ẽr-e-fō´li-um) Leaves like *Hedera* (Figure 3–58).

COMMON NAME: Baby cyclamen, ivy-leaved cyclamen, Neapolitan cyclamen.

NATIVE HABITAT: Central and southern Europe.

HARDINESS: Zone 5.

HABIT: Low, herbaceous, spreading ground cover.

SIZE: 4 inches tall, spreading to 12 inches across.

RATE: Moderate; space plants (usually as tubers) 8 to 10 inches apart.

LANDSCAPE VALUE: Usually used for facing in the foreground of a shrub or herbaceous perennial border. Also useful for accent when planted at the base of a large rock or trunk of an open tree. No foot traffic.

FOLIAGE: Basal, deciduous, cordate, often angled or lobed, to 5 1/2 inches long, marbled light green and silvery-white above, green and occasionally with deep red below, margins somewhat toothed, lasting from early fall to late spring.

FLOWERS: Solitary, on scapes to 7 inches long, individual flowers to 1 inch long, corolla tube short with 5 lobes, lobes ovate-lanceolate, rose-pink to white, effective late summer to fall.

FRUIT: 5 valved capsule, not ornamentally significant.

OF SPECIAL INTEREST: Unique in that the foliage is shed during the summer months and is present fall to spring.

CULTURE: Soil: Adaptable to most well-draining soils.
Moisture: Keep soil slightly moist, with watering as needed.
Light: Light to moderate shade.

PATHOLOGY: Diseases: Blight, bacterial tuber rot, gray mold, stunt, root rot.
Pests: Black vine weevil, aphids, thrips, cyclamen mite, broad mite, nematodes, field mice which feed on the tubers.

MAINTENANCE: Remove dead flowers and foliage as they fade.

PROPAGATION: Division: Divide tubers in fall and either replant immediately or store in cool dry place until spring.
Seed: Sow seed in midsummer to midwinter; germination occurs in 3 to 4 weeks at a temperature of 65° to 70°F; temperature should not exceed 72°F.

Cymbalaria (sim-bȧ-lā′ri-ȧ) From Latin *cymbalum*, foliage cymbal shaped. This genus is composed of about 10 species of trailing, herbaceous perennials that are native to the Old World.

FAMILY: SCROPHULARIACEAE: Figwort family.

HABIT: Herbaceous, trailing, matlike ground covers.

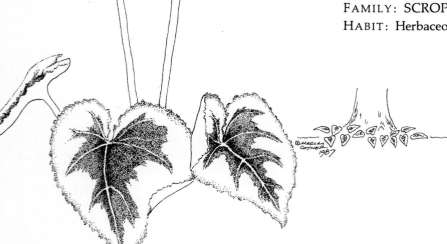

Figure 3-58 *Cyclamen hederifolium* (life-size)

RATE: Relatively fast; space either species 10 to 16 inches apart from 3 to 4 inch diameter containers.

LANDSCAPE VALUE: Best when planted as general covers for small areas. They are especially nice when planted on a steeply sloping bank or allowed to trail over a retaining wall. No foot traffic.

* * *

C. aequitriloba (ē-kwi-tri-lō´bà) Leaves with 3 equal lobes.

COMMON NAME: Toadflax basket ivy.

NATIVE HABITAT: Southern Europe.

HARDINESS: Zones 8 to 10.

SIZE: To 4 inches tall, spreading indefinitely.

FOLIAGE: Evergreen, mostly opposite, simple, pubescent, entire or slightly 3 to 5 lobed, medium green.

FLOWERS: Solitary, axillary, snapdragonlike; corolla about 1/2 inch long, lilac-blue with yellowish throat, effective throughout summer.

FRUIT: Capsules, not of ornamental significance.

* * *

C. muralis (mū-rā´lis) The word muralis in Latin means wall. Likely the reference is to the frequency with which this species is grown in conjunction with walls (Figure 3–59).

COMMON NAME: Kenilworth ivy, coliseum ivy, pennywort.

NATIVE HABITAT: Europe.

HARDINESS: Zones 5 to 9.

SIZE: 1 to 2 inches tall, spreading indefinitely.

FOLIAGE: Evergreen, mostly alternate, 3 to 7 lobed, cordate to orbicular or reniform, 1/2 to 1 inch long and wide, glabrous and glandular dotted on both sides, palmately veined, medium green.

STEMS: Trailing, somewhat striated, rooting as they touch the ground.

FLOWERS: Axillary, solitary, snapdragonlike, lilac-blue with yellowish throat, effective summer through fall.

FRUIT: Not ornamentally significant.

Generally Applicable to Both Species

CULTURE: Soil: Well-drained somewhat acidic loam with good moisture retention properties required for best growth. Tolerant of mild alkalinity.
Moisture: Not tolerant of excess heat or drought; keep ground moist with regular watering.
Light: Moderate to dense shade.

PATHOLOGY: The primary disease is damping off. Pests include snails, aphids, and mites.

Figure 3–59 *Cymbalaria muralis* (life-size)

MAINTENANCE: Requires regular light fertilization for best growth.

PROPAGATION: Cuttings: Softwood cuttings root easily.
Division: Simply divide clumps and keep moist until established.
Seed: Germination takes 1 to 4 weeks at about 55 °F.

✛✛✛✛✛✛✛✛✛✛✛✛✛✛✛

*C*yrtomium (sẽr-to´mi-um) Said to be derived from Greek, meaning arching and merging, in reference to the morphology of the venation. This genus is composed of about 10 species of rhizomatous ferns that are native to Europe.

FAMILY: POLYPODIACEAE: Polypody family.

C. falcatum (fal-kā´tum) Like a sickle, in reference to the pinnae (Figure 3–60).

COMMON NAME: Japanese holly fern, holly fern.

NATIVE HABITAT: Tropics of the Eastern hemisphere.

HARDINESS: Zone 10.

HABIT: Dense, arching fern. Coarse leaves are not typical of ferns.

SIZE: 1 to 2 feet tall, spreading to 2 feet across.

RATE: Moderate; space plants from quart- to gallon-sized containers 14 to 18 inches apart.

LANDSCAPE VALUE: Very attractive when grown in

naturalized settings for accent or facing small trees and larger shrubs, or as a border to walking paths. No foot traffic.

FOLIAGE: Stiffly erect, pinnately compound, to 2 1/2 feet long; leaflets are ovate to elliptic, to 4 inches long with margins entire or somewhat wavy, waxy, dark shiny green and leathery, evergreen, coarse textured.

PETIOLE: Densely scaly at the base.

SORI: Large, round, scattered; indusia peltate.

SELECTED CULTIVARS AND VARIETIES: 'Compactum' Dwarf selection.
'Rochfordianum' Margins of leaflets coarsely fringed.

CULTURE: Soil: Moist, organically rich, well-drained loam.
Moisture: More resistant to dry conditions than many ferns, but soil should be kept moist for best growth.
Light: Light to dense shade.

PATHOLOGY: See **Ferns: Pathology.**

MAINTENANCE: Occasionally topdress with a light application of leaf compost if they are grown in amended soils.

PROPAGATION: See **Ferns: Propagation.**

*C*ystopteris (sis-top´tĕr-is) From Greek, *kystis*, a bag, and *pteris*, a fern, alluding to the saclike covering of the sori. The genus is made up of around 18 species of delicate ferns that are native to rocky woodlands throughout the temperate and tropical zones. Commonly called bladder ferns, they have scaly, creeping rootstalks and deciduous leaves.

FAMILY: POLYPODIACEAE: Polypody family.

C. bulbifera (bulb-if´ĕr-à) Referring to the little green balls or buds underneath the fronds (Figure 3–61).

COMMON NAME: Bulblet bladder fern, berry bladder fern.

NATIVE HABITAT: Eastern North America.

HARDINESS: Zone 3.

SIZE: 2 to 3 feet tall, spreading to about 2 feet across.

HABIT: Rhizomatous, creeping fern.

RATE: Moderate; space plants from quart- to gallon-sized containers 1 1/2 to 2 feet apart.

LANDSCAPE VALUE: Fine for planting in a naturalized setting, especially in combination with low shrubs.

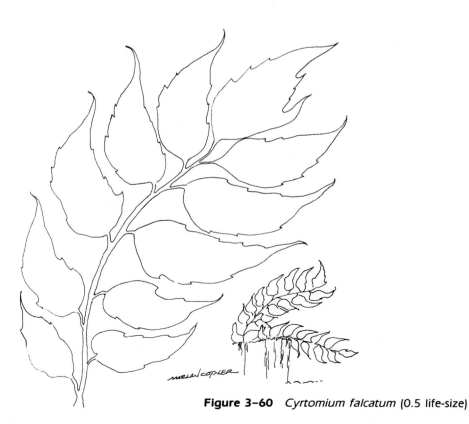

Figure 3–60 *Cyrtomium falcatum* (0.5 life-size)

Figure 3–61 *Cystopteris bulbifera* (0.16 life-size)

LEAVES: Twice pinnately compound, lanceolate, to 2 1/2 feet long by 5 inches across, twice or thrice pinnate; leaflets about 2 inches long, drooping, linear-triangular in outline; subleaflets variable, usually linear in pairs, bulblets are produced on lower surface.

PETIOLE: Usually about one-fifth the length of the leaf, yellowish with blackish colored base.

SORI: Few, scattered, indusia short, hoodlike.

CULTURE: Soil: Rich, moist, well-drained, moderately alkaline soil is best.
Moisture: Can withstand periods of drought without additional watering, but best growth and appearance are attained when soil is kept moist.
Light: Light to dense shade.

PATHOLOGY: See **Ferns: Pathology.**

MAINTENANCE: Periodic light mulching is beneficial when growing this plant in amended soils.

PROPAGATION: Collect small bulblike bodies on the fronds and sow them in summer. Also see **Ferns: Propagation.**

*C*ytisus (sit´i-sus) Greek, *kytisos*, trefoil, pertaining to the leaves of many species. This genus is composed of about 50 species of often showy, flowering, nonthorny shrubs that are commonly called brooms. Their habit varies from tall shrubs to prostrate shrublets. Stems are rushlike upon which pealike flowers are profusely born.

FAMILY: LEGUMINOSAE: Pea family.

LANDSCAPE VALUE: Good as specimens for the rock garden. They are also effective next to walls or in planters where they can trail over. Combined with relatively drought tolerant shrubs such as *Calluna, Juniperus,* and *Erica,* they make very interesting borders. Generally, they perform best in the Carolinas on the east coast and from central California to Vancouver to the west. Many locations in between are suitable, with parts of Oregon, Washington, and Canada particularly good for the growing of brooms. No foot traffic.

* * *

C. decumbens (dē-kum´benz) Decumbent growth habit (Figure 3–62).

COMMON NAME: Prostrate broom.

NATIVE HABITAT: Southern Europe.

HARDINESS: Zones 5 to 9.

HABIT: Prostrate, mat forming shrub.

SIZE: To 8 inches tall (although often much lower), spreading to over 3 feet across.

RATE: Moderate; space plants from gallon-sized containers 16 to 24 inches apart.

FOLIAGE: Deciduous, alternate, simple, singular, 1/4 to 3/4 inch long, 1/16 to 1/8 inch wide, narrowly oblong to obovate, sessile, pubescent on both sides but more below, dull, medium green.

STEMS: Decumbent to 8 inches long, 5 angled, pubescent, rooting as they touch.

INFLORESCENCE: Singular or in axillary racemes in the joints of previous years growth; flowers 1 to 3 per cluster are 1/2 to 5/8 inch long, yellow, pealike, borne in profusion in late spring and sporadically thereafter until fall.

FRUIT: 3 to 4 seeded pod, hairy and about 1 inch long, not highly ornamental.

* * *

OTHER SPECIES: *C. albus* Relatively slow spreader to 1 foot high; flowers are yellowish-white to white and are borne in terminal racemes in summer. Hardy in zones 6 to 9.
C × *beanii* Reaching 14 inches tall by 16 inches across; leaves of 1 leaflet, linear, to 1/2 inch long, hairy in youth; flowers are golden yellow, borne singularly or in threes at the nodes of previous season's growth. Hardy in zones 6 to 9.
C. × *kewensis* Reaching 10 inches tall by 4 feet across; leaves are often of 1 leaflet, but usually in threes, linear-oblong, pubescent; flowers are creamy-white or pale yellow, 1/2 inch long, borne in late spring. Hardy in zones 6 to 9.

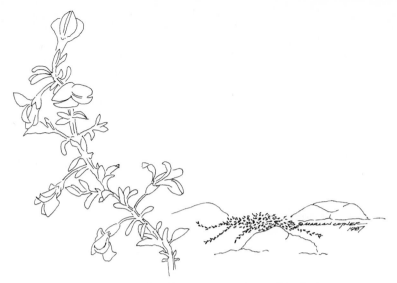

Figure 3–62 *Cytisus decumbens* (life-size)

C. procumbens To 30 inches high and over 4 feet across. Very similar to *C. decumbens,* but taller. *C. purpureus* To 18 inches tall with stems that are procumbent, nearly glabrous; leaves obovate to 1 inch long; flowers are purple, pink, or white, in late spring. Hardy in zones 6 to 9.

Generally Applicable to All Species

CULTURE: Soil: Adaptive to various well-drained soils, but generally best in light textured, freely draining loam.
Moisture: Highly resistant to drought, suitable for wind swept locations.
Light: Plant in full sun.

PATHOLOGY: Diseases: Leaf spot, blight, necrosis. Pests: Nematode.

MAINTENANCE: Prune lightly immediately after flowering to encourage branching. Do not prune heavily though, as some varieties may be slow to recover.

PROPAGATION: Cuttings: Cuttings root easily throughout the season. Mist should be used in mid-summer and a 3000 ppm/talc preparation is helpful at all times.
Grafting: Cultivars may be grafted on *C. scoparious* or *C. nigrescens* understock.
Seed: Gather upon ripening and process by soaking in concentrated sulfuric acid for 30 minutes, wash thoroughly, then sow. Seed may be stored for up to 1 year before processing.

OF SPECIAL INTEREST: Fresh seed pods are said to be poisonous.

✦✦✦✦✦✦✦✦✦✦✦✦✦✦✦✦✦

Daboecia (dab-ē´shi-à or dā-bō-ē´sē-à) From the Irish common name, St. Daboec's heath. The genus is composed of only two species of heathlike, evergreen, small shrubs that are native to Spain, Iceland, and the Azores.

FAMILY: ERICACEAE: Heath family.

D. cantabrica (kan´ti-brik-à) Meaning unknown (Figure 3–63).

COMMON NAME: Irish heath.

NATIVE HABITAT: Ireland and Spain.

HARDINESS: Zone 5.

HABIT: Low, spreading, mound shaped, woody dwarf shrub.

SIZE: 1 1/2 to 2 feet tall by 2 1/2 to 3 1/2 feet wide.

RATE: Moderate; space plants from gallon-sized containers about 2 feet apart.

LANDSCAPE VALUE: Fine specimen in rockery. Also useful in combination with various heaths and heathers of differing foliage color. As a general cover for large areas or as a facing for larger shrubs and tall growing herbaceous perennials or grasses, this plant is excellent. No foot traffic.

FOLIAGE: Evergreen, elliptic, to 1/2 inch long, apex pointed, margin revolute (rolling back), dark shiny green above and white tomentose below.

INFLORESCENCE: Nodding terminal racemes; flowers are about 1/2 inch long, calyx 4 parted, corolla 4 lobed and colored purple, effective early to late summer.

FRUIT: 4 celled capsule of no ornamental significance.

SELECTED CULTIVARS AND VARIETIES: 'Alba' Flowers white.

'Atropurpurea' Flowers rich reddish-purple.

'Bicolor' Flowers striped white and purple.

'Nana' Lower growing with smaller leaves.

'Pallida' Flowers rose-pink.

Var. *rubra* A dwarf, dark green selection with deep purple flowers.

OTHER SPECIES: *D. azorica* is less common, reaches 10 inches tall, is mounding, and is native to the Azores.

CULTURE: Soil: Prefers highly organic loam, but will tolerate sandy or even rocky soil; good drainage needed in all cases. A pH range from 4.5 to 5.5 is optimal.

Moisture: Extremely tolerant to drought and drying winds. Seldom is watering needed.

Light: Full sun to light shade, blooming best in full sun.

PATHOLOGY: No serious diseases or pests reported.

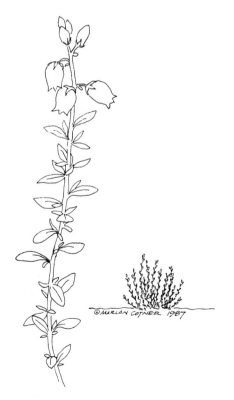

Figure 3–63 *Daboecia cantabrica* (life-size)

MAINTENANCE: Shear lightly after flowering to keep compact and neat appearing.

PROPAGATION: Cuttings: Softwood cuttings should be taken in spring.

Seed: Apparently the species can be propagated from seed, but I have not been able to locate information on specific procedures.

*D**alea** (dā´lē-à) After Dr. Samuel Dale, English botanist. This genus of around 200 species of shrubs and herbs originates primarily from dry areas of the western hemisphere.

FAMILY: FABACEAE (LEGUMINOSAE): Pea family.

D. greggii (greg´i-ī) Meaning unknown, likely a memorial (Figure 3–64).

COMMON NAME: Trailing indigo bush, trailing smoke bush.

NATIVE HABITAT: Chihuahuan Desert of Mexico.

HARDINESS: Zones 9 and 10.

HABIT: Low sprawling, woody shrub.

SIZE: To 12 inches tall (occasionally to 2 feet), spreading 4 to 8 feet across.

RATE: Relatively fast; space plants 1 1/2 to 2 feet apart.

LANDSCAPE VALUE: Best in desert areas of the Southwest as a general cover and soil retainer on moderate to large, level or sloping ground.

FOLIAGE: Evergreen, gray-green, white tomentose, compound, odd pinnate, to 1/2 inch long and 1/4 inch across; leaflets ovate to obovate, bases acute, more or less keeled, usually numbering 9.

STEMS: Trailing, rooting, about 16 inches in length, tomentose in youth, turning brown with prominent lenticels in maturity.

INFLORESCENCE: Small, purple, in dense globose heads, effective spring and summer.

FRUIT: Small legume, not of ornamental significance.

CULTURE: Soil: Tolerant of various soil types, with good drainage a necessity.

Moisture: Water frequently until established; thereafter, drought tolerance is exceptional and little or no irrigation will be needed.

Light: Full sun, tolerating light shade.

PATHOLOGY: Diseases: An unidentified fungus has caused problems with young transplants and cuttings, affecting them above the crown.

Pests: None serious.

Figure 3–64 *Dalea greggii* (life-size)

PROPAGATION: Cuttings: Stem tip cuttings in early summer generally root well when stuck in a very porous medium (at the University of Arizona in Tucson researchers have conducted studies using 50% vermiculite, 50% coarse perlite as their base medium). Kept under mist, rooting occurs in about 6 weeks. Root inducing compounds of moderate concentration have been helpful.

✦✦✦✦✦✦✦✦✦✦✦✦✦✦✦✦✦✦

D*elosperma* (dē-lō′spĕr-mā or del-ō′) From Greek, *delos*, manifest, and *perma*, seed. This genus is probably composed of about 140 species of dwarf, highly branched, succulent shrubs and subshrubs along with biennial and perennial low growing succulents. Several of these ice plants are probably worthy of growing as ground covers, but they have not yet been widely cultivated for this purpose.

FAMILY: AIZOACEAE: Carpetweed family.

*   *   *

D. cooperi (kö′pĕr-ī) After Cooper, presumably the discoverer of this species.

COMMON NAME: Trailing ice plant, mesembryanthemum, vygie.

NATIVE HABITAT: South Africa.

HARDINESS: Zones 7 to 10, mulch in areas where winter temperatures drop below 20°F.

HABIT: Low, succulent, trailing, shrubby ground cover.

SIZE: 6 to 7 inches tall, spreading to 24 inches across.

RATE: Relatively fast; space plants from pint- or quart-sized containers 18 to 24 inches apart.

LANDSCAPE VALUE: Good for soil stabilizing on steep banks and for use before fences and foundations as it is considered relatively fire resistant. Performs quite well in coastal gardens. No foot traffic.

FOLIAGE: Evergreen, opposite, spreading, incurved or recurved, succulent, 2 3/16 inch long, 1/4 inch wide, to 3/16 inch thick, nearly cylindrical, flattened upper surface that narrows toward the apex, papillate, medium green.

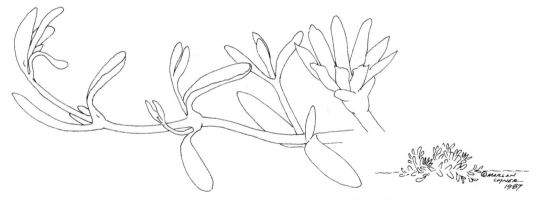

Figure 3–65 *Delosperma nubigenum* (1.25× life-size)

STEMS: Prostrate, highly branched, papillate, rooting as they touch.

FLOWERS: Solitary or grouped 3 to 7 in terminal racemose cymes, to 2 inches across, corolla of many purple petals, effective early to midsummer, followed by a second blooming period in fall.

SELECTED CULTIVARS AND VARIETIES: Several cultivars are in existence that differ primarily in the color of their flowers. Plants with white, pink, rose, purple, and mauve flowers are commonly cultivated.

FRUIT: Capsule, of no ornamental significance.

* * *

D. nubigenum (nū-bi-gēn´um) Latin, meaning born of a cloud (Figure 3-65).

COMMON NAME: Trailing ice plant, hardy ice plant, cloud-loving delosperma.

NATIVE HABITAT: South Africa, at high altitudes of the Drakensberg Mountains, where there are clouds; thus the common name cloud-loving delosperma.

HARDINESS: Reports have been made of it withstanding −25 °F. It is likely to be successful in zones 4 or 5 to 9 or 10.

HABIT: Low, mat forming, trailing, succulent ground cover.

SIZE: 1 to 2 inches tall, spreading 3 or more feet across.

RATE: Relatively fast; space plants from 3 to 4 inch diameter containers 12 to 16 inches apart.

LANDSCAPE VALUE: Good for general cover in smaller areas where a mossy carpetlike effect is desired. Its vibrant color is very eye catching. No foot traffic.

FOLIAGE: Opposite, evergreen or nearly so, often

incurving, succulent, linear, to 1 1/4 inches long, 3/16 inch across, margin entire, base truncate, apex acute, three angled in cross section with upper section flatter and broader than bottom two sections; light, glistening, shiny green. Becoming reddish in cold weather.

STEMS: Prostrate, reddish, fleshy, round, about 1/8 inch across, rooting as they creep.

FLOWERS: Solitary, many petaled, light yellow, about 3/4 inch across, effective in midspring, foul smelling.

FRUIT: Capsule, of no ornamental significance.

Generally Applicable to Both Species

CULTURE: Soil: Adaptable to almost any well-drained soil.
Moisture: Tolerant to considerable drought; infrequent thorough watering is helpful but usually not necessary during periods of hot, dry weather.
Light: Full sun.

PROPAGATION: Cuttings: Easily rooted throughout the year. Just press them into the soil surface.
Division: Simply divide plants, giving attention to soil moisture until well rooted.
Seed: Seed should be collected when ripe, and dried and stored at room temperature. It germinates in 2 to 3 weeks in high percentages.

*D*ennstaedtia (den-stet´i-à) After A. W. Dennstaedt, a German botanist. This genus is commonly called the cupferns. It is composed of some 70 species of medium to large ferns that are native to

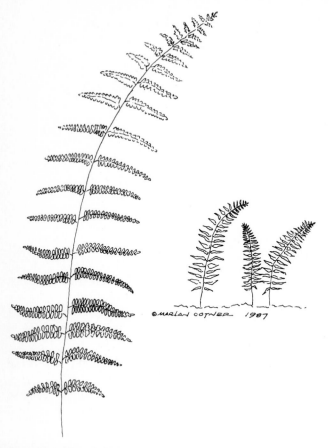

Figure 3–66 *Dennstaedtia punctilobula* (0.25 life-size)

semitropical and tropical areas of South America and Asia, with one species from North America.

FAMILY: POLYPODIACEAE: Polypody family.

D. punctilobula (punk-ta-lō´bū-là) Meaning unknown; likely referring to the small basal lobes of the pinnules, or to the dots formed by the sori on the lobes of the pinnules (Figure 3–66).

COMMON NAME: Hay-scented fern, boulder fern.

NATIVE HABITAT: Eastern North America.

HARDINESS: Zones 3 to 8.

HABIT: Rhizomatous, fern.

SIZE: 20 to 30 inches tall, spreading indefinitely.

RATE: Moderate, space bare root divisions or plants from quart- to gallon-sized containers 16 to 24 inches apart.

LANDSCAPE VALUE: A very pleasant looking plant when used alongside the base of a rock ledge or in a naturalized setting to face shrubs or trees. Tolerance to high soil moisture allows planting in swampy locations. No foot traffic.

LEAVES: Twice pinnately compound, arising singu-

larly or in small groups, broadly lanceolate, yellowish green, about 16 inches long, but occasionally reaching 2 1/2 feet; leaflets numbering about 20 pairs, compressed, somewhat staggered, bristly, hairy below, sessile; subleaflets oblong, opposite, margins indented, very fragrant.

PETIOLE: About 7 inches long, dark or blackish at base, light brown to reddish-brown above, glabrous or sparsely hairy.

SORI: Surrounded by cup-shaped indusia, located in subleaflet margins.

CULTURE: Soil: Although rich, well-drained loamy soil is best, high moisture and infertility are tolerated.

Moisture: Not markedly tolerant of drought or windy conditions. Keep soil moist with irrigation as needed.

Light: Light to moderate shade.

PATHOLOGY: See **Ferns: Pathology.**

MAINTENANCE: See **Ferns: Maintenance.**

PROPAGATION: See **Ferns: Propagation.**

*D***ianthus** (di-an´thus) From Greek, *dios,* God or divine, and *anthos,* a flower; the divine flower. This genus is made up of many hybrids and around 300 species of annual, biennial, and perennial herbs from Eurasia to South Africa. Fruit is usually 4 valved capsule, but not ornamentally significant.

FAMILY: CARYOPHYLLACEAE: Chickweed or pink family.

LANDSCAPE VALUE: Commonly used to edge borders and walks or as a general cover for smaller areas. No foot traffic.

* * *

D. arenarius (ar-ē-nā´ri-us) From Latin, *arena,* sand, sandy-pink, the flowers.

COMMON NAME: Sand pink.

NATIVE HABITAT: Finland.

HARDINESS: Zones 4 to 7.

HABIT: Low, herbaceous, tufted ground cover.

SIZE: 6 to 15 inches tall.

RATE: Moderate; space plants from 3 to 4 inch diameter containers 8 to 10 inches apart.

FOLIAGE: Evergreen, opposite, linear, quite narrow, gray-green, grassy.

FLOWERS: Single, fragrant, calyx to 1 inch long;

corolla with petals to 1/2 inch long and deeply dissected, base bearded, white, effective mid-summer.

* * *

D. deltoides (del-toi´dez) Triangular, referring to the delta on the petals (Figure 3–67).

COMMON NAME: Maiden pink.

NATIVE HABITAT: Western Europe to eastern Asia.

HARDINESS: Zones 3 or 4 to 7.

HABIT: Herbaceous, mat forming ground cover.

SIZE: 4 to 8 inches tall by 15 inches across.

RATE: Moderate; space plants from 3 to 4 inch diameter containers 8 to 10 inches apart.

FOLIAGE: Evergreen, opposite, linear-lanceolate to linear, 1/2 inch long by 1/4 inch across; margins

Figure 3–67 *Dianthus deltoides* (life-size)

pubescent, entire; glandular dotted both sides, medium green to glaucous.

FLOWERS: Solitary, regular, fragrant, terminal, 3/4 inch across, petals lavender with dark V-shaped band at the base of each, effective spring and summer, then sparsely in fall.

* * *

D. plumarius (plū-mā´ri-us) The flowers feathered or frilled.

COMMON NAME: Cottage pink, grass pink.

NATIVE HABITAT: Europe.

HARDINESS: Zones 3 or 4 to 7. Protect with mulch if snow is unreliable in northernmost areas.

HABIT: Low, herbaceous, mat forming ground cover.

SIZE: From 12 to 18 inches tall, spreading to 16 + inches across.

RATE: Moderate; space plants from 3 to 4 inch diameter containers 8 to 10 inches apart.

FOLIAGE: Lanceolate, evergreen, narrow, opposite, 1 to 4 inches long, acute, keeled, margins finely serrulate.

FLOWERS: Solitary, fragrant, calyx to 1 inch long; petals purplish, to 3/4 inch long, often fringed rose, purple, or white, effective mid-late spring.

SELECTED CULTIVARS AND VARIETIES: 'Spring Beauty' Apparently this name is (inappropriately?) applied to a mixture of seedlings from this species that have double flowers and range from white, to pink, rose, salmon, and mixtures thereof.

OTHER SPECIES AND HYBRIDS: Countless other species and hybrids may be useful as ground covers. A few of the more popular include: *D.* × 'Aqua' With double, pure white flowers. Usually from 10 to 12 inches tall.
D. × 'Doris' With salmon-pink, deep pink-eyed flowers that are relatively large, semidouble, and fragrant.
D. × *latifolius* 'Essex Witch' Flowers profuse rose-pink and fragrant; finely fringed; reaches about 5 inches tall.
D. × 'Spotti' Radiant white spotted, pinkish-red flowers with blue-green foliage to 6 inches tall.
D. × 'Zing Rose' Flowers bright rose-red throughout summer; reaching about 6 inches tall; foliage green.

Generally Applicable to All Species

CULTURE: Soil: Adaptive to a wide range of soils as long as drainage is good. A pH range of 7.0 to 8.0 is preferred.

Moisture: Not noted for tolerance to drought; keep soil slightly moist.

Light. Full sun to light shade.

MAINTENANCE: Cut plants back after flowering to encourage new growth and to get rid of dry flower heads. If grown in amended soils that are naturally acidic, an annual soil analysis is recommended to determine the need for additional lime.

PATHOLOGY: Diseases: Bacterial and fungal wilt, stem rot, leaf spot, bacterial spot, bacterial pimple, flower spot, greasy blotch, rust, bud rot, anther smut, viral streaking, mottling, ringspot, and mosaic.

Pests: Cutworm, aphids, cabbage looper, leaf roller, leaf tier, thrips, mites, root knot nematode, deer, mice, and rabbits.

PROPAGATION: Cuttings: Cuttings usually root in 2 to 4 weeks. Use a very porous medium and discontinue mist as soon as roots form. A treatment of 1000 ppm IBA/talc may facilitate the process.

Division: Divide clumps in spring or fall.

Tissue Culture: In vitro methods have been established. Refer to *Hortiscience* 10(6), pages 608–610.

Seed: Ripe seed germinates in about 10 days.

Dicentra (di-sen´trà) Derived from Greek, *di*, two, and *kentron*, spur, in reference to the dual spurs of the petals. The genus is composed of around 19 species of herbaceous perennials, many of which are tuberous or rhizomatous. They are native to North America and Asia. At least two species make excellent ground covers.

FAMILY: FUMARIACEAE: Fumitory family.

LANDSCAPE VALUE: Suitable as general ground covers for small areas in visible locations where their grace and charm can be appreciated. They are often used with excellent effect in combination with ferns. No foot traffic.

FRUIT: Capsule, of no ornamental worth.

HABIT: Rhizomatous, medium-sized, herbaceous, graceful ground covers.

RATE: Moderate; space plants from pint- to quart-sized containers 12 to 16 inches apart.

*　　*　　*

D. exima (eks-im´à) Distinguished, unusual (Figure 3–68).

Figure 3–68   *Dicentra exima* (life-size)

COMMON NAME: Fringed bleeding heart, turkey corn, staggerweed, wild bleeding heart.

NATIVE HABITAT: Eastern United States from New York to Georgia.

HARDINESS: Zones 3 to 9.

SIZE: 12 to 18 inches tall.

FOLIAGE: Deciduous, basal leaves ternately compound, stem leaves dissected, leaf segments in either case are broadly oblong to ovate, soft textured, medium gray-blue, glabrous.

INFLORESCENCE: Nodding panicles elevated on scapes that are frequently as tall as the foliage; flowers are heart shaped, corolla tapering from a cordate base into a narrow apex, colored pink to purplish, most effective early to midsummer, but the entire bloom period runs from midspring to early fall.

SELECTED CULTIVARS AND VARIETIES: 'Adrian Bloom' Crimson-red flowers, and blue-green foliage.

'Bountiful' With dark pink flowers that are most

heavily produced during midspring and then in fall, with sparse bloom in between.

'Luxuriant' With red blossoms and green foliage.

'Zestful' Flowers pink.

* * *

D. formosa (for-mō´så) Handsome, beautiful.

COMMON NAME: Pacific bleeding heart, western bleeding heart.

NATIVE HABITAT: Western North America from British Columbia to California.

HARDINESS: Zones 2 or 3.

HABIT: Same as *D. exima.*

SIZE: About 12 inches tall, spreading 10 to 18 inches across.

FOLIAGE: Deciduous leaves, long petioled all of which are basal, segments are oblong, glaucous colored below and frequently above.

INFLORESCENCE: Compound racemes; flowers with corolla of rose-purple to white, to 5/8 inch long, effective midspring to early fall.

SELECTED CULTIVARS AND VARIETIES: 'Sweetheart' with white flowers.

Generally Applicable to Both Species

CULTURE: Soil: Adaptable to various soil textures, the main requirement being good drainage. Soils rich in humus and moderately acidic are the best. Moisture: Not notably tolerant to drought; keep the soil moist with supplemental watering in summer. Light: Light to moderate shade.

PATHOLOGY: Diseases: Stem rot, wilt, downy mildew, rust.

Pests: No serious pest problems reported.

MAINTENANCE: No special maintenance is required. Shearing foliage back following the period of main bloom with *D. exima* may stimulate another period of heavy flowering in the same season.

PROPAGATION: Division: Simple division of the clumps in spring or fall is commonly practiced. Fall is the preferred time.

Cuttings: Stem cuttings should be taken after the main period of flowering. They are relatively easy to root. Root cuttings are usually cut to 3 inches long and also taken after the main blooming period.

Seed: Sow seed in late summer or fall for natural stratification outdoors, or store at 40 °F for 6 weeks and then sow.

✚✚✚✚✚✚✚✚✚✚✚✚✚✚✚

*D*ichondra (di-kon´drå) Meaning double *(di)* and grain, in reference to the shape of the fruit. This genus is composed of about nine species of low, creeping, herbaceous perennials. They are native to Mexico, South America, eastern United States, Australia, West Indies, eastern Asia, and New Zealand.

FAMILY: CONVOLVULACEAE: Morning glory family.

D. micrantha (mī-kran´thå) Meaning small flowered (Figure 3–69). Often listed as *D. repens* and *D. carolinensis. D. carolinensis* is a different species.

COMMON NAME: Dichondra, small lawn leaf.

NATIVE HABITAT: West Indies, China, Japan, Mexico, and Texas.

HARDINESS: Zone 10.

HABIT: Herbaceous, dense, mat forming, stoloniferous creeper.

SIZE: 3 to 6 inches, spreading indefinitely.

RATE: Moderate; set plugs about 12 inches apart.

LANDSCAPE VALUE: Commonly planted as a lawn substitute, dichondra lawns are dark green and lush appearing. Also useful as a filler between stepping stones or as a general cover in smaller areas. Do not walk on frozen dichondra.

FOLIAGE: Evergreen, alternate, entire, suborbicular-reniform, to 1/2 inch wide and long, venation conspicuous, sparsely appressed-pubescent below,

Figure 3–69 *Dichondra micrantha* (1.5 × life-size)

nearly glabrous above, glandular dotted both sides, dark green, petioles weak, about 1 1/2 inches long.

FLOWERS: Solitary, axillary, elevated on 1/2 inch long peduncles that are upright ascending, then becoming recurved, regular, very small to about 1/8 inch across, corolla greenish-yellow, borne in spring.

FRUIT: Single-seeded bilobed capsule, about 1/8 inch in diameter, not ornamentally significant; self-sowing.

CULTURE: Soil: Adaptable to almost any soil condition. If soil is on the organically rich side, performance is usually better.

Moisture: Dichondra is relatively sensitive to drought. Regular watering is required in hot, dry locations, especially those with porous soil. Water thoroughly and allow to dry in between.

Light: Full sun to moderate shade.

PATHOLOGY: Diseases: Rhizoctonia, sclerotium.

Pests: Dichondra flea beetle, cutworms, moths, weevils, mites, slugs, snails, and nematodes.

MAINTENANCE: Regular mowing is generally practiced to keep plantings low and compact, which helps them fight weed invasion. Usually, reel-type mowers give the best results. Fertilization is necessary on a regular basis. Frequent light applications usually are better than infrequent heavy applications. Should dichondra dry up to the point where it turns brown, it will usually recover soon after when given ample moisture.

PROPAGATION: Seed: Sow seed between the first of March and mid October. Seedlings must be kept moist with frequent syringes of water during hot or dry weather. Two pounds of seed per 1000 ft² is adequate.

Drosanthemum (drō-san´the-mum) Referring to the flowers being like those of *Drosera.* This genus is made up of about 95 species of succulent shrubs with decumbent to creeping habit. They are often wide spreading and are native to South Africa.

FAMILY: AIZOACEAE: Carpetweed family.

RATE: Moderate to relatively fast; space plants from pint- or quart-sized containers 12 to 18 inches apart.

LANDSCAPE VALUE: General covers for use around shrubs and near foundations. *D. floribundum* makes an excellent plant for erosion control on rocky slopes. No foot traffic.

*   *   *

D. floribundum (flôr-i-bun´dum) Abundant flowers (Figure 3–70).

COMMON NAME: Rosea ice plant, rosy ice plant.

NATIVE HABITAT: South Africa.

HARDINESS: Zones 9 to 10.

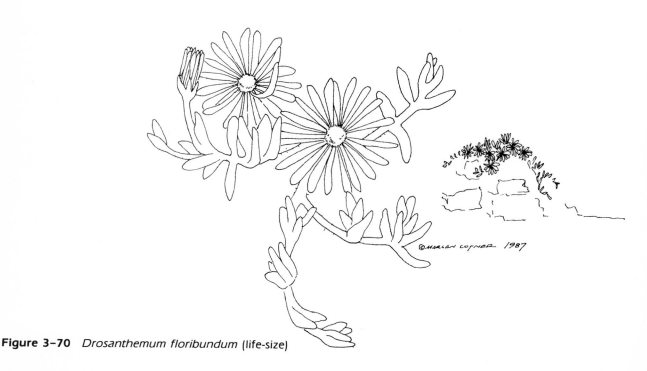

Figure 3–70 *Drosanthemum floribundum* (life-size)

HABIT: Low, succulent, matlike, trailing or decumbent ground cover.

SIZE: To 6 inches tall, spreading indefinitely.

FOLIAGE: Evergreen, opposite, whorled or fasicled, terete to 3 angled, to 1/2 inch long, 1/8 inch across, entire, glabrous, dotted with papillae, grayish.

STEMS: Terete, white tomentose, sometimes woody.

FLOWERS: Solitary or grouped in threes, regular, to 3/4 inch across, corolla with many pale pink petals, effective in spring and summer. Attractive to bees.

FRUIT: Nonornamental 4 to 5 valved capsule.

*　　*　　*

D. hispidum (his´pi-dum) With bristly hairs.

NATIVE HABITAT: South Africa.

HARDINESS: Zones 9 to 10.

HABIT: Clump forming, succulent, ground cover.

SIZE: 18 to 24 inches tall, spreading to 3 feet across.

FOLIAGE: Evergreen spreading or incurved, to 1 inch long, 1/8 inch thick, nearly cylindrical with upper sides slightly flattened, light green or reddish, with large water-soaked papillae.

FLOWERS: Solitary or paired, sometimes in threes, to 1 1/4 inch across, corolla with many deep purple and glossy petals in late spring.

FRUIT: Capsules, of no ornamental significance.

OTHER SPECIES: *D. bellum* has white or pink flowers.

D. bicolor Reaches 1 foot high, is shrubby, with very attractive bright yellow flowers and orange-tipped petals.

D. speciosum is many branched and shrubby; flowers are bright orange or cornelian red with greenish base.

Generally Applicable to All Species

CULTURE: Soil: Adaptable to a wide range of well-drained soils.

Moisture: Withstanding considerable drought once established. Watering occasionally in summer months may be beneficial.

Light: Full sun.

PATHOLOGY: No serious diseases reported.

Pests: Scales, mealybugs, aphids.

MAINTENANCE: Little or no special maintenance.

PROPAGATION: Cuttings: Easily rooted throughout the year.

Division: Division of rooted branches can be accomplished anytime provided soil is kept moist until well established.

Seed: Reported to be a common means of propagating some species.

*D**ryopteris*** (drī-op´tĕr-is) From Latin, *dryas*, a wood nymph, and *pter*, meaning wing, or possibly *drys*, oak, and, *pteris*, a fern, derived from Greek. This genus is composed of about 150 species of temperate tropical ferns. They are often leathery, evergreen, and commonly referred to as wood ferns, as many species inhabit woodlands.

FAMILY: POLYPODACEAE: Polypody family.

LANDSCAPE VALUE: Best in woodland or naturalized gardens for use as accent plants. No foot traffic.

*　　*　　*

D. intermedia (in-tĕr-mē´dē-å) Meaning unclear.

COMMON NAME: Intermediate shield fern.

NATIVE HABITAT: Eastern North America.

HARDINESS: Zone 3.

HABIT: Intermediate height fern that spreads by creeping rootstalk.

SIZE: 1 to 1 1/2 feet tall, spreading indefinitely.

RATE: Moderate; space bare root divisions or plants from quart- to gallon-sized containers 1 1/2 to 2 1/2 feet apart.

FOLIAGE: Arising in groups, semievergreen, compound, twice pinnate-pinnatifid, more or less narrow triangular in outline, leaflets 10 to 14 pairs arranged in an opposite or slightly staggered fashion, subleaflets variable, medium to dark shiny green.

PETIOLE: About 12 inches long, densely scaly, stout.

SORI: Round; indusia reniform, glandular.

*　　*　　*

D. filix-mas (fi´liks-mahs) Meaning male fern (Figure 3–71).

COMMON NAME: Male fern.

NATIVE HABITAT: England.

HARDINESS: Zone 3.

HABIT: Clump forming fern.

SIZE: Reaching from 2 to 4 feet tall by 2 1/2 feet across.

RATE: Relatively slow to spread; space bare root clumps or plants from quart- or gallon-sized containers 1 1/2 to 2 1/2 feet apart.

FOLIAGE: Semievergreen, about 18 inches long by 8 inches across, bipinnate, broadest two-thirds from apex, with about 20 pairs of leaflets; leaflets narrowly oblong-lanceolate; subleaflets in about 24 pairs, forward pointing, rounded, slightly toothed, leathery, colored yellow-green.

PETIOLE: About 4 inches long, stout, green, densely scaly; scales brown.

SORI: Large, prominent toward midvein; indusia reniform, often with hairy margins.

FRUIT: Spores, bright green, short lived, with chlorophyll, encased; spore cases stalked, relatively large, clustered, green turning dark brown.

OF SPECIAL INTEREST: Of my favorite myths associated with plants is one applied to the male fern. In times past, it was believed that if ashes derived from the burning of *D. filix-mas* were worn about one's finger in the pattern of a ring, the individual would be blessed with the ability to understand the speech and songs of birds.

* * *

D. marginalis (mär-ji-nā´lis) Sori located along subleaflet margins.

COMMON NAME: Marginal shield fern, leather wood fern.

NATIVE HABITAT: Eastern North America.

HARDINESS: Zone 3.

HABIT: Clump forming, low fern.

SIZE: 15 to 18 inches tall by 2 1/2 feet across.

RATE: Relatively slow to spread; space bare root divisions or plants from quart- to gallon-sized containers 1 1/2 feet apart.

FOLIAGE: Evergreen, about 18 inches long by 6 inches across, may reach 2 1/2 feet long, twice compound, arising in clumps; leaflets lanceolate, consisting of about 20 pairs of staggered subleaflets; subleaflets are blunt tipped, coarse, blue-green above, and green below.

PETIOLE: Stout, brittle, about one-quarter to one-third the length of the leaf, covered with brown scales that increase in density around the swollen base.

SORI: Located along the subleaflet margins, prominent, solitary or in rows, gray becoming brown; indusia reniform, prominent.

Generally Applicable to All Species

CULTURE: Soil: Best growth occurs when these ferns are planted in rich, moisture retentive, loamy soil. Adaptable to broad pH range, but moderate acidity is preferred.

Moisture: These ferns are not very tolerant of extended hot, dry conditions. Plant them in a cool, nonwindy location and keep the soil moist.

Light: Moderate to dense shade.

PATHOLOGY: See **Ferns: Pathology**

MAINTENANCE: See **Ferns: Maintenance**

PROPAGATION: See **Ferns: Propagation**

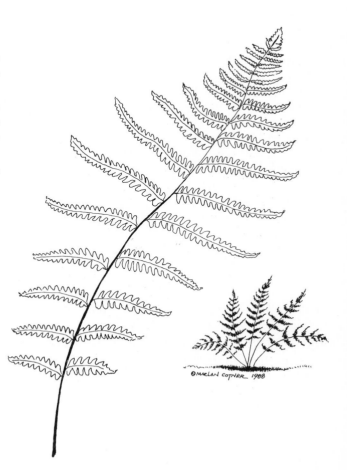

Figure 3-71 *Dryopteris filix-mas* (0.25 life-size)

✛✛✛✛✛✛✛✛✛✛✛✛✛✛✛✛

*D*uchesnea (dū-kes´nē-ȧ) After Antoine Nicolas Duchesne, an 18th century horticulturist who was an authority on true strawberries (*Fragaria*). This genus is composed of only two species of rather small, stoloniferous, herbaceous perennials. They are native to southern Asia.

FAMILY: ROSACEAE: Rose family.

D. indica (in´di-cȧ) Of india (Figure 3–72).

COMMON NAME: Mock strawberry, Indian strawberry.

HARDINESS: Zone 5.

HABIT: Low, stoloniferous, dense, mat forming herbaceous ground cover.

SIZE: About 2 inches tall, spreading indefinitely.

RATE: Relatively fast; space plants from pint- to quart-sized containers 14 to 18 inches apart.

LANDSCAPE VALUE: En masse, these plants create an excellent cover for smaller areas, especially underneath small trees and loosely growing shrubs. They are also quite valuable as soil stabilizers on steeply sloping banks. Tolerating very limited foot traffic.

FOLIAGE: Evergreen, alternate, palmately compound, each leaf with three leaflets; leaflets are ovate to obovate, 1 inch long by 1/2 inch across, coarsely toothed, pubescent on both sides along the veins, medium green with hairy petioles that are reddish at the base.

FLOWERS: Solitary, regular, 1/2 to 1 inch across; corolla 5 petaled, yellow; borne intermittently throughout the growing season.

FRUIT: Red, dry, spongy receptacle, 1/2 inch in diameter, surrounded by a large, persistent calyx, appearing strawberrylike, edible, but not very pleasing to the taste. Often eaten by birds.

CULTURE: Soil: Adaptable to most soils with the main consideration being excellent drainage.
Moisture: Able to tolerate extended dry periods, but the ground should be kept slightly moist throughout the growing season for best health and appearance.
Light: Best in light shade, but tolerant of full sun.

PATHOLOGY: Diseases: Rust.
Pests: Aphids and mites.

MAINTENANCE: Mow foliage in early spring to stimulate new growth.

PROPAGATION: Division: In spring or fall, this is the easiest and probably most common means of propagation.
Seed: Plants can be propagated by seed, but due to ease of division, seed propagation is seldom practiced.

*E*pimedium (ep-i-mē´di-um) Greek name of obscure meaning. The genus is composed of about 21 species of low growing, rhizomatous or clump forming, herbaceous perennials. They are native to Europe and Asia. In addition to the species, many hybrids are in cultivation. The species that are suitable for ground cover use are quite similar in habit, and therefore the first species, *E. alpinum*, will be described in detail and the difference of the others pointed out later.

FAMILY: BERBERIDACEAE: Barberry family.

Figure 3–72 *Duchesnia indica* (life-size)

OF SPECIAL INTEREST: One common name for epimedium, Bishop's hat, supposedly arose due to the resemblance of the four pointed flowers to the clergyman's biretta. The common name, barrenwort, was supposedly given because a root extract from the plant was believed at one time to prevent conception when given to women.

* * *

E. alpinum (al-pī´num) From Alpine regions (Figure 3–73).

COMMON NAME: Alpine epimedium.

NATIVE HABITAT: Southern and central Europe.

HARDINESS: Zones 4 to 8.

HABIT: Dense, erect, herbaceous, rhizomatous ground covers.

SIZE: From 10 to 15 inches tall, spreading indefinitely.

RATE: Moderate; space plants from pint- to quart-sized containers 9 to 12 inches apart.

LANDSCAPE VALUE: Exceptional, and some feel incomparable, ground covers for facing low trees and shrubs. Sometimes used as specimens in rock gardens of perennial borders or as edging along a walk or border planting. In any case, they contribute a strong beauty to the landscape. No foot traffic.

FOLIAGE: Basally derived, deciduous to semievergreen, biternately compound, upon erect, wiry petioles that arise from the rootstalk, leathery, heart shaped tapering to a fine point, to 5 inches long, medium green, rather coarse, but those species with smaller leaves tend to be medium textured.

INFLORESCENCE: Loose terminal racemes that are elevated on slender stems; flowers are nodding, partially obscured by foliage, about 1/2 inch across, sepals are red and arranged in whorls, corolla of 4 yellow petals, borne in midspring. Presumably self-sterile.

SELECTED CULTIVARS AND VARIETIES: Listed under the species of *A. alpinum* because it most closely resembles this parent is the hybrid *E. × rubrum* (Figure 3–73). Probably the most popular of all the epimediums, this hybrid is characterized by its persistent, deciduous leaves that are nicely edged with red in springtime. The inflorescences are held above the foliage. Flowers are showy, with bright crimson sepals and creamy white petals. The parents are *E. alpinum* and *E. grandiflorum.*

* * *

E. grandiflorum (grand-di-flō´rum) Large flowered.

COMMON NAME: Long-spur epimedium, bishop's hat.

NATIVE HABITAT: Japan, Korea, and Manchuria.

HARDINESS: Zones 3 to 8.

SIZE: 9 to 12 inches tall.

DISCUSSION: This is among the most showy of the epimediums with its smaller, more delicate, basal leaves, and leafy flower stalks. Larger flowers are elevated slightly above the foliage mass and are

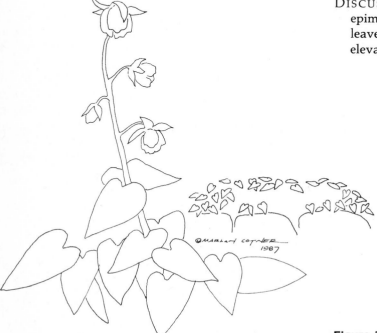

Figure 3–73 *Epimedium* × *rubrum* (0.5 life-size)

more showy than other species. They are characterized by a larger corolla of red and violet and sepals of white. Selected cultivars include: 'Album' Flowers mostly white.

'Niveum' Flowers white.

'Violaceum' Flowers light purple.

* * *

E. pinnatum (pin-nā´tum) Pinnately leaved.

COMMON NAME: Persian epimedium.

NATIVE HABITAT: Persia.

HARDINESS: Zones 5 to 8.

SIZE: 9 to 12 inches tall.

DISCUSSION: The foliage is persistent yet deciduous. Leaves are biternate with margins of leaflets spined. Flowers are brownish-red, about 5/8 inch across, and elevated to 1 1/2 feet.

SELECTED CULTIVARS AND VARIETIES: Var. *colchicum* With fewer leaflets per leaf than the species. Sometimes it is listed as *E. elegans* or *E. pinnatum elegans.*

Hybrid *E.* × *versicolor* Juvenile foliage that is mottled green and red, later turning green, and then tinged pinkish-red in fall. It has flowers that are about 1 inch wide, with sepals of rose and petals of yellow with red tinged spurs. In other respects it most closely resembles the parent *E. pinnatum.*

E × *versicolor* 'Sulphureum' Commonly cultivated, with pale yellow inner sepals and yellow corolla. Unlike *E* × *versicolor*, its spring foliage is usually not mottled.

Generally Applicable To All Species

CULTURE: Soil: Adaptive to various soil conditions. Best in rich loam that is capable of retaining moisture but does not become overly saturated. Reaction should be slightly acidic.

Moisture: Capable of weathering short periods of drought, but without supplemental water, appearance may suffer. Generally, appearance is more attractive if the site is cool and the soil kept moist.

Light: Light to moderate shade.

PATHOLOGY: Diseases: Occasionally, epimediums will become infected with mosaic virus. If this occurs, they should be dug up and destroyed, and the area replanted with a resistant plant.

Pests: Aphids are common; whether or not they are the vector for mosaic virus is not known. The physical damage they do is usually negligible.

MAINTENANCE: Remove the oldest leaves in the fall or early spring to neaten plantings.

PROPAGATION: Division: Simple division in early spring or fall is the standard means of propagation.

✝✝✝✝✝✝✝✝✝✝✝✝✝✝✝✝✝

Erica (ē-rī´kȧ or ẽr´i-kȧ) From Greek *ereike*, heath or heather. This genus is made up of about 500 species of compact, many branched, evergreen subshrubs, shrubs, and small trees. They are native primarily to South Africa, with some originating in Europe and the Mediterranean region. Many species, cultivars, and hybrids are suitable for ground cover use. Only a few of the more commonly cultivated are listed here. The book *Heathers*, by Brian Proudley and Valerie Proudley, will be invaluable to those wishing to learn more about this remarkable genus.

FAMILY: ERICACEAE: Heath family.

LANDSCAPE VALUE: Excellent in ornamental bed with ericaceous shrubs, including rhododendron, pieris, and azalea, for facing or accent. Heaths combine well with various types of heather (*Calluna*) to cover large areas and slopes as colorfully brilliant lawn substitutes. In the rock garden, they make exceptional specimens, as they do in combination with herbs in a border setting. Tolerant to salt and thus useful in coastal gardens. No foot traffic.

* * *

E. carnea (kär´nē-ȧ) Flesh colored, the flowers being pale fleshy pink (Figure 3–74).

COMMON NAME: Spring heath, snow heather.

NATIVE HABITAT: Central and southern Europe.

HARDINESS: Zones 4 or 5 to 7 or 8; snow cover is necessary in the northern range.

HABIT: Short, woody, dense, prostrate, subshrubby ground cover.

SIZE: To 1 foot tall, spreading 1 1/2 to 2 feet across.

RATE: Relatively slow; space plants from quart- to gallon-sized containers 14 to 20 inches apart.

FOLIAGE: Evergreen, whorled in fours, to 1/4 inch long, needlelike, glabrous, entire, dark shiny green.

INFLORESCENCE: One-sided racemes; flowers regular, about 1/4 inch long, corolla urn shaped, red to whitish, effective late winter to early spring.

FRUIT: 4 valved capsule, persistent but not ornamentally significant.

SELECTED CULTIVARS AND VARIETIES: There are many, a few of which include: 'Aurea' Foliage light lime-green becoming yellow-golden in winters; flowers deep pink in midwinter to spring.

'Ellen Porter' Compact habit with dark green foliage and carmine-rose flowers.

'King George' Foliage dark green and glossy; flowers

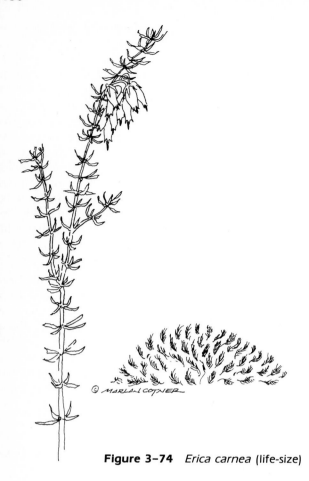

Figure 3–74 *Erica carnea* (life-size)

deep rose-pink on short inflorescences; height to about 12 inches.

'Prince of Wales' Dark green leaves with rose pink flowers; height to about 12 inches.

'Springwood Pink' Vigorously spreading with medium green foliage and flowers that become clear pink after beginning as white; effective late winter to early spring; height to about 8 inches.

'Springwood White' Vigorously spreading with bright green foliage and strikingly white flowers late winter to early spring.

'Vivellii' Compact habit; foliage dark green with bronzy winter coloration; flowers deep blood-red; height to 8 inches.

'Winter Beauty' Spreading and slow growing; foliage is dark green; flowers deep pink from early winter to early spring.

*   *   *

E. cinerea (sin-ẽr-ē´à) Gray or ashen, the underside of the leaves.

COMMON NAME: Twisted heath, Scotch heath, bell heath.

NATIVE HABITAT: Western Europe.

HARDINESS: Zones 6 or 7 to 8.

HABIT: Woody, low, spreading, shrubby ground cover.

SIZE: 12 to 24 inches tall, spreading to 2 feet across.

RATE: Relatively slow to moderate; space plants from quart-sized containers 10 to 16 inches apart.

SPECIAL MAINTENANCE: This species requires heavy shearing immediately following bloom to keep it compact and neat in appearance.

FOLIAGE: Evergreen, needlelike, in threes, to 1/4 inch long, glabrous and glossy dark green.

INFLORESCENCE: Terminal racemes or umbels; flowers whorled, corolla urn shaped, purple changing to blue, effective midsummer to early autumn.

SELECTED CULTIVARS AND VARIETIES: A few of the more common include: 'Atrorubens' With ruby red flowers.

'C. D. Easen' With vivid deep pink flowers from mid to late summer.

'Golden Drop' With yellowish foliage that changes to russet in winter; flowers pink.

'P. S. Patrick' Flowers purple.

'Violacea' Flowers purple-blue; more upright habit.

*   *   *

E. × darleyensis (där-lē-en´sis) Originating at Darley Dale nursery.

COMMON NAME: Darley heath, winter heath, hybrid heath.

HYBRID ORIGIN: *E. carnea × E. mediterranea*

HARDINESS: Zone 6.

HABIT: Woody, relatively tall, spreading, shrubby ground cover.

SIZE: To 3 feet tall, spreading 1 1/2 to 2 feet across.

RATE: Relatively fast; space plants from quart- or gallon-sized containers 1 1/2 to 2 feet apart.

GENERAL DISCUSSION: This hybrid is the result of the cross *E. mediterranea* (an upright species) and the prostrate *E. carnea*. It is exceptional as it inherited the long flowering period and bright color of *E. carnea* along with the vigorous growth and abundant flower production of *E. mediterranea*. Height is intermediate, as is hardiness. Ability to withstand adverse soil conditions is remarkable. Leaves are grouped in fours, to 1/2 inch long, needlelike, evergreen, and dark shiny green. Flowers are born in terminal leafy racemes with pink urn-shaped corolla in late autumn to early spring.

SELECTED CULTIVARS AND VARIETIES: Many exist; the two that are likely to be the most popular include: 'Arthur Johnson' Foliage light green; flowers mauve-pink on inch long inflorescences; height to about 2 feet.

'Silberschmelze' Foliage deep glossy green; flowers silvery white; height to around 1 foot.

* * *

E. tetralix (tet-trá´liks) Meaning 4 leaves that are arranged in a cross-like fashion at each node.

COMMON NAME: Crossleaf heath, bog heath, bell heath

NATIVE HABITAT: Western Europe.

HARDINESS: Zones 4 to 7.

HABIT: Medium sized, woody, upright branched, subshrubby ground cover.

SIZE: To 12 inches tall, spreading 1 to 1 1/2 feet across.

RATE: Relatively slow to moderate; space plants from pint- to quart-sized containers 12 to 16 inches apart.

FOLIAGE: Evergreen, needlelike, in 4's, to 1/8 inch long, glandular-ciliate, downy gray.

INFLORESCENCE: Dense terminal clusters; flowers urn shaped, 1/4 inch long, corolla waxy rose colored, effective early summer to early to mid fall.

SELECTED CULTIVARS AND VARIETIES: 'Alba Mollis' Compact selection with white flowers and silvery gray foliage that becomes greener with age.

'George Fraser' Flowers pale pink with bluish foliage.

Others: Many other cultivars exist; those above are two of the more noteworthy.

* * *

E. vagans (vā´ganz) Wandering, in reference to its growth habit.

COMMON NAME: Cornish heath, wandering heath.

NATIVE HABITAT: Western Europe.

HARDINESS: Zone 5.

HABIT: Low, spreading, subshrubby, woody ground cover.

SIZE: To 12 inches tall, spreading to 2 feet across.

RATE: Relatively slow to moderate; space plants from quart-sized containers 16 to 20 inches apart.

FOLIAGE: Evergreen, needlelike, in fours or fives, to 3/8 inch long, glabrous, bright shiny green.

INFLORESCENCE: Leafy racemes to 6 inches long; flowers with urn-shaped corolla to 1/8 inch long, purplish-pink, effective during late summer.

SELECTED CULTIVARS AND VARIETIES: 'Alba

Minor' Compact selection with white flowers.

'Mrs D. F. Maxwell' Foliage dark green; flowers deep red, on long spikes from early summer to fall.

'St. Keverne' With clear, bright pink terminal flowers that are effective in late summer.

Generally Applicable to All Species

CULTURE: Soil: Needing well-drained sandy loam preferably with incorporated organic matter. The pH is best when in the range from 4.5 to 5.5. *E. carnea*, *E. mediterranea*, and *E.* × *darleyensis* will take somewhat higher pH.

Moisture: Erica can withstand moderate drought conditions; however, occasional deep watering will enhance floral display and appearance of the foliage. Best in areas of high humidity and moderate temperatures.

Light: Full sun to light shade, with best floral display in full sun.

PATHOLOGY: Diseases: Wilt, powdery mildew, rust.

Pests: None serious.

Physiological: Stunting and chlorosis due to alkalinity.

MAINTENANCE: Plantings may be mowed or sheared following bloom to keep them looking neat and to maintain compact habit. A light mulch around the base is useful as it will help to moderate the temperature and moisture content of the soil.

PROPAGATION: Cuttings: Cuttings are usually made in midsummer, fall, or winter; remove lower leaves and treat with a mild rooting preparation prior to sticking in a porous medium. 1000 ppm IBA/talc is adequate.

Division: Simple division of spreading-matting types is usually performed in early spring or fall.

Seed: Species can be propagated by seed; apparently, it can be sown at any time with satisfactory result for a year or so following harvest. No stratification is needed.

✛✛✛✛✛✛✛✛✛✛✛✛✛✛✛✛✛

*E*rigeron (ĕ-rij´ĕr-on) From Greek, *eri*, early (or *ear*, meaning spring), and *geron*, old or old man, in reference to the hoary appearance of the foliage of some species in the spring. The genus is composed of about 200 species of annual, biennial, and perennial herbs. Primarily they are native to North America; however, they are found nearly everywhere in the world.

Figure 3–75 *Erigeron karvinskiianus* (life-size)

FAMILY: ASTERACEAE (COMPOSITAE): Sun-flower family.

E. karvinskiianus (kär-vin-skē-ā´nus) Memorial (Figure 3–75).

COMMON NAME: Bonytip fleabane

NATIVE HABITAT: Mexico to Venezuela.

HARDINESS: Zones 9 to 10.

HABIT: Low, herbaceous, trailing, mat forming ground cover.

SIZE: 10 to 20 inches tall.

RATE: Relatively fast to invasive; space plants from quart-sized containers 14 to 20 inches apart.

LANDSCAPE VALUE: Most frequently used as a general ground cover in small- to moderate-sized contained areas or as an underplanting around shrubs and low trees. Tolerates high salinity and is thus useful along seaside. No foot traffic.

FOLIAGE: Evergreen, alternate, mostly entire but often dentate or lobed at the apex, elliptic-lanceolate or obovate-cuneate, to 1 1/4 inch long, 1/2 inch wide, glandular dotted and pubescent on both sides, ciliate margin, medium green.

STEMS: Decumbent and trailing, multibranched, to 3 feet long.

FLOWERS: Solitary, heads 3/4 inch in diameter, regular, ray flowers white fading to pink, then reddish-purple, disc flowers yellow, effective throughout most of the growing season.

FRUIT: Achene, not of ornamental significance.

OTHER SPECIES: *E. glaucus* (seaside daisy, beach fleabane) Hardy to zone 3; reaching 10 inches tall and spreading to 18 inches across. Leaves are fleshy, semievergreen, broadly spatulate to obovate, to about 6 inches long and arranged in rosettes, colored grayish blue. Flowers are arranged in heads, colored violet, effective mid to late spring. This plant is native to the Pacific Coast of the United States from southern California to northern Oregon. A pink ray-flowered cultivar named 'Roseus' and white ray-flowered cultivar 'Alba' are available.

CULTURE: Soil: Best in well-drained loamy soil. Adaptable to a range of pH from slightly alkaline to moderately acidic.
Moisture: Relatively tolerant to drought, but best with occasional irrigation to keep soil slightly moist.
Light: Full sun.

PATHOLOGY: Diseases: Downy mildew, leaf spot, powdery mildew, rusts, aster yellows.
Pests: Aphids.

MAINTENANCE: Shear or mow plantings following bloom.

PROPAGATION: Cuttings, division, and seed are all viable means of propagation.
Seed: Seed is collected ripe and stored cool until spring; it germinates in a week or two.

Erodium (ē-rō´di-um) From Greek *erodios*, a heron, in reference to the resemblance of the style and ovaries to the head and back of a heron. This genus is commonly called heron's-bill or stork's-bill and is composed of about 60 species of annual and perennial herbs. They are native to many regions throughout the world.

FAMILY: GERANIACEAE: Geranium family.

E. chamaedryoides (kam-e-drī-oy´dez) like chamaedrys (germander) (Figure 3–76).

COMMON NAME: Alpine geranium.

NATIVE HABITAT: Mediterranean region.

HARDINESS: Zones 7 to 10.

HABIT: Herbaceous, mat forming, dense ground cover.

SIZE: To 3 inches tall, spreading 4 to 6 inches across.

RATE: Relatively slow; space plants from 3 to 4 inch diameter containers 4 to 6 inches apart.

LANDSCAPE VALUE: Small-scale general ground covers. Excellent also in rock gardens or alpine gardens. No foot traffic.

FOLIAGE: Evergreen, primarily basal, on long petioles, round-ovate, somewhat cordate, to 3/8 inch long and about the same across, margin crenate and unlobed, pubescent both sides, medium to dark green.

FLOWERS: Solitary, regular, corolla white with pinkish venation, about 1/2 inch across, effective midspring to early fall.

FRUIT: Capsule to 5/8 inch long, not of great ornamental merit.

SELECTED CULTIVARS AND VARIETIES: 'Album' With white flowers.

'Florepleno' With single and double deep pink flowers.

'Roseum' Flowers rosy pink with red venation.

OTHER SPECIES: *E. manescavii* reaches about 1 foot tall and is said to possess excellent ground covering ability. Its basal pinnate leaves are large, hairy, gray-green and coarse textured. Flowers are magenta, to 1 1/2 inch across, blooming throughout summer. Hardy to zone 6.

E. chrysanthum is hardy to zone 7, is tufted, and reaches about 6 inches tall. Its silvery basal leaves and 2 to 5 flowered inflorescences with 1/2 inch diameter blooms are a beautiful yellow in summer.

CULTURE: Soil: Adaptable to most soils, good drainage being critical.

Moisture: Not notably tolerant to drought; soil should be maintained slightly moist with periodic watering.

Light: Full sun to light shade, generally best with some shade.

PATHOLOGY: Diseases: Leaf spot, viral curly top, stem and crown rots.

Pests: Aphids, soil mealybug.

MAINTENANCE: No special requirements.

PROPAGATION: Division: Simple division of plants in spring or fall is the usual means of propagation.

Seed: The species is sometimes propagated by seed.

*E*uonymus (ū-on´i-mus) or (ū-ō´nē-mus) Named after Euonyme, the mother of the Furies in Greek mythology, or from Greek *eu*, good, and *onoma*, a name, that is, of good repute. This very popular genus is composed of 120 species of deciduous and evergreen shrubs and trees. A few climb or trail in vinelike fashion, adhering by means of aerial rootlets.

NATIVE HABITAT: Primarily native to Asia, with some species from North America, Europe, Australia, and Africa.

FAMILY: CELASTRACEAE: Bittersweet family.

E. fortunei (fôr-tū´nē-ī) After Robert Fortune, a plant collector (Figure 3–77). Once known and occasionally still listed as *E. radicans*.

COMMON NAME: Wintercreeper, tsuramaskaki (Japanese for vine spindle tree).

NATIVE HABITAT: China.

HARDINESS: Zones 5 or 6 to 9 or 10.

HABIT: The species itself is an evergreen vine that is trailing or climbing in habit. It may reach a height

Figure 3–76 *Erodium chamaedryoides* (life-size)

Figure 3–77   *Euonymus fortunei* 'Acutus' (0.75 life-size)

of 20 feet and is supported by aerial rootlets. Types for ground cover are woody, clinging vines or subshrubs.

SIZE: See individual cultivars.

RATE: Generally, moderate to relatively fast growth; dwarf forms may be slow growing; space plants 12 to 16 inches apart from 2 to 3 inch diameter containers (small leaved selections, 8 to 10 inches apart). Sometimes pint- to gallon-sized containers are used with spacing of 2 1/2 to 3 feet commonly practiced.

LANDSCAPE VALUE: Dwarf and small-leaved varieties are best used on a small scale where a dense carpet is needed. Very effective in planters, between steps, and trailing over walls. Larger-growing forms are more vigorous and make excellent general-purpose ground covers for moderate- to large-sized areas. Limited foot traffic tolerated.

INFLORESCENCE: Axillary cymes on long peduncles; 4 parted flowers, small, greenish, ornamentally insignificant. Many cultivars seldom bloom.

FRUIT: Pinkish capsules, 3 to 5 valved; effective in fall, persistent.

FOLIAGE: Evergreen, opposite, simple, elliptic, elliptic-ovate, or elliptic-obovate; length to 2 inches, with crenate-serrate margins. Dark green with prominent silvery-white venation. Variable, often sporting unique foliage that leads to new cultivars.

SELECTED CULTIVARS AND VARIETIES: 'Acutus' With dark green, narrow foliage and vigorous prostrate habit. Often this is named as var. *acuta*.
'Canadian Variegated: To about 18 inches tall and

3 feet wide; with small green, waxy leaves that are edged in white.
'Coloratus' Commonly called purple winter creeper as its fall color is purplish, deepening in winter. Rapid growth with good soil-binding ability for slopes. Stems rooting; foliage dark green and glossy, with light green veins, 1 to 2 inches long. Often listed as forma or variety *colorata*, this is by far the most frequently used selection.
'Dart's Blanket' Rapid spreading clone that originated recently in Holland. Foliage leathery, waxy dark green and 1 to 2 inches long. Fall color bronzy and purplish-red below; to about 16 inches. Reported to have good salt tolerance, and for this reason may prove useful along drives and parking lots in the north and naturally saline soils in the southwest and coastal areas.
'Emerald and Gold' Low growing, branching shrub reaching 1 1/2 to 2 feet tall; foliage smaller, dark, glossy green with yellow margins, developing reddish tints in fall.
'Golden Prince' New foliage emerging bright yellow.
'Gracilis' A variable cultivar with white, yellow, or pink coloring in the foliage.
'Kewensis' Prostrate, mat forming, with tiny leaves about 1/4 to 5/8 inch long and 1/8 to 1/4 inch wide. When allowed to climb, the upper portion becomes like the variety radicans.
'Longwood' Similar to 'Kewensis' and 'Minimus' with moderate growth rate and prostrate habit.
'Minimus' Small green leaves are 1/4 to 1/2 inch long; habit is mat forming like 'kewensis,' but taller, and leaves are somewhat larger and growth much more vigorous.
Var. *radicans* Trailing or climbing rapidly; leaves are medium to dark shiny green, usually slightly smaller and more rounded than the species, serrate, 1 1/2 to 2 inches long; flowering and fruiting sporadically.
'Tricolor' 2 to 3 feet tall; low, broad spreading shrub of dense habit, with the ability to climb. Foliage is small, variegated green, white, and cream. Developing hints of pinks and reds in autumn.
'Tustin' Reportedly dense in habit with dark green conspicuously veined foliage, under sides of leaves becoming purplish in winter.
'Vegitus' Commonly called big leaf winter creeper or evergreen bittersweet, this is a clinging vine or subshrub with rounded, 1 inch diameter, dark green leaves. Dehiscent fruit capsules reveal orange fruit in fall.

OF SPECIAL INTEREST: *Euonymus fortunei* is similar to *Hedera helix* in that the juvenile form does not flower and has a different leaf morphology than the adult stage (dimorphic). Generally, adult foliage is thicker, leathery, more rounded, and smaller than juvenile foliage.

CULTURE: Soil: Adapts to almost any well-drained soil. Intolerant of oversaturated soils. A pH range from 5.5 to 6.5 is best, but slightly more acid or alkaline conditions are tolerated.

Moisture: Tolerant to moderate periods of drought, benefiting by occasional deep watering.

Light: Tolerates range of full sun to relatively dense shade.

MAINTENANCE: All *E. fortunei* varieties and cultivars benefit from seasonal pruning; the best time is spring and a rotary mower is often the most practical method for lower growing selections.

PATHOLOGY: Diseases: Anthrachose, bacterial crown gall, dieback, powdery mildew, leaf spot, and several other fungal pathogens.

Pests: Aphids, thrips, scales (several species).

PROPAGATION: Cuttings: Cuttings root readily throughout the growing season. 1000 to 3000 ppm IBA/talc is helpful with most varieties, but not essential. Horizontal shoot cuttings tend to produce sprawling plants, and upright shoot cuttings produce more upright plants.

Layering: A viable means of propagation, but seldom used on a large scale.

Seed: Seed should be removed from fruit and stratified for 3 to 4 months at 32° to 50°F prior to sowing.

*F*erns Members of the division Pteridophyta, ferns were among the first plants to inhabit the earth. Developing after algae and mosses millions of years ago, ferns marked the transition to plants with vascular (water conducting) tissue, an important development that allowed them to flourish in areas too dry for more primitive plants. No less important was their bearing of true leaves or megaphylls. Unlike earlier leaves, the development of megaphylls marked the transition to leaves with a network of venation (rather than unbranched veins), allowing the development of longer leaves (and thus more photosynthetic area), with even greater specialization. These two developments, along with more extensive root systems, ultimately allowed the ferns to become as diverse and successful as they presently are.

HABIT: Ferns exhibit a tremendous variety of habit and form. They range from the erect-stemmed, large-leaved tree ferns of tropical rain forests, which may exceed 75 feet in height, to small-leaved aquatic organisms, and horizontally spreading, rhizomatous or stoloniferous types. Leaves are typically identifiable as compound and finely divided, but vary to simple and unlobed, from several feet long to a few inches or less. Ferns that are suitable for ground covers are characterized by compact habit and generally have mature heights of 4 feet or less. The genera discussed in this text include *Adiantum, Asplenium, Athyrium, Cyrtomium, Cystopteris, Dennstaedtia, Dryopteris, Onoclea, Osmunda, Polypodium,* and *Polystichum.*

PROPAGATION: Ferns are commonly propagated as follows:

Division: Division can be practiced throughout the year but generally is most successful in early spring while still dormant or in late fall. Success is directly related to the size of the division. Avoid overwatering and physical damage to new shoots.

Root Cuttings: Sections of unrooted rhizomes will often root if placed one-half their depth with terminal (growing point) end exposed to air. The medium should be well drained and kept moist, shaded, and warm.

Buds: Some ferns, such as *Asplenium bulbiferium*, produce new plantlets on their fronds. They arise from knots in the leaves that are called buds. The new plantlets can be removed and planted in a moist, but well-drained, medium. One variation calls for anchoring them in the soil while still attached to the mother leaf, which provides nourishment until their roots develop. Once established, they may be cut away and transplanted into individual containers.

Tissue Culture: Methods have been devised, but they are not widely practiced on a commercial scale.

Spore Culture: Ferns reproduce naturally by spores, rather than seed as do the angiosperms (flowering plants). Growing ferns by spores involves collecting spores when ripe, roguing and/or disinfection, and then sowing on sterile medium in nutrient-containing solutions or agar. With proper care, green prothalia form, which in turn are transplanted and eventually develop into mature ferns. The entire process takes about 2 years to produce a small-sized transplantable plant. Consult the book, *Fern Growers Manual,* by Barbara Hoshizaki, for details.

PATHOLOGY: Diseases: Blights, damping off, molds, leaf spots, water molds, root rots.

Other: Algae and liverworts are frequent competitors of the prothalia when spores are cultivated.

Pests: Ants, in conjunction with aphids and

mealybugs; aphids, book lice (psocids and springtails) as vectors for bacteria and fungi; fungus gnats, mealybugs, millipedes, nematodes, spider mites, scales, snails, slugs, sowbugs and pillbugs, thrips, white flies. Various larvae of moths and butterflies, along with beetles, grasshoppers, and cockroaches may feed on foliage and prothalia.

OF SPECIAL INTEREST: Those wishing to further their understanding of ferns are encouraged to consult *Fern Growers Manual*, by Barbara Hoshizaki, called "an instant classic" by the *Garden Journal*. This book deals with nearly every aspect of ferns, their culture, and usefulness. *Ferns and Fern Allies*, by John T. Mickel, is also exceptional.

Festuca (fes-tū´kà) From Latin, *festuca*, a stem or blade. This genus is composed of more than 100 species of annual and perennial grasses primarily from temperate and cold regions, but native to most areas around the world.

FAMILY: POACEAE (GRAMINEAE): Grass family.

F. ovina var. *glauca* (ō-vē´nà) Pertaining to sheep, fodder for (Figure 3–78).

COMMON NAME: Blue sheep's fescue, blue fescue.

NATIVE HABITAT: Central and southwestern Europe.

HARDINESS: Zones 4 to 9.

HABIT: Tufted, low, hummuck forming grass.

SIZE: 8 to 12 inches tall spreading to 12 to 18 inches across.

RATE: Slow to spread; space plants from pint- or quart-sized containers 10 to 16 inches apart.

LANDSCAPE VALUE: A few plants work nicely for accent in front of large rocks or boulders or as a facing to taller growing, evergreen, shrubby ground covers. Occasionally and very interestingly put to use as a border along a sunny path. Useful for arranging geometrically or as a wavy looking general cover in small- to moderate-sized areas. No foot traffic.

FOLIAGE: Evergreen, basal, involute, entire, sharply pointed, slender, stiff in youth then arching over gracefully, colored glaucous blue (of the bluest of plants), 1/2 as long as culms, sheath split.

CULM: Slender, erect, stiff.

INFLORESCENCE: Compact panicle, to 6 inches long, often one sided, on stiff stems that carry them above the foliage; spikelets mostly 4 to 5 flowered,

Figure 3-78 *Festuca ovina* var. *glauca* (0.5 life-size)

brown in maturity, not greatly ornamental in comparison to the foliage, effective early to midsummer.

FRUIT: Grain, not ornamentally significant.

OTHER CULTIVARS AND VARIETIES: Many selections have been made over the years, but few are widely distributed or vary significantly. Of the more common is a cultivar named 'Sea Urchin.' This cultivar originated from a group of seedlings at Wayside Gardens. It is more rigid and uniform in habit.

CULTURE: Soil: Best in light textured, well-drained soils.

Moisture: Highly resistant to drought, wind, and heat. Color is best, however, when soil is kept slightly moist.

Light: Best in full sun, tolerant to light shade.

PATHOLOGY: No serious diseases or pests plague this species. Crown rot will occur if soil is too moist.

PROPAGATION: Division: Selections must be propagated by division, while the variety *glauca* is commonly grown from seed.

Seed: Easily propagated by seed. Commercial supplies are available, and quick germination in high percentages can be expected.

✝✝✝✝✝✝✝✝✝✝✝✝✝✝✝✝

*F**icus** (fi´kus) Latin name for fig. This genus is very large, being composed of about 800 species of trees, shrubs, and woody clinging vines. They are native primarily to the tropics, mostly of European origin.

FAMILY: MORACEAE: Mulberry family.

F. pumila (pū´mi-là) From Latin, *pumilo*, meaning dwarf (Figure 3–79). Often this species is listed as *F. repens*.

COMMON NAME: Creeping fig, climbing fig, creeping rubber plant.

NATIVE HABITAT: China, Japan, Australia, and North Vietnam.

HARDINESS: Zones 9 to 10.

HABIT: Creeping, clinging, woody vine that clings by rootlets, yet makes a trailing ground cover when unsupported.

SIZE: 1/2 to 1 1/2 inches tall without support; spreading indefinitely.

RATE: Slow to start, but becoming moderate to fast once established; space plants from quart- to gallon-sized containers 18 to 36 inches apart.

LANDSCAPE VALUE: Use as a low ground cover in small locations where spread can easily be contained or controlled. Elevated planters are ideal as the branches and leaves look very attractive as they hang down. Tolerates minimal foot traffic.

FOLIAGE: Evergreen, cordate-ovate in youth and about 1 inch long, glossy green, base oblique; mature foliage differs from juvenile, being elliptic or oblong, 2 to 4 inches long, leathery green.

FLOWERS: Minute, unisexual, not ornamentally significant.

FRUIT: Relatively large pear-shaped figs, to 2 inches long, not edible.

STEMS: Trailing or clinging by aerial rootlets.

OF SPECIAL INTEREST: Similar to *Hedera helix* and *Euonymus fortunei* in that during the maturation process, in which a stem begins to produce flowers, the branches cease to climb and the leaf morphology changes (dimorphism).

SELECTED CULTIVARS AND VARIETIES: 'Arina' Reportedly of Denmark; with more vigorous growth and greater propensity to branch.
'Minima' With smaller leaves.
'Variegata' Foliage variegated green and white; very attractive.
unnamed Reportedly, there is an ivy-leaved selection of *F. pumila* that has recently appeared on the market. It seems that Dr. John Creech, formerly of the U.S. National Arboretum, originally collected the plant in Cape Muroto in Japan around 1960.

CULTURE: Soil: Best in well-drained, organically rich loam.
Moisture: Tolerant to short periods of drought, but best when soil is kept moist.
Light: Full sun to light shade; avoid planting in a location where southern or western exposure is subjected to reflected heat, which will cause the foliage to turn yellow.

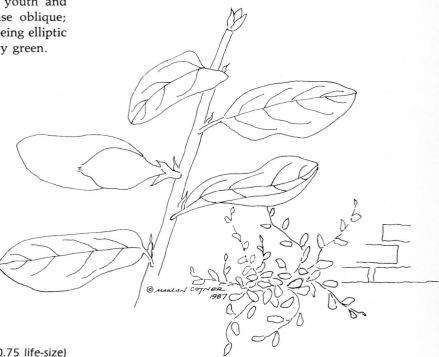

Figure 3–79 *Ficus pumila* (0.75 life-size)

PATHOLOGY: No serious diseases or pests reported.

MAINTENANCE: Occasionally clip back branches as they outgrow their bounds or attempt to climb trees or shrubs, which they can quickly overrun.

PROPAGATION: Cuttings: Cuttings root easily during summer and fall.

Layering: Rooted sections of trailing stems can be dug and transplanted during the cool seasons.

Filipendula (fil-i-pen´dū-là) From Latin *filum*, a thread, and *pendulus*, hanging, reportedly in reference to root-tubers of some species connected by threads. This genus is composed of a few species of perennial herbs that are native to various locations within the northern temperate zone.

FAMILY: ROSACEAE: Rose family.

F. vulgaris (vul-gā´ris) Common (Figure 3–80). Formerly *F. hexipetala*.

COMMON NAME: Dropwort.

NATIVE HABITAT: Europe and Siberia.

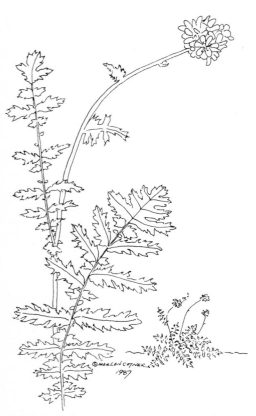

Figure 3–80 *Filipendula vulgaris* (life-size)

HARDINESS: Zone 3.

HABIT: Herbaceous, tuberous rooted, graceful ground cover.

SIZE: 1 to 1 1/2 feet tall, spreading to 1 1/2 feet across.

RATE: Moderate to relatively fast; space plants from pint- to quart-sized containers 14 to 16 inches apart.

LANDSCAPE VALUE: A fine, tough, general ground cover for use in small- or moderate-sized areas. Often it is used to face a border of herbaceous perennials. No foot traffic.

FOLIAGE: Alternate, fernlike, 4 to 10 inches long, with many pinnatifid leaflets that reach about 1 inch long, rich green with soft feathery texture.

INFLORESCENCE: Terminal corymbose panicles on leafy stems; flowers are creamy white, tinged red on the outside, about 3/4 inch wide, effective during early to midsummer.

FRUIT: Achenes, not ornamentally significant.

OF SPECIAL INTEREST: An extract of the flowers is said to be much like wintergreen oil. Roots reportedly contain tannic acid.

SELECTED CULTIVARS AND VARIETIES: 'Flore Pleno' With double white flowers; commonly called double dropwort.

'Grandiflora' Flowers creamy yellow and slightly larger than those of the species.

CULTURE: Soil: Adaptable to most soils, but showing a preference for moderately fertile soils that have good drainage.

Moisture: Tolerant to drought for a moderate period of time, but flowering is better if soil is maintained slightly moist.

Light: Full sun to moderate shade.

PATHOLOGY: No serious diseases or pests.

MAINTENANCE: Mow plantings after flowering is complete to improve appearance. Occasional division of clumps is necessary if floral display is to be optimal.

PROPAGATION: Division: Simple division of the tuberous rootstalk is commonly practiced in spring or fall.

Seed: The species is sometimes propagated by seed.

Forsythia (fôr-sith´i-à or sī´thi-à) After William Forsyth, superintendent of the Royal Gardens, Kensington (1737–1805). This genus is composed of six or seven species of erect or spreading shrubs.

FAMILY NAME: OLEACEAE: Olive family.

OF SPECIAL INTEREST: Forsythias are attractive to songbirds in that they provide good nesting sites and a food source, as their flower buds are sometimes consumed by birds in late winter (seldom to any great detriment of floral display). Hummingbirds, too, have been said to be attracted to their nectar.

* * *

F. × *intermedia* 'Arnold Dwarf' (in-tẽr-mē′dē-ȧ): Intermediate in habit with respect to parent plants (Figure 3–81).

COMMON NAME: Dwarf Arnold forsythia.

HYBRID ORIGIN: The species *F.* × *intermedia* is a hybrid formed by the cross *F. suspensa* × *F. viridissima*. It reaches 10 feet high with a wide, spreading, arching pattern to the branches. The low growing cultivar, 'Arnold Dwarf,' is the result of a seedling produced by crossing *F.* × *intermedia* with *F. japonica* var. *saxatilis* at the Arnold Arboretum in 1941.

HARDINESS: Zones 5 to 8.

HABIT: Compact, low spreading, mounding, woody shrub.

SIZE: 2 to 4 feet high by 6 to 7 feet wide.

RATE: Moderate; space plants from gallon sized containers 2 to 3 feet apart.

LANDSCAPE VALUE: Excellent for dwarf hedge or erosion controlling bank cover. Adds nice, soft, cool green texture along walkway or building entrance. No foot traffic.

STEM: Arching over and rooting as the nodes contact the soil; pith solid at the nodes, lamellate to hollow in internodes.

FOLIAGE: Deciduous, opposite, simple or three parted, ovate or oblong-ovate, to about 2 1/2 inches long, margins toothed from tip often extending below mid-section, light to medium green.

FLOWERS: Grouped 1 to 3, pale, greenish-yellow, very few produced in early to midspring.

FRUIT: Woody capsule, not ornamentally significant.

* * *

F. viridissima 'Bronxensis' (vir-i-dis′im-ȧ) Greenest, the stem.

COMMON NAME: Bronx dwarf forsythia, dwarf green-stem forsythia.

NATIVE HABITAT: Originated as a seedling of *F. viridissima* var. *koreana*. It was named in 1947 by the curator of the New York Botanical Gardens.

HARDINESS: Zones 5 to 8.

HABIT: Low, broad spreading, woody shrub.

SIZE: 1 to 3 feet tall by 2 to 4 feet across.

RATE: Moderate growing; space plants from 1 to 2 gallon sized containers 2 to 3 feet apart.

LANDSCAPE VALUE: Excellent dwarf hedge or foundation ground cover; planted en masse it makes a

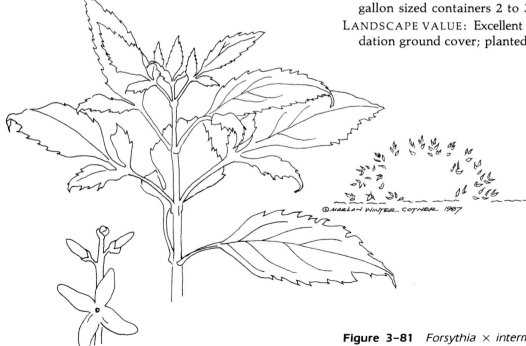

Figure 3–81 *Forsythia* × *intermedia* 'Arnold Dwarf' (life-size)

good bank stabilizer. More reliable and showy in bloom than 'Arnold Dwarf.' No foot traffic.

FOLIAGE: Deciduous, opposite, simple (unlobed or three parted), 3/4 to 1 3/4 inches long by 1/2 inch wide, narrowly oblong-ovate, gradually tapering to a point, base wedgelike, margins toothed, bright-medium green, petiole 1/2 inch long.

STEMS: Arching over, greenish in youth, later greenish-brown, pith not solid at nodes, rooting when in contact with the soil.

FLOWERS: Bright yellow, born profusely in early to midspring on old wood prior to leaf emergence.

FRUIT: Capsule, not ornamentally significant.

*   *   *

F. suspensa var. *sieboldii* (sus-pen'sà) Hanging down, the flowers.

COMMON NAME: Weeping forsythia.

NATIVE HABITAT: China.

HARDINESS: Zones 5 to 8.

HABIT: Wide, arching woody shrub.

SIZE: 4 to 5 feet tall by 10 to 15 feet across.

RATE: Moderate to relatively fast growing; space plants 4 to 6 feet apart, from 1 to 2 gallon sized containers.

LANDSCAPE VALUE: Often planted near the edge of a wall where the pendulous branches can cascade over. Also quite well suited for stabilizing steep banks. No foot traffic.

FOLIAGE: Deciduous, opposite, simple or sometimes three parted, ovate to oblong-ovate, 2 to 4 inches long, 1 to 2 inches wide, margin toothed, apex acute, base broad cuneate or rounded, dark green.

STEM: Yellowish brown, pendulous, trailing very long; pith hollow in internodes, solid at nodes.

FLOWERS: Usually solitary or grouped two or three together, 1 to 1 1/4 inches across, corolla golden yellow, effective in early to midspring.

FRUIT: Capsule, not ornamentally significant.

Generally Applicable to All Species

CULTURE: Soil: Adaptive to almost any well-drained soil, best in medium loam, with pH between 5.5 and 6.5.
Moisture: Once established, forsythias will tolerate a moderate duration of drought, but are best when soil is on the moist side.
Light: Full sun to moderate shade.

PATHOLOGY: Bacterial crown gall, leaf spots, and dieback.
Pests: Four-lined plant bug, Japanese weevil, northern root knot nematode, and spider mites.

MAINTENANCE: Pruning should be done immediately after flowering, as bud set is in the fall. No foot traffic.

PROPAGATION: Cuttings: Softwood and hardwood cuttings both root readily. Root inducing chemicals are not necessary, but dipping in a talc preparation with fungicide and 1000 to 2000 ppm IBA will increase quality and percent of take. It is not unreasonable to expect greater than 90% rootability with most cultivars.
Layering: A simple method of propagating forsythia on a small scale; rooted stem sections are best divided in spring or fall.

*F*ragaria (frà-gā'ri-à or frà-gãr'ē-à) From Latin *fraga* or *fragans*, sweet smelling, the fruit. This genus, commonly called strawberry, is made up of about 12 species of low growing, stoloniferous, herbaceous perennials. They are native to the northern temperate regions of the world. The common garden strawberry is a hybrid formed by the cross *F. chiloensis* × *F. virginiana*. It is called *F.* × *ananassa* and of course is suitable for use as a ground cover, as well as those species described here.

FAMILY: ROSACEAE: Rose family.

HABIT: Stoloniferous, herbaceous, densely carpeting, creeping ground covers.

RATE: Moderate to relatively fast; space plants from pint- or quart-sized containers 12 to 18 inches apart.

LANDSCAPE VALUE: Use as a binder for gradually sloping banks or as a general cover in small- to medium-sized areas. Naturalized in a woodland setting or as a facing to large shrubs and trees (where light is adequate), these plants are very attractive. Withstanding limited foot traffic.

*   *   *

F. chiloensis (chill-ō-en'sis) Origin, Chile.

COMMON NAME: Chiloe strawberry, beach strawberry.

NATIVE HABITAT: North America from Alaska to California, and Chile.

HARDINESS: Zones 4 or 5 to 10.

SIZE: 6 inches tall, spreading indefinitely.

FOLIAGE: Evergreen, compound; with 3 leaflets, obovate, 1 to 2 inches long, base cuneate, margin coarsely serrate, glabrous and shiny dark green above, densely silky tomentose below, turning reddish in fall; petioles 2 to 8 inches long, stout, arising from creeping rootstalk.

INFLORESCENCE: Cymose or racemose on scapes; flowers 3/4 inch wide, calyx silky, corolla with 5 petals, white, on flaccid stalks, effective in spring.

FRUIT: Small, firm, dark red, fleshy receptacle with seedlike achenes on the surface, edible but not as palatable as the garden strawberry.

* * *

F. vesca (ves´cả) Small or feeble (Figure 3–82).

COMMON NAME: European strawberry, sow-teat strawberry, woodland strawberry.

NATIVE HABITAT: European Alps.

HARDINESS: Zones 5 or 6 to 10.

SIZE: To 8 inches tall, spreading indefinitely.

GENERAL DISCUSSION: Similar in habit to *F. chiloensis*, but with leaflets to 2 1/2 inches long, glaucous below and bearing flowers and fruit throughout the season. Fruit reaches 3/8 to 3/4 inches in diameter.

* * *

F. virginiana (vẽr-jin-i-ā´nả) Of Virginia.

COMMON NAME: Wild strawberry, Virginia strawberry.

NATIVE HABITAT: Eastern Canada and the United States westward to Oklahoma.

HARDINESS: Zones 5 or 6.

SIZE: 4 to 12 inches tall, spreading indefinitely.

GENERAL DISCUSSION: Leaves are 1 to 4 inches long, blooms appear in late spring to early summer with some leaves turning red throughout summer. Fruit is small, cone shaped, and red. It is excellent for making jam.

Generally Applicable to All Species

CULTURE: Soil: Adaptable to range of soils; good drainage is quite important for success.
Moisture: Relatively tolerant to drought conditions, but best appearance is attained when soil is kept slightly moist.
Light: Full sun to light shade.

PATHOLOGY: Diseases: Anthracnose, leaf spot, blight, leaf rot, crown rot, wilt, viral diseases.
Pests: Nematodes.
Physiological: Iron-deficient chlorosis.

MAINTENANCE: Mow in early spring (well before flowering and growth of new foliage) to promote density and vigor.

PROPAGATION: Division: Division of rooted plantlets from runners (stolons) in fall or early spring is the standard means of propagation.
In Vitro: Methods have been used for propagating virus-free stock of garden strawberries, but to my knowledge have not been used to propagate any of the species mentioned here.
Seed: Seed is said to germinate readily, but is seldom used for propagating these species.

© MARLAN WINTER COTNER. 1987

Figure 3–82 *Fragaria vesca* (0.5 life-size)

✝✝✝✝✝✝✝✝✝✝✝✝✝✝✝✝✝✝

Fuchsia (few´chē-à or fū´shà) After Leonard Fuchs, a German botanist of the 14th century. The genus is made up of about 100 species of shrubs and trees. They are native from Mexico to Patagonia, along with New Zealand and Tahiti.

FAMILY: ONAGRACEAE: Evening primrose family.

F. procumbens (prō-kum´benz) Prostrate habit, without rooting (Figure 3–83).

COMMON NAME: Trailing fuchsia.

NATIVE HABITAT: New Zealand.

HARDINESS: Zone 10.

HABIT: Procumbent, herbaceous, mat forming ground cover.

SIZE: 1 to 12 inches tall (generally averaging 3 to 5 inches tall).

RATE: Moderate; space plants from pint- to quart-sized containers 10 to 14 inches apart.

LANDSCAPE VALUE: General ground cover for use in small areas. No foot traffic.

FOLIAGE: Evergreen, alternate, simple, glabrous and glandular dotted both sides, nearly round, 1/2 to 3/4 inch long, margin serrulate.

STEMS: Slender, brownish, procumbent.

FLOWERS: Solitary, axillary, 5/16 inch long, sepals green, calyx tube dark red, without petals, effective late summer to fall.

FRUIT: Bright red to purplish berry, to 3/4 inch across, persistent, effective during winter.

OF SPECIAL INTEREST: The flowers of fuchsia are said to attract hummingbirds and butterflies.

CULTURE: Soil: Organically rich, well-drained soil.
Moisture: Not markedly tolerant to drought; plant in a relatively cool location and keep the soil moist with watering as needed.
Light: Full sun to light shade.

PATHOLOGY: Diseases: Blight, root rot, rusts, wilt, dieback, leaf spot, virus.
Pests: Aphids, beetles, mealybugs, mites, scales, whitefly, thrips.

MAINTENANCE: No special maintenance required.

PROPAGATION: Cuttings: Cuttings root very easily. A treatment of 1000 ppm IBA/talc is helpful but not a necessity.
Division: Rooted stem sections can be dug up and separated in spring or fall.

✝✝✝✝✝✝✝✝✝✝✝✝✝✝✝✝✝✝

Galium (gā´li-um) From Greek *gala*, milk, the leaves of *G. verum* once having been used for the curdling of milk. This genus of about 300 species of generally thin, square-stemmed herbs is native to temperate regions throughout the world.

FAMILY: RUBIACEAE: Madder family.

G. odoratum (ō-do-rā´tum) Leaves fragrant (Figure 3–84).

COMMON NAME: Sweet woodruff, waldmeister, *musc de bois* (French, wood musk), bedstraw.

NATIVE HABITAT: Europe and Asia, temperate regions and mountains of tropical areas.

HARDINESS: Zones 4 to 8.

HABIT: Low, herbaceous, rhizomatous, upright stemmed, spreading ground cover.

SIZE: 6 to 10 inches tall, spreading indefinitely.

RATE: Relatively fast growing; space plants 8 to 12 inches apart from 3 inch diameter containers.

LANDSCAPE VALUE: Provides soft texture and uniform growth underneath shallowly rooted trees

Figure 3–83 *Fuchsia procumbens* (life-size)

Figure 3–84 *Galium odoratum* (life-size)

and shrubs. *Malus sargentii* (Sargent crab) blooms at about the same time as woodruff, and the combination of pink flowers above white is magnificent. Sweet woodruff thrives in very dense shade where few other plants can survive. Combines well with ivies, spring flowering bulbs, hosta, and ferns. Waldmeister (master of the woods, in German) is truly beautiful in a naturalized or woodland setting as an accent plant or large-scale general cover. It also finds use as an edging for perennial border or garden path. No foot traffic.

FOLIAGE: Deciduous but persisting late into fall, evergreen in warm climates, sessile, arranged in whorls of 6 to 8, glabrous and glandular dotted on both sides, obovate to oblanceolate, margins ciliate and finely serrate, to 1 1/2 inches long, with fringe of hairs at the base of whorls, fine textured, medium green, aromatic when dried.

STEMS: Four-sided, fragile, upright or decumbent, nodes prominent.

INFLORESCENCE: Axillary or terminal 1 inch diameter cymes of 1/4 inch long white flowers, in mid to late spring, mildly fragrant.

FRUIT: Two-lobed, two-seeded, indehiscent, not ornamentally significant.

OF SPECIAL INTEREST: The dried flowers and stems are strongly aromatic (much like freshly harvested hay) and are commonly used in sachets. Vegetative growth is also used as a vanillalike flavoring (following boiling in water) and in a culinary role (primarily as a garnish in wines, liquor, and other beverages). It is an essential ingredient in *Maitrank*, the German May wine that customarily is consumed the first of May. Although nonfragrant while alive, when dried the stems and leaves become very pleasantly aromatic and, in addition to having supposed insect repelling properties, are useful in potpourri, perfumes, and in the past as a stuffing for matresses; hence the common name bedstraw.

SELECTED CULTIVARS AND VARIETIES: There is a thick leaved, compact selection circulating in the trade that, to my knowledge, has not been given a name.

CULTURE: Soil: Best in rich, loamy, well-drained soil, but tolerant to sandy and heavy soils; a slightly acidic soil reaction is preferred.

Moisture: Relatively tolerant to drought, but best when planted out of strong wind in a cool location and supplied with enough water to keep the soil moist.

Light: Moderate to dense shade.

PATHOLOGY: Usually no serious pathology. Damping off can be severe when propagating by cuttings before the cells have become adequately lignified. Aphids are common in the greenhouse but not outdoors.

MAINTENANCE: Occasional weeding may be necessary as sweet woodruff outgrows its bounds. Mowing after bloom rejuvenates plantings.

PROPAGATION: Cuttings: Make cuttings in fall. Use scissors, so as not to separate the nodes, which are fragile. Proximal (toward crown) cut should be made slightly below a node. Stick at a density of one per every 2 square inches of soil area. Fall seems to be a favorable time as cells are thoroughly lignified. A talc compound with 2000 to 3000 ppm IBA works very well. Bottom heating with soil temperatures of about 60°F is helpful.

Division: Simply divide mats in spring or fall.

Seed: Commercial sources are available. Seed should be sown in the fall in a cold frame. Germination is generally slow and at a low percentage the following spring (as temperatures reach about 55°F). In sowing, freshly harvested seed usually is more productive than that which has been stored.

*G*ardenia (gär-dē´ni-à) Commemorating Dr. Alexander Garden, a botanist of South Carolina. The genus is composed of about 200 species of shrubs and small trees. They are native to the tropics and subtropics of the Old World.

FAMILY: RUBIACEAE: Madder family.

G. jasminoides 'Prostrata' (jas-min-oi´dēz) Resembling jasmine (Figure 3–85).

COMMON NAME: Creeping cape-jasmine, creeping gardenia.

NATIVE HABITAT: China

HARDINESS: Zones 9 to 10.

HABIT: Low, spreading, shrubby ground cover.

Figure 3–85 *Gardenia jasminoides* 'Prostrata' (life-size)

SIZE: 1 to 2 feet tall, spreading 2 to 4 feet across.

RATE: Relatively slow; space plants from gallon-sized containers 1 1/2 to 2 1/2 feet apart.

LANDSCAPE VALUE: Usually used as a small-to medium-scale general ground cover, especially good as a facing plant. No foot traffic.

FOLIAGE: Evergreen, opposite or whorled in threes, simple, thick, entire, somewhat revolute, 1/2 inch wide, glabrous both sides, dark glossy green often streaked with white.

STEMS: Rough, horizontal spreading, brownish in color.

FLOWERS: Solitary, single or double, corolla white, calyx 5 toothed, very fragrant, effective midsummer to early fall.

FRUIT: Ovate, orange, fleshy berry.

CULTURE: Rich, well-drained loam is best and pH in the range from 4.5 to 5.5 is optimal.
Moisture: Withstanding heat very well as long as soil is kept slightly moist. Plant high to avoid crown rot as plants are susceptible when soil does not drain adequately. North or eastern exposures are best if grown in desert regions.
Light: Full sun to moderate shade. Moderate shade is needed for best growth in inland areas.

PATHOLOGY: Diseases: Canker, leaf spots from both fungi and bacteria, bud rot, powdery mildew.
Pests: Aphids, mealybugs, scales, thrips, southern root knot nematode, white fly.
Physiological: Chlorosis when grown in alkaline soil.

MAINTENANCE: Mulch annually to conserve water.

PROPAGATION: Softwood cuttings root readily.

✚✚✚✚✚✚✚✚✚✚✚✚✚✚✚✚✚

Gaultheria (gâl-thē´ri-à) Commemorating Dr. Gaulthier, a botanist and physician of Quebec in the 18th century. The genus, commonly called wintergreen, is composed of about 100 species of evergreen, erect, or prostrate shrubs, subshrubs, and small trees. They are primarily native to the Andes of South America and to a lesser extent North America, and from Asia to Australia.

FAMILY: ERICACEAE: Heath family.

HABIT: Low, spreading, matted or procumbent, woody or herblike shrubs or subshrubs.

RATE: Relatively slow to moderate; space 4 to 6 inch diameter clumps 12 to 18 inches apart for herbaceous or semiherbaceous types. Space woody plants from gallon-sized containers 2 to 3 feet apart.

LANDSCAPE VALUE: Often used as companions to taller growing ericaceous plants such as rhododendron and azalea. Excellent in woodland or native gardens. Usually transplanted as sods or large clumps. Withstanding only very limited foot traffic.

OF SPECIAL INTEREST: Many species of *Gaultheria* possess aromatic and edible foliage and fruit. The leaves and fruit of *G. procumbens* were used in the past to produce wintergreen flavoring. Today, wintergreen flavoring is derived from *Betula lenta* or the chemistry lab.

* * *

G. hispidula (his´pi-dū-là) With hairy leaves.

COMMON NAME: Creeping pearlberry, maidenhair berry, moxie plum, creeping snowberry.

NATIVE HABITAT: Various locations of North America.

HARDINESS: Zone 3.

HABIT: Trailing, matlike, semiherbaceous ground cover.

SIZE: 2 inches tall, spreading indefinitely.

FOLIAGE: Evergreen, ovate, to 3/8 inch long, revolute, bristly below, leathery.

INFLORESCENCE: Terminal panicles; flowers white, less than 3/16 inch long, effective late spring.

FRUIT: Berrylike, china white and shiny; capsule within.

* * *

G. humifusa (hŭ-mi-fū´sà) Meaning sprawling on the ground.

COMMON NAME: Alpine wintergreen.

NATIVE HABITAT: North America.

HARDINESS: Zone 4.

HABIT: Trailing, low, matlike, semiherbaceous subshrublike ground cover.

SIZE: 4 inches tall, spreading to 12 inches across.

FOLIAGE: Evergreen, elliptic or orbicular, to 3/4 inch long, subentire to crenate, waxy.

FLOWERS: Solitary, corolla white, pink-tinged, bell shaped, 1/8 inch long, calyx glabrous, effective early summer.

FRUIT: Small, scarlet, berrylike drupe.

* * *

G. miquelina (mik-el´in-à) Commemorative.

COMMON NAME: Miquel wintergreen.

NATIVE HABITAT: Japan.

HARDINESS: Zone 5.

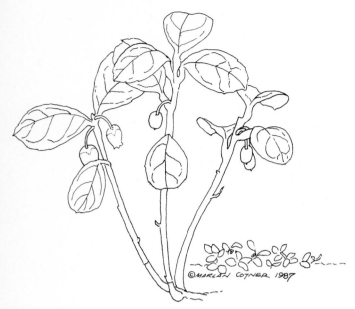

Figure 3–86 *Gaultheria procumbens* (0.75 life-size)

HABIT: Woody, low spreading shrub.

SIZE: To 12 inches tall.

FOLIAGE: Evergreen, elliptic to obovate, to 1 3/8 inch long, margin crenate-serrate, shiny green.

INFLORESCENCE: Short, nodding racemes; flowers pubescent, corolla white or pinkish, small, being less than 1/4 inch long, effective mid to late spring.

FRUIT: White or pink, small, berrylike, edible capsule, effective in fall.

*   *   *

G. ovatifolium (ō̄v-a-ti-fō ́li-um) Leaves oval shaped.

COMMON NAME: Oregon wintergreen.

NATIVE HABITAT: British Columbia to Idaho.

HARDINESS: Zone 5.

HABIT: Sprawling, shrubby ground cover.

SIZE: To 8 inches tall.

LEAVES: Evergreen, ovate to 1 1/2 inch long, leathery, toothed.

FLOWERS: Solitary, small, corolla less than 3/16 inches long, white calyx, pubescent, effective in spring.

FRUIT: Small, scarlet, berrylike, effective summer, capsule within.

*   *   *

G. procumbens (prō-cum ́benz) Procumbent growth habit (Figure 3–86).

COMMON NAME: Checkerberry, wintergreen.

NATIVE HABITAT: Eastern North America.

HARDINESS: Zone 3.

HABIT: Low, compact, creeping semiherbaceous subshrub.

SIZE: 3 to 5 inches tall, spreading indefinitely.

FOLIAGE: Evergreen, alternate, elliptic to narrowly obovate, to 2 inches long by 1 1/4 inches wide, glabrous both sides, margin slightly toothed, glandular dotted below, apex round, tip sharp, base wedge shaped below, dark glossy green above, petioles red, aromatic, bronze in fall, often variegated.

FLOWERS: Solitary, nodding, regular, tiny, white, bell-shaped corolla, effective midspring to early fall.

FRUIT: Small, fleshy, scarlet, berrylike, 2/5 inch long and slightly less wide, capsule within, effective late summer to fall, persisting often through winter, aromatic.

Generally Applicable to All Species

CULTURE: Soil: Generally best in moist, highly organic soils with a pH range from 4.5 to 5.5 Moisture: Requiring high levels of soil moisture. Soil should be kept moist with watering as needed. Light: Light to moderate shade.

MAINTENANCE: Mulch annually with acidic compost or peat if growing in highly amended soils. I have seen some of the herbaceous types respond to periodic mowing by becoming much denser than if not mowed.

PATHOLOGY: Diseases: Leaf spots, black mildew, red leaf, gall, powdery mildew. Pests: Aphids, thrips.

PROPAGATION: Cuttings: Softwood cuttings root relatively well but very slowly. They should be taken in early summer. Division: Division in early spring is the standard means of propagating herbaceous types. Seed: Reportedly, many species respond to 3 months of cold stratification at 40°F prior to sowing.

✝✝✝✝✝✝✝✝✝✝✝✝✝✝✝✝

G̶aylussacia (gā-lu-sā ́shi-ȧ) Named after the French chemist, J. L. Gay-Lussac. This genus is composed of about 40 species of shrubs that are native to South and North America.

FAMILY: ERICACEAE: Heath family.

G. brachycera (bra-kis ́e-rȧ) Short horned (Figure 3–87).

COMMON NAME: Box huckleberry.

NATIVE HABITAT: North America, including Pennsylvania, Virginia, Tennessee, and Kentucky.

HARDINESS: Zone 5.

HABIT: Low, rhizomatous, woody shrub.

SIZE: 6 to 18 inches tall, spreading indefinitely.

RATE: Slow, space plants from gallon-sized containers 2 to 3 feet apart.

LANDSCAPE VALUE: Valuable as a specimen in naturalized settings. Combining well with pines, rhododendrons, azaleas, and other ericaceous plants. Not tolerant to foot traffic.

FOLIAGE: Evergreen, alternate, simple, elliptic to ovate, medium fine texture 3/4 to 1 inch long, margins finely crenate, short petioled, glabrous, dark glossy green above, often with reddish cast when grown in full sun, paler below, winter color is bronze to reddish purple.

STEMS: Slender, rounded, pith continuous.

INFLORESCENCE: Axillary racemes; flowers perfect, corolla urceolate, campanulate, or tubular, white or pinkish, 1/4 inch long, effective mid to late spring.

FRUIT: Self-sterile, but when cross pollinated, bluish berrylike drupes are formed that are effective mid to late summer.

Figure 3–87 *Gaylussacia brachycera* (life-size)

CULTURE: Soil: Requiring well-drained, porous, organically rich soil of pH 4.5 to 5.5.
Moisture: A relatively high moisture level must be maintained at all times.
Light: Full sun to moderate shade.

PATHOLOGY: No serious diseases or pests reported.

MAINTENANCE: In amended soils it is wise to test the pH of the soil periodically and amend accordingly. Annual incorporation of peat moss will in many cases keep the pH within the acceptable range.

PROPAGATION: Cuttings: Soft and hardwood cuttings can be rooted, being helped along with 8000 ppm IBA/talc.
Seed: Sow ripe seed immediately for natural stratification or stratify first warm, then cold, for a period of 1 to 2 months each time; then sow.

*G*azania (gā-zā´ni-à) From Latin, *gaza*, treasure or riches, in reference to the large showy flowers, or possibly in commemoration of Theodore of Gaza, who translated the botanical works of Theophrastus into Latin. This popular genus is composed of about 16 species of annual and perennial rhizomatous herbs and shrubs. They are native to Africa and their leaves contain a milky juice. The species have been extensively hybridized, resulting in many outstandingly beautiful flowered selections. Unfortunately, the nomenclature has become confused in the process.

FAMILY: ASTERACEAE (COMPOSITAE): Sunflower family.

SIZE: Generally 6 to 8 inches tall, spreading indefinitely.

RATE: Moderate to relatively fast; space plants from pint-size containers 10 to 14 inches apart.

LANDSCAPE VALUE: Trailing types are excellent for edging walks and drives or in beds as facing. They can be considered versatile general ground covers for use on most medium to large, flat or sloping areas.

* * *

G. linearis (lin-ē-ā´ris) Linear or grasslike, the leaves.

COMMON NAME: Treasure flower.

HARDINESS: Zones 8 to 10.

HABIT: Stemless, low, herbaceous ground cover.

FOLIAGE: Basal rosettes, evergreen, 4 to 8 inches

long, linear to linear-lanceolate, rarely pinnatifid, ciliate and somewhat revolute margined, primarily glabrous gray-green above and white tomentose below.

FLOWERS: Regular, solitary, daisylike heads to about 2 3/4 inches across, elevated on scapes to 14 inches tall, ray flowers yellow or orange, sometimes with a dark brown basal spot, disc flowers tubular, bisexual, opening in the sun, spring to summer.

FRUIT: Achenes, not of ornamental significance.

* * *

G. rigens (rī´jenz) Rigid, stiff (Figure 3–88).

COMMON NAME: Trailing gazania.

HARDINESS: Zones 8 to 10.

HABIT: Trailing, herbaceous, rhizomatous, ground cover.

FOLIAGE: Simple, evergreen, lanceolate or obovate-lanceolate, in basal rosettes, to about 4 to 5 inches long by 1/2 inch across, entire or occasionally pinnatifid, gray-green and glabrous above, white tomentose below.

STEMS: Decumbent, glabrous or hairy, to 16 inches long.

FLOWERS: Regular, solitary, daisylike, heads to 3 inches across, on glabrous peduncles 4 to 6 inches long, ray flowers yellow to orange with dark brown white-eyed basal spot, opening in the sun in spring to summer.

FRUIT: Achenes, not ornamentally significant.

SELECTED CULTIVARS AND VARIETIES: Var. *leucoleana* Heads to 1 1/2 inch diameter with yellow ray flowers.
'Sunburst' Orange ray and black disc flowers 2 inches across; foliage gray.
'Sunglow' Entirely yellow flowered.

'Sun Gold' with 2 inch diameter butter-yellow flowers and gray foliage. Introduced in 1971 by the Los Angeles Arboreteum.
'Sunrise Yellow' Green leaved with yellow ray and black disc flowers.

OTHER INTERESTING TYPES PROBABLY OF HYBRID ORIGIN: 'Aztec Queen' Multicolored.
'Colorama' With flowers to 4 inches across in mixed colors.
'Copper King' Orange flowered with markings in various colors.
'Fiesta Red' Dark reddish-orange flowers.
'Fire Emerald' 3 to 3 1/2 inch diameter flowers in various colors with the bases of the ray flowers emerald green.
'Moon Glow' With double yellow flowers.

Generally Applicable to Both Species

CULTURE: Soil: Tolerant of most soils, even those low in fertility, provided drainage is good.
Moisture: Requiring only moderate amounts of additional water during dry periods. Water only as often as necessary to maintain turgor in foliage.
Light: Full sun.

PATHOLOGY: Diseases: Crown rot.
Pests: Soil mealybug, thrips.

MAINTENANCE: Mow after flowering to keep plantings dense and to stimulate new growth.

PROPAGATION: Division: Clones must be propagated by division, usually in early spring, late fall, or winter.
Seed: Many hybrid groups are propagated by seed. Germination of spring sown seeds takes about 1 to 2 weeks at 70 to 80 °F; covering to exclude light is helpful.

Gelsemium (jel-sē´mi-um) Latinized version of *gelsomino*, Italian for jasmine. The genus is composed of only three species of twining shrubs that are native to eastern Asia and eastern North America.

FAMILY: LOGANIACEAE: Logania family.

G. sempervirens (sem-pēr-vī´renz) From Latin, *semper*, always, and *vireo*, green (i.e., evergreen) (Figure 3–89).

COMMON NAME: Carolina jessamine, Carolina yellow jasmine, yellow jessamine, evening trumpet flower, Carolina wild woodbine, yellow false jessamine.

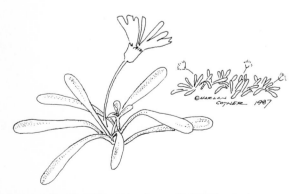

Figure 3–88 *Gazania rigens* (0.25 life-size)

NATIVE HABITAT: Woodlands of eastern North America from Virginia southward.

HARDINESS: Zones 7 to 9.

HABIT: Shrubby twining vine that will climb with support; unsupported, it is a useful, sprawling ground cover.

SIZE: To 3 feet without support, spreading 20 to 30 feet across.

RATE: Moderate; space plants from gallon-sized containers 3 to 4 feet apart.

LANDSCAPE VALUE: Best as a general cover for large areas where it will not interfere with shrubs and trees. Next to a retaining wall, the stems cascade over gracefully. No foot traffic.

FOLIAGE: Evergreen, or semievergreen in northern range, opposite, simple, entire, lanceolate or oblong-lanceolate, rarely ovate, 1 to 4 inches long, acute or acuminate apex, short petioled, narrowing at base, glossy dark green above, paler below, purplish in winter.

INFLORESCENCE: Cymes or solitary; flowers fragrant, perfect, to 1 1/2 inch long, corolla bright yellow and shaped like a funnel with 5 imbricate lobes, effective in spring, and often sparsely again in fall.

FRUIT: Compressed 3/4 to 1 1/2 inch long capsule of little ornamental significance.

SELECTED CULTIVARS AND VARIETIES: 'Pride of Augusta' ('Plena') A double flowering selection; bloom period extends into summer.

OF SPECIAL INTEREST: Vegetative growth and flowers are poisonous when ingested, which should be considered when selecting a site for this plant. Children should be made aware of the potential danger and it should be eradicated from livestock pastures. Reportedly, the toxic substances are gelsemine, gelsemicine, and other related alkaloids. Ingestion results in headache, dizziness, visual disturbances, and in severe cases muscular disfunction.

G. sempervirens is the official flower of South Carolina.

CULTURE: Soil: Adaptable to most soils, but best in rich loam of pH 5.0 to 6.0
Moisture: Tolerant of short periods of drought, but appearing best when moisture is maintained with watering as needed.

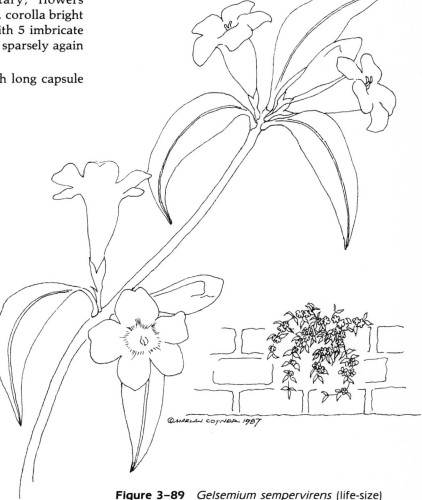

Figure 3–89 *Gelsemium sempervirens* (life-size)

Light: Full sun to light shade, with flowers more profuse in full sun.

PROPAGATION: Cuttings: Cuttings root well when taken in mid to late summer and treated with a root inducing preparation of moderate concentration such as 3000 ppm IBA/talc.

Seed: Seed is also occasionally used; collect seed as it ripens, store cool and dry; then sow the following spring. Give temperatures of 65° to 75°F.

Genista (je-nis'tà) Ancient Latin name for the broom plant. This genus, like *Cytisus* (a closely related genus), is commonly called broom. It is composed of from 75 to 90 species of usually small deciduous, sometimes armed shrubs. Leaves are simple or trifoliate, small, or sometimes nearly non-existent. Branches are green and compose a good share of the photosynthetic surface. They are native to Europe, north Africa, and western Asia.

FAMILY: FABACEAE: Pea family.

DESCRIPTION: Generally, the various ground covering species can be described as follows:

HABIT: Low to medium height, clumplike, woody based, shrubby ground covers.

RATE: Moderate; spacing of 1 1/2 to 2 1/4 feet is often adequate for plants from quart- to gallon-sized containers.

LANDSCAPE VALUE: Sometimes used as general cover or even as a lawn substitute in moderate- or smaller-sized areas. Combines well with species of *Erica* and *Calluna* and makes interesting specimens for the rock garden. Withstands planting on slopes in any event, but not tolerant to foot traffic.

FOLIAGE: Usually alternate, simple, or with leaves in threes, from 1/2 to 1 inch long, most often deciduous.

STEMS: Green, occasionally spiny, pith rounded and continuous.

INFLORESCENCE: Terminal racemes or heads; flowers papilionaceous, usually yellow, calyx 2-lipped, claws on wing and keel petals united to staminal tube, generally quite showy.

FRUIT: Flat legume.

GROUND COVERS: The most useful ground covering species are further described as follows:

* * *

G. germanica (ger-man'i-kà) German.

COMMON NAME: German woadwaxen, German broom.

NATIVE HABITAT: Central and southern Europe.

HARDINESS: Zone 5 to 7.

SIZE: To 12 inches tall, spreading to 12 inches across.

DISCUSSION: A dwarf, spiny species that has simple leaves that are elliptic-oblong; flowers are born profusely on short racemes in early summer and are colored brilliant yellow.

* * *

G. hispanica (his-pan'i-kà) Spanish.

COMMON NAME: Spanish gorse, Spanish broom.

NATIVE HABITAT: Spain and northern Italy.

HARDINESS: Zones 7 to 9.

SIZE: To 12 inches tall, spreading to 2 feet across.

DISCUSSION: This species has simple leaves that are elliptic-oblong; stems spiny; flowers many, colored yellow, born late spring to early summer.

* * *

G. horrida (hor-i'dà) In Latin *horrida* means shaggy, rough, or bristly. Here, the reference is to the spiny branches.

COMMON NAME: Cushion woadwaxen.

NATIVE HABITAT: Spain and southern France.

HARDINESS: Zone 7.

SIZE: To 1 1/2 feet tall.

DISCUSSION: This species is densely branched, spiny, with silvery-gray leaves that are grouped in threes; flowers appear in small terminal heads from midsummer to early fall, colored yellow.

* * *

G. lydia (lid'i-à) From Lydia, an ancient maritime province of western Asia Minor.

NATIVE HABITAT: Eastern Europe.

HARDINESS: Zone 7.

SIZE: To 1 1/2 feet tall.

DISCUSSION: An excellent dwarf shrub with slender, pendulous branchlets; flowers golden-yellow, born in profusion in late spring to early summer.

* * *

C. pilosa (pi-lō'sà) Downy.

COMMON NAME: Silky-leaf woadwaxen.

NATIVE HABITAT: Europe.

HARDINESS: Zones 5 to 8.

SIZE: To 12 inches tall, spreading 2 to 3 feet across.

DISCUSSION: This species is unarmed, with obovate to oblong, downy, simple leaves; flowers are many, yellow, in terminal racemes in early summer.

SELECTED CULTIVARS AND VARIETIES: 'Vancouver Gold' A low growing (4 to 6 inches tall), broad spreading (to 3 feet across) cultivar that was selected by E. H. Lohbrunner in Victoria, British Columbia. Named in 1983 at the University of British Columbia Botanical Garden, this selection flowers profusely, golden-yellow, in late spring and early summer. Introduced into the trade in 1985, 'Vancouver Gold' seems to be worthy of use as specimen atop a rock ledge where it can cascade over, or as a general cover and soil stabilizer on banks and large, open sites.

* * *

G. sagittalis (saj-i-tā´lis) Like an arrow, the winged branchlets.

COMMON NAME: Arrow broom.

NATIVE HABITAT: Central Europe.

HARDINESS: Zones 3 to 4.

SIZE: Reaching 6 to 12 inches tall, spreading to 2 feet across.

DISCUSSION: This species has branches that are broadly winged and prostrate; leaves are singular, simple, ovate to oblong, 3/4 inch long, villous; flowers yellow, arranged in terminal racemes, effective early summer.

* * *

G. sylvestris (sil-ves´tris) Of woods or forests.

COMMON NAME: Forest broom.

NATIVE HABITAT: Yugoslavia.

HARDINESS: Zone 6.

SIZE: To 6 inches tall.

DISCUSSION: This species has leaves that are singular, linear to narrowly lanceolate; stems spiny; flowers bright yellow in terminal racemes from early to midsummer. The var. *pungens* is more spiny than the species.

* * *

G. tinctoria (tink-tō´ri-à) Of dyers (Figure 3–90).

COMMON NAME: Dyer's greenwood, dyer's woad-waxen.

NATIVE HABITAT: Europe and western Asia.

HARDINESS: Zone 5.

SIZE: 2 to 3 feet tall.

DISCUSSION: Characterized by striped branches,

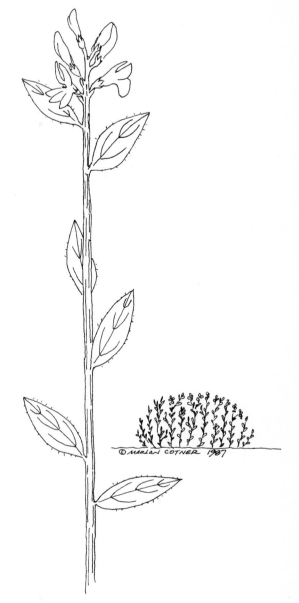

Figure 3–90 *Genista tinctoria* (life-size)

leaves that are singular, bright green, elliptic-oblong, to 1 inch long; flowers are bright yellow on terminal racemes from early summer to early fall.

Generally Applicable to All Species

CULTURE: Soil: Best growth obtained when grown in infertile, well-drained, sandy or rocky soils. Moisture: Able to withstand considerable drought. Occasional deep watering in summer may be beneficial, but too much watering will make plants become open and less attractive. Light: Full sun to light shade.

PATHOLOGY: Diseases: Dieback, powdery mildew, rust.

Pests: Bean aphid, scales, genista caterpillar.

MAINTENANCE: Prune very lightly after flowering to increase density of branches and to promote compactness.

PROPAGATION: Cuttings: Both softwood and hardwood cuttings root readily. Treatment with 3000 to 8000 ppm IBA/talc is recommended by many authorities.

Seed: Seed may be processed and sown at once upon ripening, or may be stored cool and dry for a year. Processing consists of soaking in sulfuric acid for 30 minutes. Following acid treatment, a soak in hot water (190 °F) for 18 hours may increase germination percentages.

*G*eranium (je-rā′ni-um) From Greek, *geranos*, a crane, the fruit resembling the head and beak of a crane. This genus is composed of more than 300 species of annual and perennial herbs and shrubs that originated in temperate and montane tropic regions the world over. Plants of this genus are not the same as the florist's geranium, which is commonly grown on porches and patios. Geraniums of that sort belong to genus *Pelargonium*. Many true geraniums are suitable for ground covers; however, their introduction into commercial horticulture in the United States has been slow. One of the more common species, *G. incanum*, is described here, and the unique features of the others are discussed later.

FAMILY: GERANIACEAE: Geranium family.

HABIT: Herbaceous, rhizomatous, upright, spreading by trailing, occasionally clump forming ground covers.

RATE: Relatively fast; space plants from quart- or gallon-sized containers approximately two to three times the expected mature height of the plants apart from each other.

LANDSCAPE VALUE: Excellent for edging pathways and facing shrubs in a border setting. Good also as specimens in rock garden or as general covers (sometimes as a lawn substitute) or soil binder on slopes in areas of moderate size.

* * *

G. incanum (in-kā′num) Hoary-grayish-white.

COMMON NAME: Carpet geranium.

HARDINESS: Zones 9 to 10.

SIZE: 6 to 10 inches tall.

FOLIAGE: Mostly evergreen, opposite or whorled, palmately dissected, with segments linear, 1/4 inch long by 3/16 inch across, revolute, sparsely hairy and medium green above, lighter and tomentose below, basal leaves long petioled, upper short petioled.

FLOWERS: Solitary, regular; corolla 5 petaled, imbricated, to 1 inch across, rosy purple to magenta pink, effective spring to fall.

FRUIT: Long-beaked stiff carpels attached to styles, not of great ornamental importance. Fruit is interesting for the fact that at maturity they disperse seed by explosively erupting.

* * *

G. cinereum (sin-ēr′i-um) Meaning unclear.

NATIVE HABITAT: Pyrenees Mountains.

HARDINESS: Zone 5 or 6.

SIZE: To 6 inches tall and mounding.

DISCUSSION: This species has leaves that are 5 to 7 lobed; summer borne flowers are axillary or from crown, 1 inch across, petals pink with dark stripes. Cultivars and varieties include 'Album' (white flowered) and var. *subcaulescens* which has hairs on the sepals that are spreading not adpressed.

* * *

G. endressii (en-dres′i-i) After Endress.

COMMON NAME: Endress cranesbill.

NATIVE HABITAT: Pyrenees Mountains.

HARDINESS: Zones 5 or 6.

SIZE: To 15 inches tall, spreading to 2 feet across.

DISCUSSION: This species has leaves that are basal, light green; flowers magenta-pink, fragrant, effective midsummer. The cultivar 'Wargrave Pink' has salmon-pink flowers.

* * *

G. himalayense (him-à-lā′en-sē) Himalayan.

COMMON NAME: Himalayan cranesbill.

NATIVE HABITAT: Turkestan, south of northern Tibet.

HARDINESS: Zone 4.

SIZE: 12 inches tall, spreading to 2 feet across.

DISCUSSION: This species has leaves that are deeply cut, 5 to 7 parted, to 3 or 4 inches long; petals lilac with purple venation, entire flowers in pairs on axillary peduncles, 1 1/2 inch across, born midsummer. The cultivar 'Plenum' had double, purple-blue flowers. The variety *alpinum* (sometimes listed as *G. grandiflorum*) has wrinkled, deep blue flowers.

*   *   *

G. macrorrhizum (mak-rō-rhiz´um) Large rooted (Figure 3–91).
COMMON NAME: Bigroot geranium.
NATIVE HABITAT: Southern Europe.
HARDINESS: Zone 3.
SIZE: 12 to 18 inches tall, spreading to 2 feet across.
DISCUSSION: This species has dark green, 5 to 7 lobed leaves and terminal flowers that are colored magenta, which appear throughout summer.

*   *   *

G. pylzowianum (pil-zō-ē-ān´um) After Pylzow.
NATIVE HABITAT: China.
HARDINESS: Zone 7.
SIZE: 12 inches tall.

DISCUSSION: This species has 5 lobed leaves that are deeply cut; floweres on axillary peduncles, corolla purple, borne early summer.

*   *   *

G. sanguineum (san-gwin´ē-um) Meaning blood-red, color of the flowers.
COMMON NAME: Bloody cranesbill.
NATIVE HABITAT: Europe and Asia.
HARDINESS: Zones 3 to 8.
SIZE: To 12 inches tall.
DISCUSSION: This species has leaves that are 5 to 7 lobed and deeply cut, turning scarlet in autumn. Its flowers are born on axillary peduncles, corolla is reddish purple to magenta, effective midspring to late summer. The cultivar 'Album' has white flowers, while the variety *Prostratum* is dwarf and more compact. The latter may occasionally be listed as variety *haematodes*. 'Lancastriense' (Prostratum') is a low mounding selection with reddish-pink flowers with darker venation.

*   *   *

G. wallichianum (wâl-litsch-ē-ān´um) For Dr. Wallich, a director of the Botanical Gardens, Calcutta, India.
COMMON NAME: Wallich geranium.
NATIVE HABITAT: Himalayas.
HARDINESS: Zone 6.
SIZE: 2 to 4 inches tall, prostrate.

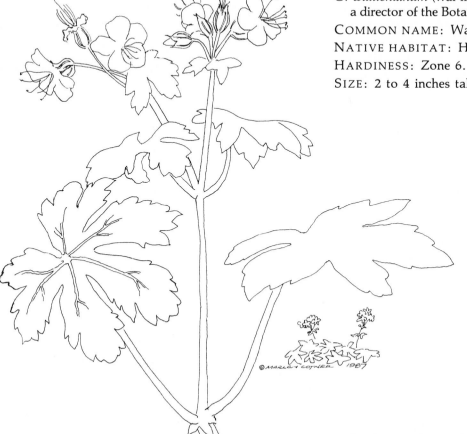

Figure 3–91 *Geranium macrorrhizum* (0.66 life-size)

DISCUSSION: This species has leaves that are 3 to 5 lobed, colored soft green, and velvety to the touch; purple flowers are borne on axillary peduncles in midsummer. The cultivar 'Buxton's Blue' has lavender-blue flowers.

Generally Applicable to All Species

CULTURE: Soil: Adaptable to most soils with good drainage and a pH range of 6.0 to 8.0.
Moisture: Somewhat tolerant to drought, but best growth is attained when soil is maintained slightly moist with irrigation as needed during summer.
Light: Full sun to light shade, blooming best in full sun.

PATHOLOGY: Diseases: Leaf spots, both bacterial and fungal; rusts, downy mildew, leaf gall, viral mosaic.
Pests: Four-lined plant bug, aphids.

MAINTENANCE: Shearing growth back about halfway through bloom may initiate another heavy flowering period later and will help keep the plants compact. Seed is generally viable and self-sowing, but hybrid selections may not come true from seed. In such cases, it is advisable to shear off fading flowers before fruit has a chance to ripen. Over time, plants in mass plantings may show deterioration. In such cases they should be thinned.

PROPAGATION: Cuttings: Root cuttings are most effective in late fall after the roots have been thoroughly frozen. They should be prepared by cutting into 1 to 2 inch long sections and laying horizontally on the medium. Cover with 1/4 to 3/8 inch of medium and keep slightly moist. Moderate bottom heat will speed the formation of shoots, while overwatering should be avoided as rotting will occur. No growth regulators are needed. Terminal stem cuttings should be made from pathogen-free stock and stuck in a porous, sterile medium. They can be rooted year round.
Seed: Commercial sources are available. Germination occurs rapidly at temperatures around 70 °F.

✚✚✚✚✚✚✚✚✚✚✚✚✚✚✚✚✚✚

*G*lechoma (glē-kō′mȧ) From Greek, *glechon,* a sort of thyme or pennyroyal; old generic name for ground ivy. This genus is composed of about ten species of low growing, creeping herbs that are native to Europe and Asia.

FAMILY: LABIATAE: Mint family.

G. hederaceae (hed-ẽr-ā-cē′ȧ) Hedera (ivy)-like, its habit (Figure 3–92).

COMMON NAME: Ground ivy, gill-over-the-ground, runaway robin, field balm, alehoof.

NATIVE HABITAT: Europe and Asia.

HARDINESS: Zone 3.

HABIT: Creeping, herbaceous, matlike, rhizomatous, low ground cover.

SIZE: 3 to 6 inches tall, spreading indefinitely.

RATE: Fast to invasive; space plants from 3 to 4 inch diameter containers 12 to 14 inches apart.

LANDSCAPE VALUE: Plant in smaller confined areas as a general ground cover. These plants often invade lawn areas but are easily controlled with broadleaf herbicides. They withstand limited foot traffic.

LEAVES: Evergreen, opposite, simple, 1/4 to 1 inch long by about the same width, orbicular or reniform (kidney shaped) in outline, long petioled, conspicuously palmately veined, glandular dotted and somewhat hairy both sides, margins crenate and sparsely ciliate, dull dark green.

STEMS: Squarish, revolute.

INFLORESCENCE: Loose axillary verticillasters, each with a few flowers; corolla bluish, to 1/2 inch long, 2-lipped, borne spring and summer.

Figure 3–92 *Glecoma hederacea* (life-size)

FRUIT: Of 4 glabrous nutlets, not of ornamental significance.

SELECTED CULTIVARS AND VARIETIES: Var. *variegata* Leaves with white or pinkish margins.

CULTURE: Rich, fertile, well-drained loam is best, but poorer infertile soils may be desirable as they will slow growth.
Moisture: Best appearance and most lush growth is realized when the soil is kept moist with frequent irrigation.
Light: Full sun to moderate shade.

PATHOLOGY: No serious diseases reported, but aphids and mites are common pests.

MAINTENANCE: Mow occasionally to keep plantings looking uniform if uniformity is desirable.

PROPAGATION: Division: Dividing is effective anytime as long as soil is kept moist until roots establish.
Seed: Seed is reported to germinate readily.

Gypsophila (jip-sof´i-là) From Greek, *gypsos*, chalk, and *phileo*, to love, in reference to the plant's love for chalky soil. This genus is composed of about 125 species of annual, biennial, and perennial herbs and subshrubs. They are mainly native to Eurasia, with a few from Egypt and Australia.

FAMILY: CARYOPHYLLACEAE: Chickweed or pink family.

LANDSCAPE VALUE: Ground covering types of gypsophila are excellent when grown next to the top of a rock ledge where they can billow over. As a specimen in a rock garden, they lend a unique softening character, as they do when put to use for edging walks and accenting building entrances. No foot traffic.

* * *

G. paniculata (pan-ik-ū-lā´tà) Panicled, the inflorescences.

COMMON NAME: Baby's breath.

NATIVE HABITAT: Caucasus to central Asia.

HARDINESS: Zones 3 to 8.

HABIT: Compact, low, graceful, rhizomatous, herbaceous ground cover.

SIZE: 2 to 3 feet tall, spreading 3 to 4 feet across.

RATE: Moderate; space plants from pint- to quart-sized containers 18 to 24 inches apart.

FOLIAGE: Opposite to fasicled or whorled, linear to lanceolate, to 3 inches long, acute, glaucous, margins ciliate, midrib conspicuous, rarely pubescent.

STEMS: Highly branched, glaucous, glabrous, rarely pubescent.

INFLORESCENCE: Loose, profusely branched panicles, pedicles to 1/4 inch long; flowers to 1000 per panicle, petals white or pinkish to reddish about 1/8 inch long, 1/4 inch wide, effective midsummer, then sometimes again, but sparsely, in late summer to midautumn.

FRUIT: 4 valved capsule, not ornamentally significant.

SELECTED CULTIVARS AND VARIETIES: 'Alba' Flowers white.
'Compacta' Dense growing.
'Compacta Plena' With double pink flowers from late spring to midsummer, reaching 15 inches tall.
'Flore Pleno' Double flowers.
'Grandiflora' Flowers larger.
'Perfect' With large double white flowers.
'Pink Fairy' Compact habit; flowers pink.

* * *

G. repens (rē´penz) Creeping (Figure 3–93).

COMMON NAME: Creeping gypsophila, creeping baby's breath.

NATIVE HABITAT: Mountains of northwest Spain to the Carpathians.

HARDINESS: Zone 3.

HABIT: Prostrate, herbaceous, mat forming ground cover.

SIZE: To 6 inches tall, spreading 18 to 24 inches across.

RATE: Moderate; space plants from pint- or quart-sized containers 14 to 18 inches apart.

FOLIAGE: Opposite, simple, linear, curved, to 1 inch long, gray-green.

STEMS: Many, glabrous, decumbent, to 18 inches long.

INFLORESCENCE: Panicles, almost corymbose; flowers 5 to 30 per panicle, pedicles to 5/16 inch long, petals 1/4 inch long, corolla about 1/3 inch wide, white, lilac, or pale purplish, born in profusion from early to midsummer.

FRUIT: 1 celled, dehiscent capsule that is of no significance ornamentally.

SELECTED CULTIVARS AND VARIETIES: 'Alba' With white flowers from early summer to frost; reaching 4 inches tall.
'Bodgeri' See *G. paniculata* 'Compacta Plena.'

'Rosea' With rose pink flowers; reaching 6 inches tall.

'Rosy Veil' With double pinkish to white flowers from early to late summer; reaching 18 inches tall.

Generally Applicable to All Species

CULTURE: Soil: Best in light textured or gravelly soils with excellent drainage and a pH of 7.0 to 8.0. Moisture: Although well suited to resisting drought, periodic deep watering in summer is beneficial.
Light: Full sun.

PATHOLOGY: Diseases: Crown gall, blight, damping off, aster yellows, fasciation.
Pests: Aphids, slugs, snails.

MAINTENANCE: Cut stems back following bloom to encourage new growth and to eliminate self-sowing. If soil is naturally acidic, an occasional application of lime may be needed.

PROPAGATION: Cuttings: Cuttings are usually taken in midsummer and are easily rooted in a very porous medium.
Grafting: Occasionally, cultivars are grafted on the species.

Seed: Germination occurs in 2 to 3 weeks at temperatures between 65° and 70°F. Commercial supplies are available.

*H*ebe (hē′bē) Presumably named after Hebe, the mythological Greek goddess of youth and spring, who was cupbearer to Zeus and wife of Hercules. This genus is made up of from 70 to 88 species of glabrous or hairy shrubs, subshrubs, and trees. They are native to South America, New Guinea, and primarily to New Zealand. Many of the prostrate growing species are useful for ground covering. Some that are more frequently used are described here.

FAMILY: SCROPHULARIACEAE: Figwort family.

HABIT: Woody, low growing, spreading shrubs.

RATE: Moderate to relatively fast; space plants from gallon-sized containers 2 to 3 feet apart.

LANDSCAPE VALUE: General cover for small- to medium-sized areas. Also, commonly used in the rock garden as specimens, or in front of taller shrubs as facing in a shrub border. Tolerant to salt in the air and therefore useful in coastal areas. Generally, plantings have not been very successful east of the Rockies. No foot traffic.

*   *   *

H. chathamica (chat′am-i-kȧ) Origin.

NATIVE HABITAT: Chatham Islands of South Pacific.

HARDINESS: 8 or 9 to 10.

SIZE: To 18 inches tall, spreading to 3 feet across.

FOLIAGE: Evergreen, opposite, simple, 4 ranked, elliptic to oblong, to 1 inch long, dark green.

Figure 3–93 *Gypsophila repens* (life-size)

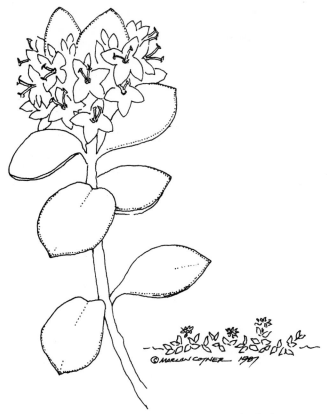

Figure 3–94   *Hebe decumbens* (life-size)

INFLORESCENCE: Racemes to 1 inch long; flowers with calyx and corolla both 4 lobed, petals lavender, effective in summer.

FRUIT: Dehiscent capsules, not of ornamental significance in this species or the others.

*   *   *

H. decumbens (dē-kum´benz) Description of the habit (Figure 3–94).

COMMON NAME: Ground hebe.

NATIVE HABITAT: New Zealand.

HARDINESS: Zone 5.

SIZE: To 14 inches tall.

FOLIAGE: Succulent, evergreen, opposite, simple, 4 ranked, short petioled, grayish, margin red.

INFLORESCENCE: Racemes to 1 inch long; flowers 1/4 inch across, corolla white, effective in spring.

*   *   *

H. pinguifolia 'Pagei' (ping-gwi-fō´li-a) Fat or thick leaved.

NATIVE HABITAT: New Zealand, South Island (the species).

HARDINESS: 8 or 9 to 10.

SIZE: To 9 inches tall, spreading to 5 feet across.

FOLIAGE: Evergreen, fleshy, simple, opposite, elliptic-obovate, to 3/4 inch long, sessile, glaucous-gray, leathery textured.

INFLORESCENCE: Dense spikes; flowers about 3/8 inch across, corolla white, borne profusely in late summer.

OTHER SPECIES: *H. albicans* has small gray-green leaves; mounding habit to about 10 inches tall; flowers white in late spring and early summer.

Generally Applicable to All Species

CULTURE: Soil: Tolerant to various soil types, good drainage being the most important factor.
Moisture: Watering regularly is required for optimal growth throughout the summer.
Light: Full sun in cool areas, light shade in warm climate.

PATHOLOGY: No serious diseases or pests reported.

MAINTENANCE: Lightly shearing on an annual basis is useful as it keeps plants neat and compact.

PROPAGATION: Cuttings: Cuttings should be taken in mid to late summer from mature wood.

✦✦✦✦✦✦✦✦✦✦✦✦✦✦✦✦

*H*edera (hed´ĕr-a) The ancient Latin name for ivy. The genus *Hedera* is composed of five or six species of evergreen, woody vines, or shrublets native to Europe, North Africa, and Western Asia. Ivies are interesting in that they display dimorphism, meaning that the habits of the juvenile and adult stage are different. In this case, juvenile growth is vinelike and trailing. The foliage is generally lobed. Upon climbing to a height at which support is lacking or to which the plant can no longer climb as genetically determined, the branches no longer cling but project outward. In addition, the leaves (most often lobed in youth) change shape, becoming unlobed and often elliptic-lanceolate, darker, thicker, and more leathery in texture. Furthermore, these branches soon begin to bear small green flowers arranged in panicles. Fruit follows as a green-turning black berry in fall.

Cuttings from adult stage ivies will root (helped immensely with bottom heat) and grow not as a vine, but rather as a broad shrub that may reach 6 or more feet in height. These tree ivies, as they are commonly referred to, are then given the name *H. arborescens*, meaning treelike. To keep matters as unconfused as possible, if may be wise to name these forms as standard for the juvenile state: namely genus, species, variety or cultivar, followed by the designation *arborescens*. As you will learn, there are hundreds of varieties and cultivars of *Hedera.* Most belong to the species *H. helix.*

When examining a planting of a single clone of ivy, it is common to view a good deal of variability in leaf morphology. Ivy is, in fact, quite prone to mutation, and once a cutting is taken from a section that appears different, the new growth that occurs will likely be atypical also. This is how many of the ivies have originated. These sports (or mutations) may look quite similar to the parent and reversion is common; thus there is a certain amount of confusion in nomenclature and erroneous identification and sale in the horticultural community. Identification is often difficult at best.

Those desiring to study ivy in detail are encouraged to read *The Ivy Book*, by Suzanne Warner Pierot, *Ivies*, by Peter Q. Rose, and various other material available through membership in the American Ivy Society, National Center for American Horticulture, Mt. Vernon, Virginia 22121.

Many ivies have ground covering merits. A small percentage are hardy in the North and many have not been tested to the point where their northern range is known. Some of the more popular are listed here, but several others are worthy of trial.

FAMILY: ARALIACEAE: Ginseng family.

HABIT: Trailing, clinging vines in the juvenile form. Woody, low, spreading shrub as adult form. Adult may reach several feet tall.

SIZE: 2 to 6 inches without support, spreading indefinitely.

RATE: Most are slow to start, then moderate to relatively fast growing; plant spacing of 10 to 14 inches is common; slow growing types should be planted 6 to 10 inches apart. Generally, they are available in 2 to 3 inch diameter pots.

LANDSCAPE VALUE: Very versatile ground covers for large- or moderate-sized areas. Excellent along building foundations where soil is often alkaline. Plant next to ledge and allow them to trail over or near tree or wall where they will climb. In any event, they make exceptionally rich looking evergreen covers. Excellent as soil stabilizers on banks and steep slopes. Will not tolerate regular foot traffic, but not harmed by infrequent traffic.

OF SPECIAL INTEREST: The flowers of adult ivies are said to contain great quantities of nectar and thus are attractive to bees. Birds occasionally will build their nests in the relative security of ivy branches. Both the foliage and fruit of ivies are reported to contain a saponin called hederin, which is mildly poisonous when ingested. Supposedly, it will cause a burning sensation in the throat and, in severe cases, gastroenteritis.

* * *

H. canariensis (ka-nãr-ē-en´sis) Of the Canary Islands.

COMMON NAME: Canary Island ivy, Algerian ivy, Madeira ivy.

NATIVE HABITAT: Canary Islands and North Africa.

HARDINESS: Zone 7 to 9 or 10.

LEAVES: Evergreen, alternate, simple; juvenile leaves ovate, entire, or shallowly 3 to 7 lobed, cordate in outline, to 6 inches wide by 6 inches long; glossy above, light to medium green, glabrous; ovate to lanceolate and darker in maturity.

STEMS: Burgundy red hairs; scalelike, 10 to 15 rays.

INFLORESCENCE: Panacled umbels; flowers small, regular, petals green, borne in fall, originating from adult stems only.

FRUIT: Black drupe, late fall.

SELECTED CULTIVARS AND VARIETIES: 'Azorica' Vigorous; with leaves 3 to 6 inches across; 5 to 7 lobed, vivid green, tomentose in youth.
'Gloire de Marengo' Leaves edged in silvery white; sometimes listed as 'Variegata.'
'Margino-Maculata' Leaf margins cream colored and flecked or spotted green.
'Ravensholst' Vigorous form with larger leaves than the species.
'Striata' With leaf margins green and mid-vein section streaked light green to ivory.

* * *

H. colchica (kōl´chi-kȧ) From Colchis, a country on the Black Sea (Figure 3–95).

COMMON NAME: Persian ivy, Colchis ivy, fragrant ivy.

NATIVE HABITAT: Asia and southeastern Europe.

HARDINESS: Zone 5.

LEAVES: Evergreen, alternate; juvenile leaves heart shaped, entire, or slightly lobed, thick, leathery to 10 inches long, dark-dull green, veins thick and recessed in upper surface, protruding below. Petioles green, smelling somewhat like celery when crushed, hence the common name fragrant ivy.

STEMS: Green; internodes 1 1/2 to 2 1/2 inches long, hairs scalelike, 15 to 20 rays.

FRUIT: Black drupe in late fall.

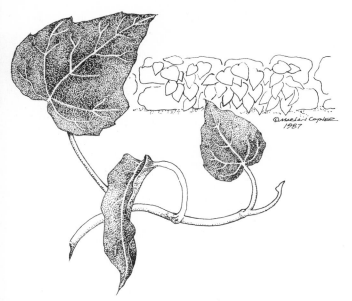

Figure 3–95 *Hedera colchica* (0.33 life-size)

SELECTED CULTIVARS AND VARIETIES: 'Dentata' Thought by many to be among the most versatile ground covers; foliage is large, to 10 inches long by 8 inches wide and glossy green, leaf margins finely toothed.
'Dentata Variegata' Leaves similar to 'Dentata,' but margins variegated irregularly creamy-yellow.
'Gold Leaf' See 'Sulfur Heart.'
'Paddy's Pride' See 'Sulfur Heart.'
'Sulfur Heart' Sometimes listed as var. *dentata aureo striata*, cv. 'Gold Leaf' or cv. 'Paddy's Pride.' Leaves are light green with irregular splashes of yellow or lighter green; veins are slightly lighter green or yellowish, margins shallowly toothed.

* * *

H. helix (he´licks) Spiral or twisted. Referring to the arrangement of the petioles on the stem (Figures 3–96 and 3–97).

COMMON NAME: English ivy.

NATIVE HABITAT: Europe.

HARDINESS: Zones 5, 6 or 7 to 9 or 10. A few hardy northward to zone 4. It is wise to consult an ivy specialist prior to trying a new type in northern or hot summer areas.

LEAVES: Evergreen, alternate, simple, juvenile leaves quite variable, but ordinarily 3 to 5 lobed, usually 2 to 3 inches long, but ranging from 1/2 to 5 inches long, margins nearly entire, triangular to ovate in outline, dark green to medium green, shiny glabrous above, often with whitish venation, lighter green to yellowish below; mature leaves are elliptic to ovate in outline, are unlobed and paler green than juvenile leaves. Petioles burgundy red in any event.

STEMS: Burgundy red in youth, green turning woody and brown with age, hairs with 4 to 6 rays.

INFLORESCENCE: Panicled umbels; flowers perfect, small, pale-green from early to midautumn.

FRUIT: Black drupes, about 1/4 inch across, with 2 to 5 seeds, reported to be poisonous, effective in fall.

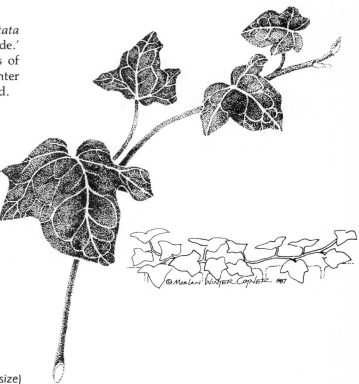

Figure 3–96 *Hedera helix* 'Baltica' (0.75 life-size)

Figure 3-97 *Hedera helix* 'Fan' (0.75 life-size)

SELECTED CULTIVARS AND VARIETIES: At least 100 cultivars of *H. helix* are grown in the United States. Some plants are considered ramulose type (branching freely) and are designated here with the letter R. Others branch very little, generally producing only one or two lateral branches in response to pruning. Those that are as hardy as the species are designated with the letter H. Others are to be assumed only hardy in climates of mild winters and should not be grown in northern regions unless recommended by an experienced horticulturist from that area. Cultivars of all ivies and particularly *H. helix* are commonly grouped into general categories in accord with the shape of their leaves. They are as follows:

Typical: Ivies with leaves shaped typically like the species *H. helix.*
Bird's foot: With elongated terminal lobe.
Fan shaped: With broad leaves reminiscent of the shape of a fan.
Curlies: With leaves that curl, overlap, and ripple.
Heart shaped: Shaped as the name describes.

'Albany' R (typical) Leaves sharply five lobed, purplish-green above, purplish below.
'Arborescens' A designation for the adult stage of a species or cultivar.
'Argenteo Variegata' (typical) Leaves variegated white.

'Aureo Variegata' (typical) Leaves variegated creamy yellow.
'Baltica' H (typical) Similar to the species with leaves somewhat smaller and venation more commonly lighter than the species. This is a useful cultivar in the Midwest and East.
'Brocamp' R&H Leaves mostly entire, to about 2 inches long and 1/2 inch wide. Lanceolate, sometimes with asymmetrical lateral lobe, light to medium green, sometimes listed as 'salicifolia.'
'Bulgaria' H (typical) A fine selection with broader, more leathery, dark green leaves. Reported to be quite drought resistant; often purplish in fall.
'Buttercup' H (typical) With small leaves that are yellow in youth if grown in sun.
'Caenwoodiana' (bird's foot) Leaves smaller, three lobed, dark green, raised white veins and elongate terminal lobe.
'Cavendishii' (bird's foot) Three lobed, gray-green, with margins variegated creamy yellow.
'Digitata' (fan) Leaves 5 lobed, digitate, center lobe slightly longer then laterals, dark green with whitish venation.
'Fan' R (fan) With 5 to 7 short lobed, broad, light green leaves from 3/4 to 2 inches wide.
'Galaxy' R&H (bird's foot) Leaves star shaped with long terminal lobe. Foliage medium green, new growth lighter and glossy.
'Glacier' (typical) Leaves small, 5 lobed, with margins pinkish and creamy white.

'Green Feather' R Oddity. Leaves five narrowly laciniate, 1/2 to 1 1/2 inches long; bright green, veins lighter, feathery textured. Lateral lobes spoonlike.

'Goldheart' (typical) Leaves smaller, 3 to 5 lobed, green with yellow and cream-colored center; petioles reddish.

'Hahn's' or 'Hahn's Self-Branching' R (typical) Leaves smaller, light green; common in the south.

'Harrison' H (typical) Similar to 'Baltica,' dark green, sometimes reddish purple in winter.

'Hibernica' H (typical) Larger leaves that are shiny in youth, 3 to 5 inches long and wide. More vigorous than the species. Thought by some to be the most commonly cultivated form of *H. helix* in America. Often sold as *H. helix* and *H. helix* 'Baltica.' It is said to be a tetraploid.

'Maculata' (typical) Leaves 5 lobed, variegated yellow-green, dark green, cream, and white.

'Marmorata' (typical) Sometimes listed as 'Discolor.' Characterized by small leaves, mottled white on dark green.

'Minima' (typical) Leaves small, 3 to 5 lobed, undulating, somewhat cordate, dull dark green, with grayish veins and petioles, becoming purplish in fall (sometimes listed as 'Atropurpurea'), reported hardy in New York City.

'Needlepoint' R (bird's foot) Popular in the South. With small leaves, three lobed, terminal lobe elongate. May be listed as 'Manda.'

'Pedata' (bird's foot) Leaves small and 5 lobed, lobes lanceolate; commonly called bird's-foot ivy.

'Pittsburg' R&H (typical) Bushy, small to medium sized. Leaves with pronounced terminal lobe; venation whiter than the species.

'Poetica' Rare. Juvenile leaves are typical but squarish as basal lobes are reduced. Usually seen in adult form because of its unique dull-orange fruit.

'Rumania' Similar to and possibly identical to 'Bulgaria.'

'Sagitifolia' (bird's foot) Small leaved, usually about 1 inch long, flat, delicate, usually 5 lobed, terminal lobe elongated, all lobes coming to a sharp point, grayish green.

'Sagitifolia Variegata' R (bird's foot) Unlike the implications of the name, this selection is not from 'Sagitifolia.' Leaves are small, three lobed, two basal lobes vestigial and backward pointing where present, terminal lobe elongate.

'Thorndale' H (typical) Leaves slightly smaller than the species with pronounced terminal lobe and attractive creamy-white venation.

'238th Street' H (heart shaped) Leaves average 1 1/2 inches long and wide; leathery, dark green, shaped the same in adult stage; veins yellowish, unique in

that it easily reaches the adult stage in the north and that cuttings from the adult form produce trailing, vining shoots, rather than being shrubby. Frequently listed as 'Tom Boy.'

'Watsonian' H (typical) Leaves somewhat smaller than the species, cusped, glossy, dark green with purplish or light green venation. Originating in Grand Rapids, Michigan.

'Wilson' H (typical) Leaves small, medium green with white venation, much like a miniature 'Thorndale' as the terminal lobe is quite pronounced. discovered by Dr. Chadwick of Ohio State University in wes'.rn Pennsylvania.

'Woerner' H (typical) Vigorous, with dull green new leaves on long petioles; lobes tipped with white spines, turning purple in winter.

Generally Applicable to All Species

CULTURE: Soil: Quite adaptive to various soil types as long as good drainage exists. Faster spreading in rich, organic soils. pH is optimal between 6.0 and 7.0; however, most ivies will tolerate slightly alkaline conditions.

Moisture requirement: Established ivy is generally well suited to withstand periods with little rain and high temperatures; however, growth is more lush when supplemented with periodic deep watering.

Light: Full sun to dense shade; variegated forms realize their best color in full sun or light shade.

PATHOLOGY: Bacterial leaf spot and stem canker, powdery mildew, sooty mold

Pests: Aphids, puss caterpillar, cabbage looper, eight spotted forester, omnivorous looper, mealybugs, scabs, two spotted mite, parasitic plant dodder. *H. canariensis* and *H. colchica* seem less susceptible to pathogens than *H. helix*.

MAINTENANCE: Little or no maintenance required. Clip back stems as they outgrow their bounds.

PROPAGATION: Cuttings: Make stem cuttings from new growth, placing one or two nodes below the medium surface. Treating with a root inducing preparation of 1000 to 3000 ppm IBA/talc or solution is beneficial. Arborescent forms root slowly and are greatly helped by bottom heat of 85 °F. Cuttings also tend to root better if taken from stock plants that have been grown in the shade rather than the sun. Root cuttings also can be used, but are generally not as efficient as stem cuttings. Root cuttings from adult forms produce juvenile plants. Seed: Sow when ripe or store dry and cool for up to 1 year.

Grafting: Ivy can be grafted upon *Fatshedera lizei* (*Fatsia japonica* × *Hedera helix*) as a rootstock for standard forms. *Acanthopanax sieboldianus* and *H. helix* are also used as an understock. Usually only adult forms are grafted.

Division: Division of rooted leaf/node sections that result from natural layering is best performed in fall or early spring.

***H**edyotis* (hay-dē-ō´tis) Formerly *Houstonia*, from Greek *hedys*, sweet, and *otos*, ear. Reference unclear. This genus is composed of about 400 species of weak stemmed shrubs and herbs. They are tropical, subtropical, and temperate in origin.

FAMILY: RUBIACEAE: Madder family.

H. caerulea (sē-rū´li-à) Cerulean, dark blue, the flowers (Figure 3–98).

COMMON NAME: Bluets, Quaker ladies, innocence, bright eyes, Venus's pride, angel eyes, star of Bethlehem, wild forget-me-not, Quaker bonnets, little washerwoman.

NATIVE HABITAT: Eastern North America.

HARDINESS: Zone 3.

HABIT: Herbaceous, broad, carpeting ground cover.

SIZE: 4 to 6 inches tall, spreading 12 + inches across.

RATE: Moderate; space plants from pint-sized containers 12 to 14 inches apart.

LANDSCAPE VALUE: General cover or accent plant in wet soil areas. Attractive when naturalized in wooded setting. No foot traffic.

FOLIAGE: Evergreen, deciduous in colder limits, opposite, simple, oblanceolate, to 3/8 inches long, about 1/8 inch across, glabrous and glandular dotted both sides, light to medium green.

STEMS: Squarish, glabrous, and glandular dotted.

FLOWERS: Solitary, regular, corolla salverform, 1/2 inch long, about 1/2 inch across, 4 petaled, violet, blue, or white with yellow eye, effective spring through summer.

FRUIT: 1/8 inch diameter, many seeded capsule, not ornamentally significant.

CULTURE: Soil: Rich, moist, acidic soil is preferred. Moisture: Not greatly tolerant to drought; keep soil moistened with regular watering.
Light: Light to moderate shade.

PATHOLOGY: Diseases: Downy mildew, leaf spots, rusts.
Pests: No serious pests reported.

MAINTENANCE: Little or no special maintenance needed.

PROPAGATION: Division: Division is usually performed in spring.

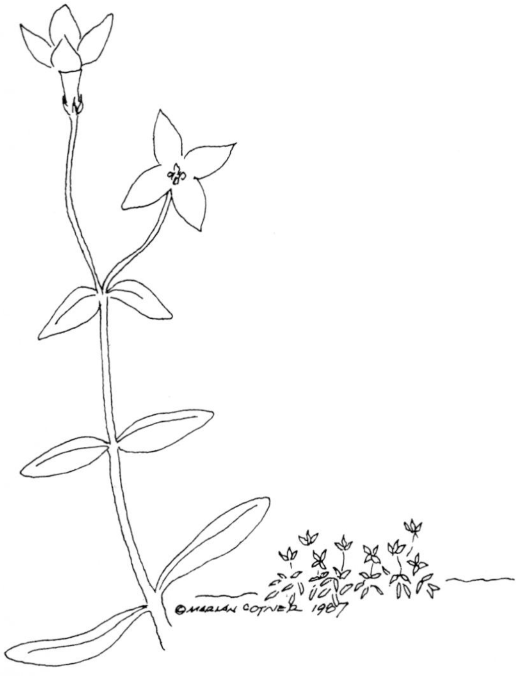

***H**elianthemum* (hē-li-an´the-mum) From Greek, *helios*, the sun, and *anthemon*, a flower, the sun rose. This genus is composed of about 110 species of subshrubs and herbs. They are native to Europe and North and South America.

FAMILY: ROSACEAE: Rose family.

H. nummularium (num-mul-ăr´ē-um) Coin shaped, reference unclear (Figure 3–99).

COMMON NAME: Common sun rose.

NATIVE HABITAT: Mediterranean region.

HARDINESS: Zones 5 or 6 to 10.

HABIT: Low, trailing, somewhat loose, woody shrub.

SIZE: 6 to 12 inches tall, spreading to 3 feet across.

RATE: Moderate once established; space gallon-sized container plants 2 to 2 1/2 feet apart.

LANDSCAPE VALUE: Fine for rock garden specimen or for general cover when planted en masse. No foot traffic.

Figure 3–98 *Hedyotis caerulea* (life-size)

Figure 3–99 *Helianthemum nummularium* (life-size)

FOLIAGE: Evergreen, mostly opposite, ovate to lanceolate, entire, margins flat or slightly revolute, to 2 inches long by 1/4 inch wide, gray tomentose below, colored gray-green above.

STEMS: Ascending to procumbent, white, hairy.

INFLORESCENCE: 1 to 2 flowered, one sided cymes; flowers usually about 1 inch across, regular; corolla of 5 petals, pink or dusty rose fading to apricot, lasting only one day but showy as new flowers open continually, effective early to midsummer.

FRUIT: Ovoid, 3 valved capsule, not ornamentally significant.

SELECTED CULTIVARS AND VARIETIES: Var. *multiplex* With small leaves and double coppery colored flowers.
H. × *hybridus* Many hybrids are available. Frequently, they are sold under the common name sun rose, and their ancestry is often not well recorded. Generally, they are from 8 to 12 inches tall, spreading to twice their height. They come in a variety of colors that ranges from white, red, pink, cream to yellow or combinations of these. Double flowered forms are also cultivated.

CULTURE: Soil: Light, well-drained, somewhat gravelly soil is best. Intolerant to saturated soils and best if pH is slightly alkaline.
Moisture: Very tolerant to drought, with infrequent summer watering to some benefit.
Light: Full sun.

PATHOLOGY: Diseases: Leaf spot.
Pests: None serious.

MAINTENANCE: Occasionally shear lightly to promote compact habit. In northern areas where snow cover is unreliable, a mulch of evergreen boughs is recommended to shade foliage and screen it from drying winds.

PROPAGATION: Cuttings: Take stem cuttings in summer. Treatment with 1000 ppm IBA/talc may result in greater root formation but is not essential.

Hemerocallis (hem-ẽr-ō-kal´is) From Greek, *hemeros*, a day, and *kallos*, beauty, meaning beautiful for a day. This genus is composed of about 15 species of clump forming herbaceous perennials. They are native from central Europe to China, and frequently Japan. Spread is by rhizomes, and roots are more or less tuberous and fleshy. Tight, compact clumps result if rhizomes are short, while more open and loose forms result from those with longer rhizomes. Daylilies, as they are commonly called, have been in cultivation for ages and only recently have begun to become very popular (and often quite expensive). Currently, a mind boggling assortment of hybrids and polyploids are available in an endless range of colors and growth habits.
The genus is generally described here and is followed by brief descriptions of the unique features of the more useful ground covering species and hybrids.

FAMILY: LILIACEAE: Lily family.

HABIT: Herbaceous, lilylike, rhizomatous ground covers.

RATE: Moderate to relatively fast; spacing of 18 inches to 3 feet is usually acceptable for plants from quart- or gallon-sized containers; vary spacing in accord with rate of spread. Most often bare root divisions are sold.

LANDSCAPE VALUE: As a general ground cover, daylilies can be useful in both small and moderate to large areas. They are often excellent as specimens when planted next to the top of a retaining wall or in a rockery. Frequently, they are planted as edging to a walkway or garden path. Near an entrance or in the foreground of a large rock or boulder, the effect as an accent plant can be spectacular. At the edge of a pond, stream, ditch, or on slopes of moderate incline, daylilies serve well for the purpose of soil retention and are often seen along railroad beds, where they have been used for this purpose. No foot traffic.

OF SPECIAL INTEREST: Flowers of daylilies are attractive to both hummingbirds and butterflies.

FOLIAGE: Evergreen to deciduous, basal, linear, entire, with sharp or blunt pointed apex, keeled, grasslike, 2 ranked, from 6 to 14 inches long on varieties that are considered relatively low, and up to 2 or even sometimes 4 feet long on taller forms, arching over as graceful fans, colored light to medium green, grayish or glaucous.

INFLORESCENCE: Flowers in clusters on scapes that elevate them above the mass of foliage; flowers are regular, 6 parted in two whorls of three, 3 outer segments being petallike sepals alternating with 3 inward petals, 6 stamened, 1 pistil; corolla usually short lived, hence the meaning of the common name, yet effective over a period of time as new flowers continually open, giving an average bloom period of a month or so; colors range from yellow, orange, reddish, to purplish and in unlimited combinations; some species open at night; most species predominantly bloom in spring or summer.

FRUIT: 3 chambered loculicidal, dehiscent, few-seeded capsule, not of ornamental significance.

* * *

H. aurantiaca (â´ran-ti-ā-kà) Golden-orange, the flowers.

COMMON NAME: Orange daylily.

NATIVE HABITAT: Japan.

HARDINESS: Zone 6.

SIZE: To 3 feet tall, spreading 3 to 5 feet across.

DISCUSSION: This species is characterized by 2 to 3 foot long leaves that are about 1 inch wide, evergreen, and coarse textured. Flowers are orange, often flushed with purple and effective midsummer. Var. *littorea* has narrow gray-yellow petals. Var. *major* has larger red-orange flowers.

* * *

H. fulva (ful´và) Tawny.

COMMON NAME: Orange daylily, fulvous daylily, tawny daylily.

NATIVE HABITAT: Europe and Asia.

HARDINESS: Zones 2 or 3.

SIZE: To 2 feet tall, spreading 3 to 5 feet across.

DISCUSSION: Rapidly spreading, with leaves 1 1/2 to 2 feet long by 1 3/8 inch across; flowers 3 1/2 inch wide by 5 inches long, nonfragrant, brownish to orangish-red, on tall (to 6 feet) scapes, effective midsummer. Quite common, this species is often used as a parent in hybridizing. Selections include:
'Cypriana' With many brownish, 4 1/2 inch flowers; leaves glossy.
'Europea' Reportedly similar to the species, but a triploid. It is common along railroad tracks and frequently the pollen parent in crosses that have resulted in many red-flowered cultivars.
'Kwanso' Sometimes listed as 'Flore Pleno,' this cultivar is common. Leaves are broad and coarse, sometimes striped white. Flowers are double, borne slightly later than the species.
'Rosea' With rosy red flowers, used extensively in breeding as a source for red flower pigmentation.

* * *

H. lilioasphodelus (lil-ē-ō-as´fō-dell-us) Like lily and *Asphodelus*.

COMMON NAME: Lemon daylily, yellow daylily, lemon lily.

NATIVE HABITAT: Eastern Asia.

HARDINESS: Zone 3.

SIZE: To 3 feet tall, spreading 3 feet across.

DISCUSSION: This species has fragrant 4 inch lemon-yellow flowers on arching scapes in spring.

* * *

H. middendorffii (mid-den-dorf´i-ī) After Middendorff, a botonist.

COMMON NAME: Middendorff daylily.

NATIVE HABITAT: Siberia.

HARDINESS: Zone 3.

SIZE: To 1 foot tall.

DISCUSSION: A small species with pale orange flowers in late spring and early summer.

* * *

H. × *hybridus* (Figure 3–100) Cultivars of hemerocallis that have arisen from hybridization may number in the thousands and continue to grow at a rapid pace. Those of commercial merit usually are noteworthy for their floral display or decreased vigor, yet when selecting cultivars for ground-cover purposes one must also pay attention to foliage color and texture.
Flowers range in size from 3 to 8 inches across and the colors are in unlimited combinations. Foliage comes in shades of green, gray, and bluish green.

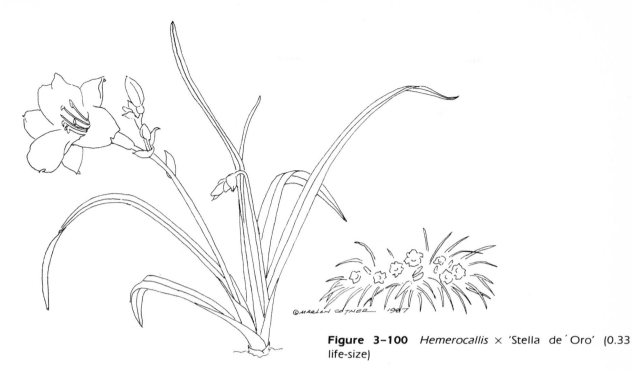

Figure 3–100 *Hemerocallis* × 'Stella de´ Oro' (0.33 life-size)

The cultivar 'Stella de' Oro' (Figure 3–100) is an excellent hybrid that reaches 2 feet tall and has an extended bloom season running from midspring to early fall. Flowers are golden yellow and 2 1/2 inches across.

Generally Applicable to All Species

CULTURE: Soil: Adaptive to almost any soil, moist, rich, well-drained loam being preferred.

Moisture: Because of their fleshy, water-storing rhizomes and roots, little supplemental water is needed once they become well established. Occasional deep watering during the summer, however, may benefit appearance, especially for those that bloom later in the season. In the event that watering is practiced, it is wise to let the soil become dry between waterings.

Light: Full sun to moderate shade. In general, dark flowered selections are at their best with some shade during the hottest part of the day.

PATHOLOGY: Diseases: Leaf spot, russet spot, root and crown rot.

Pests: Flower thrips, long horned weevil, Japanese beetle, grasshoppers, mites, slugs, wasps, southern root knot nematode.

MAINTENANCE: For most attractive appearance, remove flower stalks once blooming is finished. Should bloom diminish over time, division will bring it back. Annual mulching will furnish nutrients and improve soil, but is not mandatory.

PROPAGATION: Cuttings: Some types produce leafy side proliferations that will readily root under mist in summer.

Division: Clumps are usually divided in fall or spring. Each division should have a good root structure and have at least two or three offshoots.

In vitro: Methods have been developed and utilize flower sepals or petals as explants.

Seed: Species can be propagated by seed. Ripe seed should be chilled moist for 6 weeks prior to planting and usually germination is good. In most cases, seed is used as a means of establishing new cultivars rather than for large-scale propagation of a species.

Herniaria (hĕr-ni-ā´ri-à) From Latin, *hernia*, rupture, for which the plant was supposed to be a remedy. The genus *Herniaria* is composed of about 35 species of annual and perennial, many branched herbs.

FAMILY: CARYOPHYLLACEAE: Chickweed or pink family.

H. glabra (glā´brà) Smooth, hairless, the foliage (Figure 3–101).

COMMON NAME: Common burstwort, rupturewort.

NATIVE HABITAT: Europe and northern and western Asia, Africa, and Turkey.

Figure 3–101 *Herniaria glabra* (life-size)

HARDINESS: Zones 5 to 10.

HABIT: Herbaceous, creeping, stoloniferous, mosslike ground cover.

SIZE: To 4 or 6 inches tall, often shorter; spreading indefinitely.

RATE: Moderate; space plants from 2 to 4 inch diameter containers 6 to 12 inches apart.

LANDSCAPE VALUE: Valuable for planting between stepping stones or patio cracks. Often nice combined with spring flowering bulbs when used as a general cover for smaller areas. Tolerates limited foot traffic.

FOLIAGE: Evergreen, opposite becoming alternate toward the apex, simple, ovate to elliptic, 1/8 to 3/8 inch long, 1/8 inch wide, entire, ciliate, occasionally cuspidate, glabrous and glandular dotted both sides, bright glossy green turning reddish in cold weather.

STEMS: Hairy to glabrescent, glandular dotted, rooting as they creep.

INFLORESCENCE: Axillary clusters on short branches; flowers regular, sepals ovate and usually glabrous, petals greenish white and minute, not ornamentally significant.

FRUIT: Indehiscent single-seeded nutlet, not ornamentally significant.

CULTURE: Soil: Tolerant to most well-drained soils. Moisture: Shallow rooted, burstwort is benefited by regular watering during the summer months. Light: Full sun to light shade.

PATHOLOGY: Pests: Aphids.

MAINTENANCE: Fertilize lightly in spring and fall; otherwise, no special maintenance.

PROPAGATION: Simple division is the standard means of propagation. Usually it is carried out spring or fall, but can be successful anytime provided soil is kept moist until reestablished.

*H*euchera (hū´kĕr-a̓ or hoy´kĕr-a̓) Named after J. H. Heucher, a German botanist. This genus, commonly called alumroot, is made up of around 35 to 50 species of herbaceous, basal leaved perennials that are native mainly to western North America.

FAMILY: SAXIFRAGACEAE: Saxifrage family.

H. sanguinea (san-gwin´ē-a̓) Blood-red, the flowers (Figure 3–102).

COMMON NAME: Coral bells.

NATIVE HABITAT: Mexico and Arizona.

HARDINESS: Zones 3 to 9.

SIZE: 12 to 24 inches tall, spreading 2 to 3 feet across.

HABIT: Herbaceous, clump forming ground cover.

RATE: Moderate to relatively slow; space plants from pint-sized containers 12 to 14 inches apart.

LANDSCAPE VALUE: Used commonly as a general ground cover for small areas. Very well suited for edging a perennial border or walkway. No foot traffic.

FOLIAGE: Evergreen, basal, on hairy petioles to 5 inches long, tufted, orbicular to cordate or ovate, palmately 5 to 9 lobed, 1 to 2 inches long and wide, veins palmate and conspicuous, glandular dotted and pubescent on both sides on the veins, dark green turning reddish in winter.

INFLORESCENCE: Cymose panicles on slender scapes to 20 inches tall, elevated above foliage; flowers regular, 5 petaled, bright red to pink, about 1/2 inch long by 1/4 inch wide, effective midspring to early fall.

OF SPECIAL INTEREST: Flowers of coral bells are popular with hummingbirds.

FRUIT: Dehiscent capsule, not ornamentally significant.

SELECTED CULTIVARS AND VARIETIES: There are many cultivars and hybrids. A few of the more common are listed here. 'Alba' Flowers white; reported to come true from seed.

'Bressingham Hybrids' Hybrid group of mixed colors that range from white and pink to coral red.

'Chatterbox' With deep rose-pink flowers.

'Chartreuse' Flowers chartreuse; habit compact.

'Garnet' Flowers deep rose-pink.

'Green Ivory' Vigorous, with many white flowers.

'June Bride' Flowers white on 15-inch stems.

'Matin Bells' Coral-red flowers from early summer to early fall.

'Queen of Hearts' Flowers red and large.

'Rhapsody' Flowers clear rose-pink.

'Scarlet Sentinel' With scarlet flowers; very vigorous.

'Snowflakes' Excellent white flowered selection.

'Splendens' Flowers dark crimson.

'Virginalis' Flowers white.

CULTURE: Soil: Well-drained loamy soil is best.
Moisture: Drought tolerance is good, but appearance and flowers are best when soil is kept moistened.
Light: Full sun to moderate shade.

PATHOLOGY: Diseases: Leaf spots, powdery mildew, stem rot, leaf and stem smut.
Pests: Strawberry root weevil, foliar nematode, mealy bugs, four-lined plant bug.

MAINTENANCE: Relatively maintenance free; dead flower stalks may be sheared after bloom to enhance appearance.

PROPAGATION: Cuttings: Leaf cuttings are usually made in midsummer. They should contain the entire leaf blade, petiole, and a small section of stem.
Division: Simple division of the clumps in spring or fall is a common means of propagation.
Seed: Collect when ripe, store cool, and sow the following spring. Primarily seed should be used as a means of locating new cultivars and not a means of propagating existing ones.

✦✦✦✦✦✦✦✦✦✦✦✦✦✦✦✦✦✦

*H*ippocrepis (hip-ō-kre'pis) Derived from Greek, *hippos*, a horse, and *crepis*, a shoe, in reference to the shape of the seed pod. This genus is composed of about 12 species of alternate leaved herbs and shrubs that are native to the Mediterranean region.

FAMILY: LEGUMINOSAE: Pea family.

H. comosa (kō-mō'sà) Hairy, in tufts (Figure 3–103).

COMMON NAME: Lady's fingers.

NATIVE HABITAT: Mediterranean region.

HARDINESS: Zones 6 to 10.

Figure 3–102 *Heuchera sanguinea* (life-size)

HABIT: Herbaceous, low, spreading ground cover.

SIZE: 1 to 2 feet tall, often lower, spreading to 3 feet across.

RATE: Relatively fast; space plants from gallon-sized containers 1 1/2 to 2 1/2 feet apart.

LANDSCAPE VALUE: Ordinarily used as a bank cover, as roots bind soil well. Sometimes used as a lawn substitute on a small scale. Tolerates light foot traffic.

FOLIAGE: Alternate, odd-pinnate, with many small leaflets; leaflets arranged in pairs of 4 to 7, oblong, medium green.

INFLORESCENCE: Long peduncled umbels or headlike racemes; flowers small, golden yellow, effective in spring and somewhat in summer.

FRUIT: Many flat 1 inch long legumes, covered with reddish glands.

CULTURE: Soil: best in rich, well-drained loam, but tolerant to infertile and rocky soils. Supposedly adaptable to pH.
Moisture: Quite tolerant to drought, but best growth is attained when watered regularly during summer months when weather is hot and dry.
Light: Full sun.

PATHOLOGY: No serious diseases or pests reported.

MAINTENANCE: Mow or shear plantings after bloom to enhance appearance and promote compactness.

PROPAGATION: Cuttings: Heel cuttings from preflowering shoots should be taken in late spring and can be transplanted later in the same year.
Seed: Seed germinates readily, but usually cuttings are the preferred method. If seed is used, it should be sown in early summer.

✛✛✛✛✛✛✛✛✛✛✛✛✛✛✛✛✛✛

Hosta (hos´tà) After Nicolous Thomas Host and Joseph Host, Austrian botanists. Sometimes still called *Funkia*, (fung-ē´à), after H. Funk, a German botanist; however, the proper genus name is *Hosta*.

FAMILY: LILLIACEAE: Lily family.

The genus **Hosta** is comprised of about 40 species of herbaceous, clump forming, monocotyledonous perennials with short rhizomes and basal leaves. These plants are quite popular among herbaceous perennial growers. Hybrids, many species, cultivars, and varieties are available. New types seem to be introduced constantly, much of which can be attributed to in vitro propagation, which speeds up multiplication and thus the occurrence of spontaneous mutation. A unique characteristic of this genus is the propensity to form variegated mutants, probably more so than any other single genus. The expected nomenclatural confusion is also extensive. The American Hosta Society should be consulted

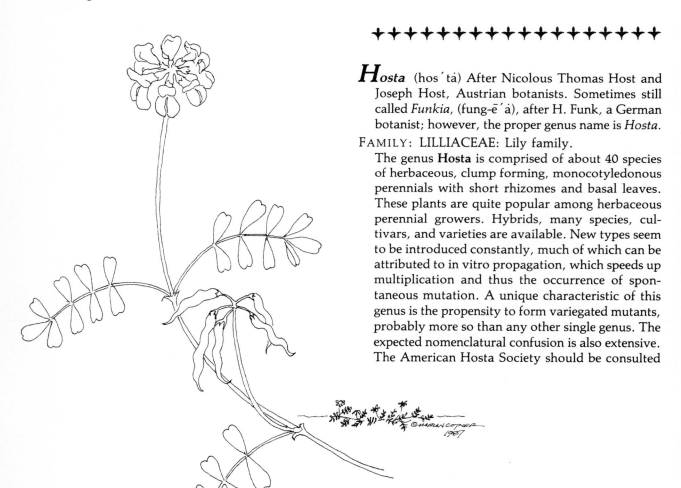

Figure 3-103 *Hippocrepis comosa* (life-size)

whenever there is doubt as to classification: American Hosta Society, 9448 Mayfield Road, Chesterland, Ohio 44026.

Hostas are generally described next, followed by details of individual species, cultivars, and varieties.

NATIVE HABITAT: Japan, with some species from China and Korea.

COMMON NAME: Plantain lily, day lily, hosta, funkia.

HABIT: Low, broad, clump forming herbaceous ground covers.

RATE: Generally moderate to relatively fast growing, except for some dwarf forms, which spread at a slow pace. Spacing of 15 to 20 inches is usually acceptable for plants from pint- to gallon-sized containers.

FOLIAGE: Deciduous, basal, tufted, petioled, entire, quite often conspicuously veined.

INFLORESCENCE: Scapose, bracted, one-sided racemes, usually elevated well above foliage; flowers are 6 segmented, corolla funnelform. Both hummingbirds and bees are attracted to the nectar.

FRUIT: Many seeded, three valved, loculicidal capsule, not ornamentally significant.

OF SPECIAL INTEREST: The number of hybrid hostas is staggering. The nomenclature is often confusing, and identical plants are commonly listed by many different names. Be sure to read catalog descriptions thoroughly prior to ordering. Reportedly, the Japanese sometime eat the leaves as a vegetable.

LANDSCAPE VALUE: Widely used as an edging plant along walks and perennial borders. Highly useful for accent among ornamental beds containing shrubs and low-growing trees. Combining well with astilbes, ferns, and day lilies. A dramatic lush appearance is obtained when massed to cover large bed or bank. Intolerant of foot traffic.

* * *

H. decorata (dek-ō-rā´tȧ) Decorative. Sometimes listed a *H.* 'Thomas Hogg.'

COMMON NAME: Blunt plantain lily.

NATIVE HABITAT: Japan.

HARDINESS: Zones 3 to 9.

FOLIAGE: Ovate to elliptic, four or five nerved on each side of midrib, to 6 inches long, blunt or abruptly pointed, green with prominent white margin.

SIZE: 12 to 24 inches tall, spreading to 2 feet across.

FLOWERS: Narrowly campanulate, dark violet in late summer.

* * *

H. fortunei (fôr-tū´-nē-ī) After Robert Fortune, plant collector.

COMMON NAME: Fortune's Plantain lily.

NATIVE HABITAT: Japan.

HARDINESS: Zones 3 to 9.

SIZE: 2 feet tall, spreading to 2 feet across.

FOLIAGE: Cordate, ovate to 5 inches long by about 3 inches wide, 8 to 10 nerves on either side of midrib, glaucous, pale green in color.

FLOWERS: Pale lilac to violet, on scapes to 3 feet in midsummer.

SELECTED CULTIVARS AND VARIETIES: 'Hyacinthina' Leaves green to blue-gray.
'Marginato Alba' With large leaves (to 11 inches long) that are broadly edged in white.

* * *

H. lancifolia (lan-si-fō´li-ȧ) Lance leaved.

COMMON NAME: Narrow-leaved plantain lily.

NATIVE HABITAT: Japan.

HARDINESS: Zones 3 to 9.

SIZE: 1 1/2 to 2 feet tall, spreading to 18 inches across.

FOLIAGE: Ovate-lanceolate, 5 to 7 inches long by 2 1/2 to 3 1/2 inches wide, 5 to 6 nerves on either side of midrib, often long, pointed, dark glossy green; petiole slender.

FLOWERS: Pale lavender flowers are borne in late summer to early fall, elevated on 2 foot tall scapes.

* * *

H. plantaginea (plan-ta´gin-ē-ȧ) Plantain like.

COMMON NAME: Fragrant plantain lily.

NATIVE HABITAT: Japan.

HARDINESS: Zones 3 to 9.

SIZE: 10 inches tall, spreading to 3 feet across.

FOLIAGE: Glossy, yellowish green, to 10 inches long and 7 inches wide, ovate to cordate-ovate in outline, 7 to 9 nerves on either side of midrib.

FLOWERS: White, held horizontally, blooming from late summer to early fall. More valuable in flower than many other hostas. Cultivar 'Grandiflora' has longer foliage; flowers longer and narrower than the species.

* * *

H. sieboldiana (sē-bold-ē-ā´nȧ) After Siebold (Figure 3–104). Syn. *H. glauca.*

COMMON NAME: Siebold plantain lily.

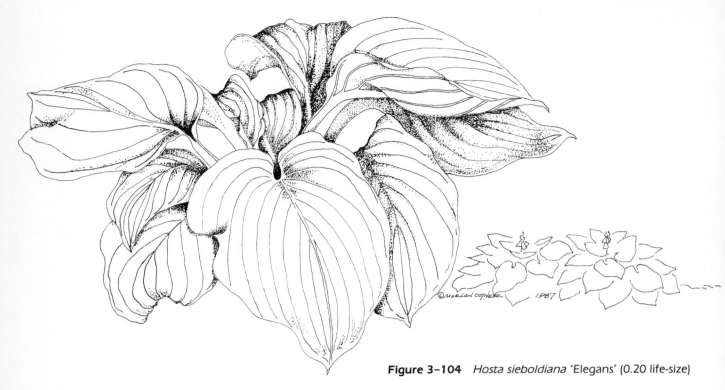

Figure 3–104 *Hosta sieboldiana* 'Elegans' (0.20 life-size)

NATIVE HABITAT: Japan.

HARDINESS: Zones 3 to 9.

SIZE: 18 inches tall, spreading to 3 feet across.

FOLIAGE: Ovate with cordate base, 10 to 15 inches long by 6 to 10 inches wide, 12 or more nerves on either side of midrib create a crinkled or wrinkled appearance.

FLOWERS: Pale lilac, effective mid to late summer on 2 foot scapes, sometimes obscured by the foliage.

SELECTED CULTIVARS AND VARIETIES: 'Aurora Borealis' Similar to 'Frances Williams' but colors more intense; growth more vigorous, to 4 feet tall and 5 feet wide.
'Elegans' Leaves broad, rounded, glaucous.
'Frances Williams' Blue-green with gold margined leaves; to 3 feet tall and 4 feet across.
'Golden Sunburst' With golden foliage.

*   *   *

H. undulata (un-dū-lā´tà) Wavy, the leaves.

COMMON NAME: Wavy-leaved plantain lily.

NATIVE HABITAT: Japan.

HARDINESS: Zones 3 to 9.

OF SPECIAL INTEREST: Unique in that variegation is better in partial shade than sun. Probably a hybrid, self-sterile, sometimes offered as *H. variegata* or *H. mediopicta*.

SIZE: 2 to 3 feet tall, spreading to 2 feet across.

FOLIAGE: Small, elliptic to ovate, to 6 inches long by 3 inches wide, apex pointed, about 10 nerves on either side of midrib, margin undulating, variegated green with cream or white middle, petiole long and winged.

FLOWERS: 2 inches long, pale lavender on 3 foot scapes in summer.

SELECTED CULTIVARS AND VARIETIES: 'Albo Marginata' Leaves edged in white.
'Erromena' Slightly larger and more robust.
'Univittata' Leaves with narrow central stripe and margin less wavy.

*   *   *

H. ventricosa (ven-tri-kō´sà) Irregularly swollen.

COMMON NAME: Blue plantain lily.

NATIVE HABITAT: Japan and Siberia.

HARDINESS: Zones 3 to 9.

SIZE: 3 feet tall, spreading to 3 feet across.

FOLIAGE: Ovate-cordate, large, to 9 inches long by 8 inches wide, with slightly twisted, short, pointed apex; 7 to 9 nerves on either side of midrib, dark, shiny green; petiole narrowly winged.

FLOWERS: Dark violet to near blue, to 2 inches long on 3 foot scapes. They are abundant, effective in late summer. Seeds are fertile and fading; flowers should be removed to prevent self-sowing.

SELECTED CULTIVARS AND VARIETIES: 'Aureo

Maculata' Leaves mottled green and yellow in spring.

'Aureo-marginata' Yellow margined foliage.

Generally Applicable to All Species

CULTURE: Soil: Sandy or light-loam soil that is relatively high in organic matter with good drainage is preferred. A pH range from 5.5 to 6.5 is best. Moisture: Lush foliage is not markedly drought tolerant. Plants should be kept out of windy location and soil should be kept moist. Light: Light to dense shade, flowering best in light shade. Will take full sun if in cool location with high humidity.

PATHOLOGY: Diseases: Leaf spots, crown rot. Pests: Snails and slugs feed on leaves, as do earwigs, cutworms, inchworms, weevils, and deer. Rodents, too, do a wonderful job of eating the crowns, often in woodland settings.

MAINTENANCE: Requiring practically no maintenance. Annual additions of organic matter may be beneficial when grown in amended soils.

PROPAGATION: Division: Simply divide clumps in spring or fall. In vitro: Methods utilizing transverse stem sections as explants have been developed. Seed: Propagation from seed is relatively simple. Cold stratification does not seem to be needed. Germination usually occurs in 2 to 3 weeks at 70°F.

✛✛✛✛✛✛✛✛✛✛✛✛✛✛✛✛✛

*H*ydrangea (hi-dran´ jē-ȧ) From Greek, *hydor*, water, and *aggeion*, a vessel, or vase, in reference to the shape of the seed capsule. This genus is composed of about 20 species of erect or climbing, op- posite leaved, deciduous or evergreen shrubs. They are native to North and South America, and Asia.

FAMILY: HYDRANGEACEAE: Hydrangea family.

H. anomala ssp. *petiolaris* (ȧ-nom´a-lȧ pet-i-ō-lā´ris) Long petioled (Figure 3–105).

COMMON NAME: Climbing hydrangea.

NATIVE HABITAT: China and Japan.

HARDINESS: Zone 4 to 7 or 8.

SIZE: 3 to 6 inches tall without support, spreading indefinitely.

HABIT: Woody climbing vine with footlike holdfasts, sprawling as a ground cover when unsupported.

RATE: Slow for first few years then moderate; space plants from gallon-sized containers 3 to 4 feet apart.

LANDSCAPE VALUE: A fine cover for large open area, whether flat or sloping. Large showy flowers are outstanding. No foot traffic.

FOLIAGE: Dediucous, opposite, simple, glabrous or nearly so, broadly ovate to ovate, apex acute or acuminate, base cordate or rounded, 2 to 4 inches long, slightly less across, margins serrate, dark green and shiny above, petioles 1/2 to 4 inches long.

Figure 3–105 *Hydrangea anomala* ssp. *petiolaris* (life-size)

STEM: Brown, exfoliating with age, new growth stout and light green, rootlike holdfasts along internodes.

INFLORESCENCE: Large 6 to 10 inch diameter flat-topped corymbs on 1 to 1 1/2 inch stalk; outer flowers sterile and showy white (1 to 1 3/4 inches across); inner flowers fertile, dull white, not nearly as showy, fragrant, 15 to 22 stamened; outstanding overall effect in early summer.

FRUIT: Capsules, not ornamentally significant.

OF SPECIAL INTEREST: The species *H. anomala* can be distinguished from the more common subspecies *petiolaris* in that the leaves of the species are elliptic-ovate, 3 to 6 inches long, 1 to 3 inches across; inflorescences are smaller, 6 to 8 inches across, with fewer showy sterile flowers, 9 to 15 stamens. Although not as showy as the subspecies and less often cultivated, this plant does have excellent floral display as well.

CULTURE: Soil: Best in fertile, well-drained loam, but tolerant of a wide range of soil types. the optimal pH range is 6.0 to 7.0.
Moisture: Quite tolerant to drought, but floral display is at its finest when ground is kept moist.
Light: Tolerant of range from full sun to dense shade; however larger, more showy flowers are produced when grown in light shade.

PATHOLOGY: Generally free of diseases and pests but sometimes plagued by the following: Diseases: Stem rot, when cuttings or small plants are potted into poorly draining media; bud blight, leaf spots, rust, powdery mildew.
Pests: Two spotted mite, rose chafer, aphids, leaf tier, scales, nematodes.
Physiological: Alkaline induced iron chlorosis.

MAINTENANCE: Relatively maintenance free; just cut back shoots as they outgrow their bounds.

PROPAGATION: Cuttings: Cuttings root marginally in mid to late spring and are aided by the use of a 4000 ppm IBA/talc preparation. It is important to stick them in a very porous medium and to discontinue mist as soon as rooted.
Seed: Commercial sources are available and seed is said to germinate readily after 2 to 3 months of cold stratification.

++++++++++++++++++

*H*ypericum (hī-pĕr´i-kum or hī-pĕr´ik-um) Greek name of obscure meaning, possibly from *hyper*, over, and *ereike*, a heath, in reference to the natural habitat of some species. This genus commonly is thought to contain about 300 species of evergreen and deciduous subshrubs. They are native to temperate regions throughout the world; usually with showy yellow flowers and insignificant fruit.

FAMILY: HYPERICACEAE: Hypericum family.

LANDSCAPE VALUE: Low, trailing forms of St. John's-wort are all excellent general covers for medium to large areas. The rhizomatous species are additionally good on slopes and retain soils well. Each are used for edging walks or driveways where spread can be controlled or there is room to spread. No foot traffic.

OF SPECIAL INTEREST: The flowers of St. John's-wort are important to bees not in that they produce nectar, but rather they provide copious amounts of pollen, the source of protein for bees.

*   *   *

H. buckleyi (buk-lē´ĭ) Proper name.

COMMON NAME: Blue Ridge St. John's-wort.

NATIVE HABITAT: North Carolina and Georgia.

HARDINESS: Zones 5 to 8.

HABIT: Semiwoody, low, matlike, decumbent, small, shrubby ground cover.

SIZE: To 12 inches tall.

RATE: Fast but easily controlled; space plants from quart- or gallon-sized containers 1 1/2 to 2 1/4 feet apart. Branches do not root readily.

FOLIAGE: Opposite, simple, entire, obovate, 1/2 to 3/4 inch long, deciduous.

FLOWERS: In bundles of three or solitary, to 1 inch across, yellow, effective early summer.

*   *   *

H. calycinum (kal-is´e-num or kal-i-sī´num) From Latin, *calyx*, a cup, probably in reference to the large, cup-shaped calyx (Figure 3–106).

COMMON NAME: Aaronsbeard, rose of sharon, gold flower, creeping St. John's-wort.

NATIVE HABITAT: Southeastern Europe and Asia Minor.

HARDINESS: Zones 6 to 10, occasionally winter killed to the ground, but roots survive and recovery is swift.

HABIT: Dense, trailing, semiwoody, spreading by rhizomes, stems freely rooting.

SIZE: 12 to 18 inches tall, spreading indefinitely.

RATE: Fast; space plants from pint- to quart-sized containers 2 1/2 to 3 feet apart.

FOLIAGE: Semievergreen, opposite or whorled, 4 ranked, simple, entire, ovate to oblong, to 4 inches long by 1 inch across, glabrous and glandular dotted both sides, conspicuously net veined below, medium to dark green above and glaucous below, color becoming somewhat purplish in fall.

STEMS: Glabrous, somewhat angular, woody near base.

INFLORESCENCE: Terminal cymes or solitary flowers; flowers regular, to 3 inches across, very showy, corolla 5 petaled, colored brilliant yellow, petals spreading, stamens numerous and erect spreading, anthers reddish, effective early summer to early fall, borne on new growth.

* * *

H. ellipticum (ē-lip´te-kum) Elliptic, the shape of the leaves.

COMMON NAME: Pale St. John's-wort.

NATIVE HABITAT: Southeast Canada and the eastern seaboard states of the United States as far south as Virginia.

HARDINESS: Zone 3.

HABIT: Herbaceous, rhizomatous, dense ground cover.

SIZE: To 12 inches tall.

RATE: Moderate; space plants from quart- or pint-sized containers 1 to 1 1/2 feet apart.

FOLIAGE: Simple, opposite, elliptic, to 1 1/4 inch long, semievergreen.

INFLORESCENCE: Few to many flowered terminal cymes; flowers yellow, small, effective mid to late summer.

* * *

H. olympicum (ō-lim´pi-kum) Of Mt. Olympus, applied to several mountains in southeast Europe, western Asia, and Turkey.

COMMON NAME: Olympic St. John's-wort.

NATIVE HABITAT: Southeastern Europe.

HARDINESS: Zone 6.

HABIT: Woody, procumbent, low, shrubby ground cover.

SIZE: To 12 inches tall.

RATE: Moderate; space plant from quart- to gallon-sized containers 2 1/2 to 3 feet apart.

FOLIAGE: Simple, opposite, oblong-lanceolate to elliptic-oblong, to 1 1/2 inches long, sessile, translucent-dotted, gray-green.

Figure 3–106 *Hypericum calycinum* (0.5 life-size)

INFLORESCENCE: Terminal cymes; flowers to 2 1/2 inches across, showy, golden-yellow, effective midsummer.

* * *

H. patulum (pat´ū-lum) Spreading, the habit.

COMMON NAME: Goldencup St. John's-wort.

NATIVE HABITAT: China.

HARDINESS: Zone 6.

HABIT: Broad spreading shrub, laxly upright branched.

SIZE: To 3 feet tall, spreading 4 to 5 feet across.

RATE: Moderate; space plants from gallon-sized containers 2 1/2 to 3 feet apart.

FOLIAGE: Semievergreen, opposite, ovate to ovate-oblong or lanceolate-oblong, medium green above and paler below.

INFLORESCENCE: Cymes or solitary flowers; flowers with sepals broadly ovate to rounded and overlapping, 5 petaled corolla of yellow, to 2 inches across, effective in midsummer.

SELECTED CULTIVARS AND VARIETIES: Var. *henryi* More vigorous than the species; with larger flowers to 2 1/2 inches across.
'Hidcote' Lower, to 18 inches tall and wide, with 2 inch diameter yellow flowers from early summer to midfall.

* * *

H. reptans (rep´tanz) Creeping.

COMMON NAME: Creeping St. John's-wort.

NATIVE HABITAT: Western China and the Himalayas.

HARDINESS: Zone 7.

HABIT: Woody, prostrate growing, shrubby ground cover that roots as it spreads.

SIZE: 4 to 6 inches tall, spreading indefinitely.

RATE: Moderate, space plants 1 to 2 feet apart from quart-sized containers.

FOLIAGE: Simple, opposite, elliptic-oblong, 1/2 inch long, medium green.

FLOWERS: Solitary, to 1 3/4 inch across, yellow, effective summer to fall.

Generally Applicable to All Species

CULTURE: Soil: Widely adaptable, but best in fertile, loamy soil with good drainage and a pH from 5.5 to 6.5.
Moisture: Able to survive moderate periods of drought; however, best when supplied with occasional deep irrigation so as to maintain soil moisture.

Light: Full sun to partial shade.

PATHOLOGY: Diseases: Leaf spots, powdery mildew, rust, fungal induced root rot.
Pests: Nematode that causes roots to rot.

MAINTENANCE: Shear in early spring to rejuvenate growth and to stimulate branching.

PROPAGATION: Cuttings: Both soft and hardwood cuttings root readily and are aided by the use of a mild root inducing hormone application of 1000 to 2000 ppm IBA/talc.
Division: Simple division of clumps in early spring or fall is an effective means of propagating small quantities.
Seed: Ripe seed may be sown upon harvest or stored cool and dry for up to 1 year with good germination. Germination does not take long; however, plantlets may be rather slow to mature. They may be started in plugs and transplanted once roots are well established. Commercial supplies of seed are available for some species.

*I*beris (ī-bē´ris) From Iberia, the ancient name of Spain, where many species originated. This genus is commonly called candytuft and is composed of small, annual and perennial herbs. Usually glabrous, often with woody base (suffrutescent), they are native to central Europe and the Mediterranean region.

FAMILY: CRUCIFERAE: Mustard family.

LANDSCAPE VALUE: Commonly used as small area general covers. They often lend themselves nicely to specimen planting in rock gardens and borders and may be used to face evergreen foundation plantings. No foot traffic.

FRUIT: Silicle, not ornamentally significant.

* * *

I. gibraltarica (ji-brawl-tãr´i-kà) Of Gibraltar.

COMMON NAME: Gibraltar candytuft.

NATIVE HABITAT: Spain and Morocco.

HARDINESS: Zones 6 to 9.

HABIT: Low, sprawling subshrub.

SIZE: To 12 inches tall, spreading 3 to 4 feet across.

RATE: Relatively slow; space plants from pint- or quart-sized containers 10 to 16 inches apart.

FOLIAGE: Evergreen, simple, oblong-spatulate, to 1 inch long, margin toothed, glabrous, dark green.

INFLORESCENCE: Flat, umbelliform clusters to 2 inches across; flowers 4 petaled, lilac or light purple, effective in late spring.

* * *

I. saxatilis (saks-at´i-lis) Inhabiting rocks.

COMMON NAME: Rock candytuft.

NATIVE HABITAT: Southern Europe.

HARDINESS: Zones 3 to 7 or 8.

HABIT: Low, sprawling, subshrub.

SIZE: To 6 inches tall, spreading to 4 feet across.

RATE: Moderate; space plants from quart-sized containers 2 to 2 1/2 inches apart.

FOLIAGE: Evergreen, linear, semicylindrical on vegetative stems, flat on flowering stems, small, colored midnight green.

INFLORESCENCE: Terminal corymbs; flowers small, white, effective mid to late spring.

* * *

I. sempervirens (sem-pẽr-vī´rens) Always green (Figure 3–107).

COMMON NAME: Evergreen candytuft, edging candytuft.

NATIVE HABITAT: Europe and Asia.

HARDINESS: Zones 3 to 8.

HABIT: Low, herbaceous, dense, upright branched, spreading subshrub.

SIZE: To 12 inches tall, spreading 3 to 4 feet across.

RATE: Relatively slow, space plants from pint-sized containers 1 1/2 to 2 feet apart.

FOLIAGE: Evergreen, alternate, simple, entire, linear to oblanceolate, 1 to 1 1/2 inches long, 1/8 to 1/4 inch across, glabrous, dark glossy green.

STEMS: Stiff, semiwoody, slightly reddish.

INFLORESCENCE: Lateral racemes to 2 inches across; many flowered, 4 petaled, white corolla, effective and very showy late spring to early summer.

SELECTED CULTIVARS AND VARIETIES: 'Alexander's White' Compact habit; early bloom.
'Christmas Snow' With flowers in spring and again in early fall.
'Little Gem' Compact habit, to 5 inches tall, and about 8 to 10 inches across.
'October Glory' ('Autumn Beauty') Blooming spring and fall.
'Purity' Flowers whiter, more numerous and longer lasting.
'Snowflake' With larger waxy flowers in broad clusters; foliage longer, broader and darker green.

Generally Applicable to All Species

CULTURE: Soil: Adaptable to various soils, with good drainage being the key factor for success.
Moisture: Not markedly resistant to drought; watering regularly in summer to maintain soil in a slightly moistened condition is recommended.
Light: Full sun to light shade; foliage and flowers at their best in full sun. The tips of the leaves may desiccate somewhat in hot summer areas.

PATHOLOGY: Club root, damping off, downy mildew, powdery mildew, white root.
Pests: Oystershell scale, diamond back moth, southern root knot nematode.

MAINTENANCE: Clip or mow plants halfway to the ground after flowering to keep them compact and to rejuvenate growth.

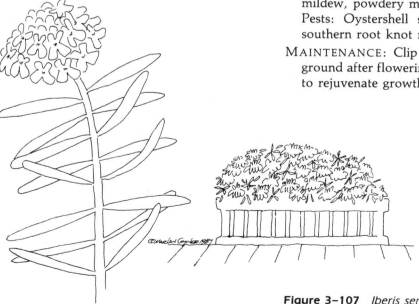

Figure 3–107 *Iberis sempervirens* (life-size)

PROPAGATION: Cuttings: Cuttings are rooted easily in mid to late summer. They may be helped with the use of a rooting compound of 3000 ppm IBA/talc; however, it is not necessary. The medium should be very porous.

Division: Simple division into small clumps in spring or fall is effective.

Seed: Collect when ripe and sow immediately (species only). Indoors seed can be germinated in 1 to 2 weeks at 70° to 85°F and may be helped by light.

Ilex (i′leks) From the Old Latin name *ilex*, an evergreen (Holm oak) which members of the genus *Ilex* were supposed to resemble. This genus is commonly called holly and is composed of primarily evergreen, dioecious or polygamodioecious shrubs and trees. They are native to temperate and tropical areas of North and South America and Asia. Several low, broad spreading cultivars are useful for ground cover.

FAMILY: AQUIFOLIACEAE: Holly family.

HABIT: Ground covering types are low, spreading, compact, woody shrubs.

RATE: Generally relatively slow to moderate; 1 or 2 gallon sized container plants should be spaced according to their expected mature spread. In most cases, a distance of two-thirds the expected mature spread is acceptable for spacing.

LANDSCAPE VALUE: Hollies make superb edgings along walks and drives. They are also good as dwarf hedges (can be sheared for a formal effect). Sometimes used to good effect for foundation planting. No foot traffic.

*   *   *

I. cornuta (kôr-new′ta) Horned, in reference to the leaf margins, which are spiny (Figure 3–108).

COMMON NAME: Chinese holly, horned holly.

NATIVE HABITAT: China.

HARDINESS: Zone 7.

I. cornuta is a densely branched, evergreen shrub or small tree, not itself useful as a ground cover. Many low growing selections, however, are excellent for this purpose. It is generally described as follows:

FOLIAGE: Evergreen, simple, alternate, short petioled, cordate or oblong-entire to quadrangular, sinuately toothed 1 to 3 per side, thick, leathery, dark glossy green.

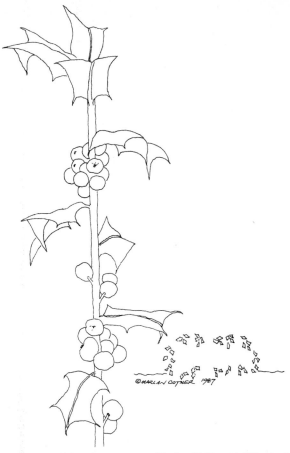

Figure 3–108   *Ilex cornuta* 'Bufordii Nana' (life-size)

FLOWERS: Inconspicuous, dioecious, numerous, solitary, dull white, small; male flowers more fragrant than female; borne late spring to early summer.

FRUIT: Abundant, bright red, persistent (through winter), globose drupe.

SELECTED CULTIVARS AND VARIETIES: 'Carissa' A dense, dwarf selection that reaches 3 to 4 feet tall and spreads to 4 to 6 feet across. Foliage is ovate, 2 to 3 inches long by 1 1/2 inches across, entire, leathery, waxy, dark green; nonfruiting.

'Bufordii Nana' Low, compact selection that usually is below 3 feet tall, but may reach 5 to 6 feet if not pruned. Foliage is lustrous dark green, usually with single terminal spine; fruiting profusely and without pollination. Spreads 5 to 6 feet across.

'Berries Jubilee' Plant patent 3168, slow growing, compact dwarf, mounding selection with heavy production of large red fruit.

'Rotunda' (dwarf Chinese holly) Compact, dwarf, mounding selection; reaches 3 to 4 feet tall by 6 to

8 feet across; foliage is 5 spined, dark glossy green, nonfruiting.

* * *

I. crenata (krē-nā´tà) Leaves crenulate, that is, with rounded teeth about the margins.

COMMON NAME: Japanese holly, box-leaved holly.

NATIVE HABITAT: Japan.

HARDINESS: Zone 6.

I. crenata, as with *I. cornuta*, is not itself useful as a ground cover. However, cultivars are. They are briefly described as follows:

FOLIAGE: Evergreen, simple, alternate, crowded, short petioled, elliptic, oblong or narrowly obovate, to 1 1/4 inch long, to 5/8 inch wide, margins crenate or serrate, leathery, glabrous, shiny, deep green above, pale below.

FLOWERS: Male flowers in few flowered fasicles, white; female flowers solitary, dull white; small and ornamentally insignificant in either case, blooming late spring to early summer.

FRUIT: 1/4 inch diameter, globose, black drupe, residing underneath the foliage, effective in fall.

SELECTED CULTIVARS AND VARIETIES: 'Compacta' Low, many branched, compact globose habit, usually maintained below 3 feet, but reaching 5 to 6 if unpruned; spreading 3 to 6 feet across.
'Convexa' Similar to 'Compacta,' but upper leaf surfaces convex.
'Green Island' Loose, broad-spreading shape, to 3 feet tall by 6 feet across.
'Helleri' Dwarf, mounded, compact shape; leaves small; reaching little more than 3 feet tall.
'Hetzi' Similar to 'Compacta' but leaves larger; faster growing.
'Kingsville' Loose habit, usually 3 feet tall by 6 feet across.
'Kingsville Green Cushion' Very dense, low growing selection that may reach 8 to 10 inches tall and spread to about 3 feet across.

* * *

I. vomitoria (vom-i-tō´ri-à) Emetic, the fruit poisonous and causing nausea.

COMMON NAME: Youpon holly, Cassina, Cassena, Cassine.

NATIVE HABITAT: Southeastern United States.

HARDINESS: Zone 7.

I. vomitoria, like the other species discussed, is not itself a ground cover, but selections of it serve the purpose. The following description is applicable to them.

FOLIAGE: Simple, alternate, evergreen, ovate to oblong or elliptic, to 1 1/2 inches long, 1/2 to 3/4 inch across, margins shallowly toothed, leathery, dark green and glossy above.

INFLORESCENCE: Flowers small, greenish white; male flowers in axillary clusters on 1 year old wood; female flowers solitary or paired, neither ornamentally significant, borne in midspring.

FRUIT: Numerous 1/4 inch diameter scarlet drupes, often persisting through winter.

SELECTED CULTIVARS AND VARIETIES: 'Nana' Dwarf, compact; with smaller foliage; reaching 3 to 5 feet tall by 8 feet across.
'Shelling's Dwarf' (Stoke's Dwarf) Similar to 'Nana' but leaves darker, habit more compact, and not producing fruit as it is a male clone.

Generally Applicable to All Species

CULTURE: Soil: Best in fertile, well-drained loam with a pH of 5.2 to 6.4.
Moisture: Tolerance to drought is only moderate; it is wise to water deeply to keep soil moist during periods of hot, dry weather.
Light: Full sun to light shade.

PATHOLOGY: Diseases: Bacterial blight, leaf, twig, and flower blights, canker, leaf spots, mildew, root rot, rust, sooty mold.
Pests: Nematodes and mites are the most frequently encountered pests.

MAINTENANCE: Generally a light shearing in spring is all that is needed. In northern areas where snow cover is questionable, it is wise to wrap plants with burlap in late fall to prevent dessication.

PROPAGATION: Cuttings: Cuttings taken in fall root well when given bottom heat of around 70 °F. Treating with a root inducing preparation of 3000 ppm IBA/talc is beneficial.

✦✦✦✦✦✦✦✦✦✦✦✦✦✦✦

*I*ndigofera (in-di-gof´ẽr-à, or in-di-gō´fẽr-à) Indigo, the blue dye, from Latin, *indicus*, Indian, where it comes from, and *fero*, to produce. More simply known as indigo, this genus is composed of around 700 to 800 species of herbaceous perennials and shrubs. They are native to warm climates throughout the world.

FAMILY: FABACEAE (LEGUMINOSAE): Pea family.

HABIT: Low growing, dense, suckering shrubs.

RATE: Relatively fast; space plants from 1 to 2 gallon sized containers 3 to 4 feet apart.

LANDSCAPE VALUE: Best used in larger areas for general cover and on banks where their extensive root systems make excellent soil stabilizers. Severe winters may kill top growth, but it usually comes right back. No foot traffic.

* * *

I. incarnata (in-kär-nā´ta) Meaning flesh colored, the flowers (Figure 3–109).

COMMON NAME: White Chinese indigo.

NATIVE HABITAT: China and Japan.

HARDINESS: Zone 5.

SIZE: To 18 inches tall, spreading 6 to 8 feet across.

FOLIAGE: Semievergreen, odd-pinnately compound, leaflets in 3 to 6 pairs, elliptic, to 2 1/2 inches long, soft medium green.

INFLORESCENCE: Racemes to 8 inches long; flowers pink, pealike, to 3/4 inch long, effective midsummer.

FRUIT: Legume.

SELECTED CULTIVARS AND VARIETIES: Var. *alba* With white flowers.

* * *

I. kirilowii (kir-i-lō´i-ī) Memorial.

COMMON NAME: Kirilow indigo.

NATIVE HABITAT: China and Korea.

HARDINESS: Zones 4 to 7.

SIZE: To 3 feet tall.

FOLIAGE: Semievergreen, odd-pinnately compound, leaflets in 3 to 5 pairs, to 1 1/2 inches long.

INFLORESCENCE: Racemes to 5 inches long; flowers bright rose, 3/4 inch long, midsummer.

Generally Applicable to Both Species

CULTURE: Soil: Best in well-drained medium loam with a pH of 4.5 to 7.5.
Moisture: Not highly tolerant to drought; supplemental water should be provided often enough to keep the ground slightly moistened.
Light: Full sun.

PATHOLOGY: No serious diseases or pests.

MAINTENANCE: Shear in spring, removing any winter-killed shoots.

PROPAGATION: Cuttings: Softwood cuttings and root cuttings are both viable means of propagation. Stem cuttings should be treated with 4000 ppm IBA/talc.
Division: Division of rooted suckers is best done in fall or spring.
Seed: Seed is processed by soaking in hot water (190 °F) for 12 hours, then sowing.

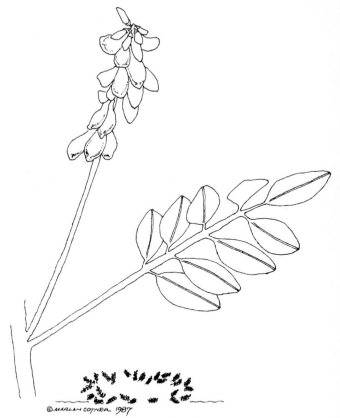

Figure 3–109 *Indigofera incarnata* (0.5 life-size)

✦✦✦✦✦✦✦✦✦✦✦✦✦✦✦✦

*I*ris (ī´ris) From Greek, *iris*, a rainbow, presumably in reference to the many colors of flowers. In Greek mythology, Iris was Juno's messenger and was supposed to have traveled the rainbow between heaven and earth, and was later transformed into a rainbow herself. This genus is composed of about 200 species of monocotyledonous, usually rhizomatous (a few bulbous) herbs. Primarily, they are native to the northern temperate zone. Irises are very popular and have been extensively hybridized to create a tremendous variety of flower characteristics. Colors range from tints and shades of pink, blue, violet, and purple to white, brown, yellow, orange, and black. Combinations of all these are also cultivated.

Irises are divided into different categories as determined by their flower morphology and root structure. The rhizomatous division includes three groups. They are as follows:

Bearded irises: The corolla of these Irises have a 'beard' pattern of hairs on the basal one-half of the falls (drooping sepals).

Crested irises: Instead of a beard, a central cockscomblike crest resides along the basal one-half of the falls.

Beardless Irises: Having smooth falls; this is the most extensive group.

A bulbous division also exists with several species included, and another is in existence that is neither bulbous nor entirely rhizomatous. Various subgroups further break down the above three groups. For those interested in gaining an in-depth knowledge of this genus, there are several books available. Membership in the American Iris Society is also encouraged. Its address is 6518 Beachy Avenue, Wichita, Kansas 67206.

Three species that are suited to ground covering are generally described next; a brief description of the unique features of the various cultivars and varieties follows.

FAMILY: IRIDALES: Iris family.

HABIT: Herbaceous, leafy, rhizomatous or bulbous ground covers.

LANDSCAPE VALUE: Fine for mass planting where their swordlike leaves are visible. Planting around walks and drives as edging is often very attractive, as is planting in a rock garden as a specimen. On a gradually sloping bank or in a ditch, their profuse root systems bind soil quite adequately. No foot traffic.

* * *

I. cristata (kris-tā´tȧ) Crested.

COMMON NAME: Dwarf crested iris, crested iris, crested dwarf iris.

NATIVE HABITAT: North America from Maryland to Georgia and Missouri.

HARDINESS: Zones 3 to 9.

SIZE: 3 to 4 inches tall, spreading indefinitely.

RATE: Relatively fast; space plants from pint- or quart-sized containers, 16 to 24 inches apart.

FOLIAGE: Deciduous, sword shaped, 4 to 9 inches long by 3/4 inch across, oriented vertically, colored green.

FLOWERS: Small, 1 to 2 flowered spathes, perianth tube longer than spath; falls obovate, to 1 1/2 inches long, pale-lilac, white, or yellow crested, effective in spring.

FRUIT: Capsule, not ornamentally significant.

SELECTED CULTIVARS AND VARIETIES: 'Alba' With white flowers.

* * *

I. pumila (pū´mi-lȧ) Dwarf or diminuitive.

COMMON NAME: Dwarf bearded iris.

NATIVE HABITAT: Europe and Asia Minor.

HARDINESS: Zone 4.

SIZE: 4 to 5 inches tall, spreading indefinitely.

RATE: Moderate; space plants from pint- or quart-sized containers 14 to 16 inches apart.

FOLIAGE: Sword shaped, 4 to 8 inches long, green, deciduous.

FLOWERS: Spathes 1 to 2 flowered; perianth tube 2 to 3 inches long, color varying from yellow to deep lilac; fall reflexed, oblong, about 2 inches long, bearded bluish, white, or yellow. Effective in spring.

FRUIT: Capsule, of no ornamental significance.

SELECTED CULTIVARS AND VARIETIES: Many are available, a few of the more popular include 'Alba' Flowers cream colored.
'Atropurpurea' Flowers purple.
'Aurea' Flowers clear bright yellow.
'Cyanea' Flowers deep blue.
'Lutea' Flowers golden yellow.

* * *

I. tectorum (tek-tō´rum) On roofs. This iris was grown on thatched roofs in Japan (Figure 3–110).

COMMON NAME: Roof iris, wall iris.

NATIVE HABITAT: China and Japan.

HARDINESS: Zone 5 to 9.

SIZE: To 10 inches tall, spreading indefinitely.

RATE: Space plants from pint- to quart-sized containers 14 to 16 inches apart.

FOLIAGE: Sword-shaped, to 12 inches long, thin, conspicuously nerved.

FLOWERS: Spathes 2 to 3 flowered, perianth tube about 1 inch long, segments deep lilac to bluish-purple; falls nearly orbicular to 2 inches long, crest deeply cut, margin wavy, colored white with brownish-violet streaks, standards (inner upright petals) colored lilac. Effective in late spring.

FRUIT: Similar to other species.

SELECTED CULTIVARS AND VARIETIES: 'Alba' Flowers white with yellow crest.

Generally Applicable to All Species

CULTURE: Soil: Best in fertile loam, but quite adaptable to others.
Moisture: Able to withstand hot, dry periods, but generally best when soil is kept moist with periodic irrigation as needed.

Figure 3-110 *Iris tectorum* (0.75 life-size)

Light: Full sun to moderate shade.

PATHOLOGY: Diseases: Numerous. Bacterial soft rot and leaf spots, blossom blight, leaf spot, rust, rhizome rot, crown rot, black rot, ink spot, fusarium basal rot, blue mold, iris mosaic virus. Pests: Aphids, tulip bulb aphid, lesser bulb fly, iris borer, Florida red scale, iris thrips, verbena bud moth, iris weevil, zebra caterpillar, bulb mite, stem and bulb nematodes, grubs, Deman's meadow nematode, southern root knot nematode.

MAINTENANCE: Relatively little maintenance required. Plants can be divided once they become crowded to improve the bloom if so desired. Fall shearing of the foliage (back to ground level) is recommended to keep plantings thick and to reduce the incidence of disease.

PROPAGATION: Division: Divide or cut rhizomes apart in late summer. Separate offsets on bulbous varieties.

*J*asminum (jas′mi-num or jas′min-um) Said to be derived from *ysmym*, the Arabic name for jasmine. This genus is composed of about 200 species of deciduous and evergreen shrubs or climbing, vinelike shrubs. They are native to tropical and subtropical regions of eastern and southern Asia, Malaya, Africa, and Australia.

FAMILY: OLEACEAE: Olive family.

HABIT: Woody, scrambling, often pubescent, climbing or nonclimbing, vinelike shrubby ground covers.

RATE: Relatively fast; space plants from gallon-sized containers 3 to 4 feet apart.

LANDSCAPE VALUE: Often used to good effect as bank covers or near the top of rock walls or ledges where the trailing branches can cascade gracefully over. No foot traffic.

*   *   *

J. floridum (flôr′i-dum or flō′ri-dum) Likely in reference to the floriferous (abundantly flowering) nature of this species.

COMMON NAME: Showy jasmine.

NATIVE HABITAT: China.

HARDINESS: Zones 8 to 10.

SIZE: 3 to 4 feet tall, spreading 5 to 7 feet across.

FOLIAGE: Semievergreen, alternate, pinnately compound; leaflets 3 or rarely 5, elliptic to oblong in outline, to 1 1/2 inches long.

STEMS: Dark green, glabrous, erect-arching, quadrangular.

INFLORESCENCE: Many flowered cymes; flowers with calyx teeth subulate; corolla salverform, tube to 3/4 inch long, lobes ovate, acute, half as long as tube, yellow, effective in early summer.

FRUIT: Black berry.

*   *   *

J. mesnyi (mez-nē′ĭ) After William Mesny.

COMMON NAME: Primrose jasmine.

NATIVE HABITAT: China.

HARDINESS: Zone 8.

SIZE: Mounding 5 to 6 feet tall, spreading 6 to 8 feet across.

FOLIAGE: Evergreen, compound, leaves opposite; leaflets in threes, oblong to lanceolate, to 3 inches long by 1/3 to 3/4 inch wide, shiny dark green.

STEMS: Erect, arching over and trailing to 10 feet, quadrangular, green, glabrous.

FLOWERS: Solitary, calyx with leafy teeth; corolla often double, salverform, of 6 to 10 divisions usually longer than the tube, yellow, with darker center; effective in spring.

FRUIT: Black berry, seldom produced in cultivation.

* * *

J. nudiflorum (nū-di-flō´rum) Naked flowered, the shrub blooming when the branches are leafless (Figure 3–111).

COMMON NAME: Winter jasmine.

NATIVE HABITAT: China.

HARDINESS: Zones 5 to 10.

SIZE: To 2 or 3 feet tall, spreading 4 to 7 feet across.

FOLIAGE: Deciduous, opposite, pinnately compound; leaflets in threes, ovate to oblong-ovate, to 1 1/4 inches long by 1/2 inch wide, entire, narrowed at ends, shiny green, short petioled.

STEMS: Arching and trailing to 15 feet, quadrangular, nonclimbing, green, glabrous.

FLOWERS: Solitary, axillary, perfect; corolla tube salverform and slender, to 2 inches long and 1 inch across, 6 lobed, bright yellow, opening in early spring prior to leaf emergence.

FRUIT: 2 lobed black berry seldom produced in cultivation.

OF SPECIAL INTEREST: Flowers are savored by hummingbirds.

Generally Applicable to All Species

CULTURE: Soil: Adaptable to most well-drained soils with acidic pH.
Moisture: Moderately tolerant to drought; periodic thorough deep watering is needed during the summer months.
Light: Full sun to moderate shade.

PATHOLOGY: Diseases: Bacterial crown gall, blossom blight, leaf spot, root rot.
Pests: Nematodes, scales, citrus whitefly.

MAINTENANCE: Cutting plants to within 6 inches of the ground every few years will rejuvenate growth. With some species, a lawn mower works well.

PROPAGATION: Cuttings: Leafy, semihardwood cuttings taken in late summer are easily rooted under mist. Application of 4000 ppm IBA/talc is beneficial.
Layering and division: Suckers and rooted nodal segments can be dug up and transplanted for small-scale propagation.
Seed: Sow as soon as ripe or store dry and cool until the following spring.

Juniperus (jŏ-nĭp´ĕr-us) Old Latin name for the juniper tree. This widely cultivated genus is com-

Figure 3–111 *Jasminum nudiflorum* (life-size)

posed of some 70 species of monecious or dioecious, evergreen, coniferous shrubs and trees. They are almost universally used in the landscape for purposes ranging from specimen plantings, foundation concealment, hedging, screening, massing as ground covers, accent and whatever else the imagination can conjure. Available in a wide range of textures, shapes, and sizes, along with practically the whole spectrum of colors, they are surely America's most serviceable plants.

HABIT: Junipers that are especially useful as ground covers are characterized by broad spreading, procumbent or decumbent habits of growth, with compactness that is capable of discouraging weed growth. There are *many* selections that fit this description.

FAMILY: CUPRESSACEAE: Cypress family.

LANDSCAPE VALUE: Excellent en masse as general covers for large or moderate open areas. On slopes for erosion control, they often perform quite well despite modest root systems. Many forms are excellent as a facing to open shrubs and trees where light is adequate. Next to a rock ledge where branches can hang over, they are graceful and eye-catching. For foundation planting and rock garden specimens, ground covering junipers are exceptional. They tolerate little or no foot traffic.

OF SPECIAL INTEREST: The fruit of many juniper species is consumed by several different types of birds and other wildlife. Even though many of the ground covering selections fruit less heavily, they too may be attractive to wildlife. The Pfitzer juniper is said to attract cedar waxwings, robins, and mockingbirds. In addition, most nonprostrate selections (and even some of them) serve well as nesting sites for a variety of birds.

* * *

J. chinensis (chi-nen´sis) Of China.

COMMON NAME: Chinese juniper.

NATIVE HABITAT: China and Japan.

HARDINESS: Zones 4 to 9.

J. chinensis itself is shrubby or treelike and may reach 60 feet tall. Features that it exhibits that generally apply to ground covering selections are as follows:

RATE: Slow to moderate; spacing plants from 1 or 2 gallon-sized containers should be based on mature width of cultivars and generally will range from 3 to 4 feet apart.

FOLIAGE: Evergreen, of two forms: juvenile leaves are awl shaped, 1/3 inch long, ending in a spiny point, spreading, arranged in whorls of three or in opposite pairs, upper surface with two glaucous bands and concave with green midrib running nearly to the apex, convex below. Adult foliage is 4 ranked, in opposite pairs, overlapping, to about 1/16 inch long, blunt tipped, convex below, with pale green margin.

FLOWERS: Usually dioecious, female flowers small and yellowish; male flowers profuse, small and also yellowish. Neither is of ornamental significance.

CONES: Male cones numerous, catkinlike, yellowish or brownish, about 1/25 inch across, primarily on adult branches. Female cones (fruit) powdery blue, ripening to brown during the second year, 1/3 inch across, 4 to 8 scaled, fleshy, 2 to 5 seeded.

SELECTED CULTIVARS AND VARIETIES: 'Fruitlandii' Compact, low-growing, maybe 1 1/2 to 2 feet tall, foliage dull green, rather coarse textured, possibly to 6 feet across. This cultivar arose as a sport from the cultivar 'Pfitzeriana Compacta.'
'Gold Coast' Plant patent 2491, graceful, compact, moderate growing, low spreading clone with golden yellow new foliage that intensifies with cold weather in fall.
'Old Gold' Similar to 'Pfitzeriana Compacta' but with golden or bronzy golden new foliage, and somewhat more delicate and refined habit than 'Pfitzeriana Compacta.'
'Pfitzeraina' Broad spreading shrub, usually not exceeding 4 to 5 feet tall and 10 feet across. Foliage is light green, with new branches somewhat pendulous.
'Pfitzeriana Compacta' Broad spreading, 12 to 18 inches tall and 6 feet across; colored gray-green.
'Pfitzeriana Aurea' With branches light golden yellow in youth; similar to 'Pfitzeriana' in habit, but somewhat smaller and flatter growing.
'Plumosa' Reaching 3 to 4 feet tall and spreading broadly. Foliage is rather dull gray-green in summer becoming purplish in fall. Both white and gold variegated versions are available.
'San Jose' Reaching 12 to 18 inches tall and spreading to 8 feet across, this lower, dull green selection is popular for its coarse texture and adaptability to harsh environments and sandy soil.
Var. *sargentii* A popular blue-green, scaly leaved, low and broad spreading variety, it will generally reach from 12 to 24 inches tall and may spread to 9 feet across.
Var. *sargentii* 'Compacta' Similar to var. *sargentii* but more compact in habit.
Var. *sargentii* 'Glauca' Lower than var. *sargentii* (only reaching 18 inches tall), with attractive, soft textured, bluish foliage.
Var. *sargentii* 'Viridis' Similar to var. *sargentii* but foliage is light green.
'Sea Spray' Low spreading sport of 'Pfitzeriana Glauca,' to 12 inches tall, reported to be very hardy and disease resitant.

CULTURE: Soil: Best in well-drained loamy soil with a pH of 7.0 to more acidic.
Moisture: Relatively drought tolerant once established. Occasional deep watering in summer will be beneficial.
Light: Full sun.

* * *

J. communis (kom-mū´nis) Common, that is, in groups or communities (communal).

COMMON NAME: Common juniper, English juniper.

NATIVE HABITAT: Europe, Asia, and North America.

HARDINESS: Zones 2 to 6; does not tolerate southern climates.

J. communis itself is a morphologically variable species with a habit that ranges from a straggly spreading shrub to a medium-sized tree. Size ranges from 2 to 35 feet tall. The species itself is seldom cultivated; however, many selections are popular. Those useful in ground covering are treated later, while the species is generally described as follows:

RATE: Relatively slow; space plants from 1 to 2 gallon sized containers 2 1/2 to 3 1/2 feet apart.

FOLIAGE: Evergreen, prickly, narrowly ternate, persistent for about 3 years, sessile, commonly adpressed on older branches, tapering to a sharp point, approximately 3/5 inch long, concave above with broad white longitudinal stripe, occasionally with green midrib at base, keeled below.

FLOWERS: Dioecious, small male and female flowers, both yellow, ornamentally insignificant.

CONES: Male are catkinlike, yellow, solitary, about 1/3 inch long, cylindrical. Female cones (fruit) are berrylike, solitary, about 1/2 inch long, green in youth, ripening to blue or black in 2 to 3 years.

SELECTED CULTIVARS AND VARIETIES: Var. *depressa* Low spreading or prostrate shrub rarely reaching 4 feet tall and 8 feet across; light yellowish or brownish-green, becoming bronzy in winter.
'Depressa Aurea' Similar to the variety *depressa* in habit, but foliage is yellow in youth.
'Depressed Star' A uniformly shaped selection of the variety *depressa* which ranges from about 1 to 1 1/2 feet tall and spreads to 8+ feet across.
'Effusa' Very low, prostrate selection. Leaves are dull green above, silvery white below, and color is retained throughout the year. Height is generally 2 to 4 inches tall, and spread may reach 4 or more feet across.
'Nana' Slow growing, dark green prostrate selection. Sometimes listed as 'Saxatilis' or 'Siberica'; to 2 or 3 feet tall.

CULTURE: Soil: Adaptable to most soils, including those that are infertile and sandy. Preferred pH range from 7.0 to more acidic.
Moisture: Very tolerant to drought; do not hesitate to plant in windy, open exposures. Occasional deep watering may be of benefit in dry, hot periods.
Light: Full sun.

* * *

J. conferta (kon-fĕr´tȧ) From Latin, *confercio*, compact, the habit of growth, or crowded, the leaves.

COMMON NAME: Shore juniper.

NATIVE HABITAT: Seacoast of Japan.

HARDINESS: Zones 6 to 8.

RATE: Relatively slow; space plants from 1 to 2 gallon sized containers 3 to 4 feet apart.

J. conferta is itself a dense, procumbent, wide spreading species that is useful as a ground cover. It may reach 1 to 1 1/2 feet tall and spread to over 6 feet across.

FOLIAGE: Evergreen, linear, awl-shaped, 1/4 to 5/8 inch long, to 1/16 inch across, ternate, spiny pointed, crowded and overlapping, grooved above, convex below.

FLOWERS: Dioecious, of no ornamental significance.

CONES: 1/4 to 1/2 inch wide, globe shaped, both male and female; female with 3-angled ovate seeds, bloomy in maturity.

SELECTED CULTIVARS AND VARIETIES: 'Blue Pacific' Blue-green in color; low trailing habit; about 6 to 12 inches tall.
'Boulevard' Foliage glaucous-green, with main branches oriented horizontally, prostrate habit.
'Compacta' More compact and prostrate than the species.
'Emerald Sea' Foliage blue-green; of low, prostrate habit, reaching 6 to 12 inches tall.

CULTURE: Soil: Tolerant of a wide range of well-drained, preferably loamy soils, but adaptable to sandy soils. Withstanding infertility and high salinity.
Moisture: Considerably tolerant to drought, but intolerant of excess moisture. Only infrequent watering is recommended, if at all, once plants are established.
Light: Full sun.

* * *

J. × *davurica* (dā-vūr´i-kȧ) Meaning unknown.

COMMON NAME: Davurian juniper.

HYBRID ORIGIN: *J. sabina* × *sheppardii*.

HARDINESS: Zones 5 to 9.

RATE: Slow to moderate; spacing of 3 to 4 feet is usually acceptable for plants from 1 to 2 gallon sized containers.
Cultivars of this supposed hybrid are commonly listed as members of the species *J. chinensis*, yet are unique in that their mixed adult and juvenile foliage is carried upon relatively stout, twisted, cordlike branches. I have not seen the plant *J* × *davurica*, but some of the cultivars are remarkable and excellent for ground covering and surely worthy of more attention than presently receiving.

SELECTED CULTIVARS AND VARIETIES: 'Expansa' (frequently listed as 'Parsoni' or *J. squamata* var. *expansa* 'Parsoni') is commonly cultivated and is a rather attractive, low, mounding shrub that reaches 3 feet high and as much as 9 feet across. Foliage is dark green, arranged in dense clusters on filiform, rigid branches that are elevated and oriented horizontally above the ground. Reportedly introduced by Parson's Nursery in New York, as a Japanese import.
'Expansa Variegata' Similar to 'Expansa,' but with scattered white leaved branches among the green.
'Expansa Aureospicata' Less vigorous, with occasional yellow leaved branches.

CULTURE: Same as *J. chinensis.*

*   *   *

J. horizontalis (hor-i-zon-tā´lis) Horizontal habit of growth (Figure 3–112).

COMMON NAME: Creeping juniper, creeping cedar, creeping savin.

NATIVE HABITAT: Northeastern United States.

HARDINESS: Zones 3 to 9.

HABIT: Procumbent to decumbent, mat forming, woody ground cover.

RATE: Slow to moderate; space 1 to 2 gallon sized plants 3 to 4 feet apart.

SIZE: 1 to 2 feet tall, spreading 4 to 6 or slightly more feet across.

FOLIAGE: Evergreen, mature leaves are mostly scalelike, entire, about 1/16 inch long, 4 ranked, adpressed, acutely pointed, with glandular depression on back. Juvenile leaves are awl shaped, in opposite pairs; both usually glaucous, becoming purplish in fall.

FLOWERS: Dioecious, small and usually not very ornamental.

SELECTED CULTIVARS AND VARIETIES: 'Admirabilis' Blue-green, primarily juvenile foliaged selection (awl shaped); height to about 12 inches, and secondary branches angled upward. This clone is male, and reportedly was introduced by Plumfield Nurseries.

'Adpressa' A green selection with glaucous new growth. Habit is dense; to about 10 inches tall.

'Alpina' Rather erect selection that exhibits dwarf characteristics; reaching 1 to 2 feet tall by only 5 feet across.

'Aurea' New foliage colored golden yellow.

'Bar Harbor' Low spreading selection of excellent landscape merits; height to 10 inches tall, spreading to 6 or more feet across, steel blue, mostly awl foliaged, turning purplish-blue in autumn.

'Blue Acres' Prostrate, spreading clone with new growth bluish and older foliage blue-green.

'Blue Chip' An excellent silvery-blue selection with mounding habit to 8 or 10 inches tall.

'Blue Horizon' Supposedly very similar to 'Wiltoni,' but more prostrate, not mounding in the center.

'Blue Mat' Blue-green; dense, prostrate habit.

'Douglasii' (Waukegon juniper) A fairly popular steel-blue trailing clone that turns purplish in autumn; reaching 1 to 1 1/2 feet tall.

'Dunvegan Blue' Steel-blue, low, spreading clone.

'Emerald Isle' Low, spreading, compact, green, feathery foliaged selection.

'Emerald Spreader' Feathery textured, very prostrate selection; introduced and patented by Monrovia Nursery.

'Emerald' Low growing, blue foliaged selection that reaches about 6 inches tall and turns purplish in winter.

'Gray Carpet' Gray-green, feathery foliaged selection that reaches about 1 foot tall.

'Green Acres' Similar to 'Blue Acres,' differing in a predictable fashion.

'Hughes' Blue-green, low growing selection that reaches about 12 inches tall, and becomes slightly purplish in winter.

'Huntington Blue' Densely branched, wide spreading selection with blue-gray foliage that becomes purplish in winter.

'Jade Spreader' Compact, very low, jade-green selection.

'Livida' Dense, mat forming selection with dark green leaves and grayish bloom.

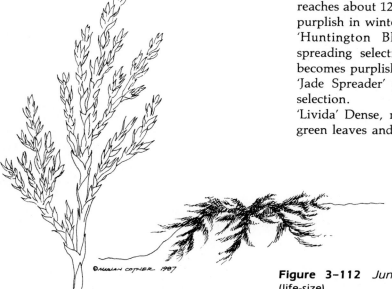

Figure 3–112 *Juniperus horizontalis* 'Bar Harbor' (life-size)

'Livingston' Foliage scalelike, steel-blue (blue-green in winter), with procumbent habit to 8 inches tall.

'Marcell' Low growing with procumbent habit to only about 6 inches tall.

'Petraea' Blue-green, juvenile foliaged selection to 10 inches tall.

'Plumosa' A rather dense, wide spreading selection with foliage that is scalelike grayish-green, rather coarse, colored purplish in winter.

'Plumosa Aunt Jamima' Blue-green, low spreading clone to about 8 inches tall, flat topped, with interesting name.

'Plumosa Compacta' More dense version of the cultivar 'Plumosa.'

'Plumosa Compacta Youngstown' This selection is similar to 'Plumosa Compacta,' but with less tendency to turn purplish in autumn.

'Plumosa Fountain' Rapid growing, flat topped selection that is similar to 'Plumosa Aunt Jamima,' but to 16 inches inches tall. *There is a strong need for introduction of more cultivars that begin with the word plumosa, for the nomenclature has not yet reached the point where it is totally confusing.* The search should begin immediately!

'Prince of Wales' Dense, procumbent, bright green foliaged selection that reaches 4 to 6 inches tall and becomes purplish-brown in winter.

'Pulchella' A selection with gray-green leaves and height of 4 to 6 inches.

'Sun Spot' Habit similar to 'Douglasii,' but with occasional yellow leaved branches.

'Turquoise Spreader' A Monrovia Nursery introduction with low, spreading habit and dense turquoise-green, feathery foliage.

'Variegata' Foliage variegated creamy-white, vigorous and prostrate.

'Webberi' Low spreading habit to 1 foot high; foliage finely textured, bluish-green.

'Wiltoni' ('Blue Rug') Extremely popular, low, trailing clone that may reach from 3 to 6 inches tall. Foliage is silvery-blue and color is maintained well throughout winter with usually only a hint of purple. Introduced by South Wilton Nurseries of Wilton, Connecticut in early 1900s.

'Winter Blue' Much like 'Plumosa,' but with light green summer foliage that becomes bluish in winter.

'Yukon Belle' Low, silvery-blue selection; reportedly more hardy than the species.

CULTURE: Soil: Adaptable to most soils with moderate drainage and a pH in the range from 4.9 to 5.8.

Moisture: Quite drought tolerant and easily withstanding locations that receive heat and wind. Occasional watering in summer may be useful, but usually is not needed.

Light: Full sun.

*   *   *

J. procumbens (prō-kum´benz) Procumbent habit of growth. Frequently listed as variety of *J. chinensis.*

COMMON NAME: Japanese garden juniper.

NATIVE HABITAT: Japan.

HARDINESS: Zones 4 to 9.

HABIT: Procumbent, woody, mat forming shrub.

SIZE: Usually 10 to 12 inches tall, occasionally somewhat taller; spreading to 10 or more feet across.

RATE: Relatively slow; spacing of plants from 1 to 2 gallon sized containers is usually adequate in the range of 2 1/2 to 3 1/2 feet for cultivars and 3 to 4 feet for the species.

FOLIAGE: Evergreen, similar to *J. chinensis,* about 1/3 inch long, concave above, spiny pointed, glaucous with green midrib, convex below, bluish-gray.

FLOWERS: Dioecious, male flowers small, yellow; female flowers small, greenish. Neither of ornamental significance.

FRUIT: 3 seeded, subglobose, about 1/3 inch across, not commonly witnessed in cultivation.

SELECTED CULTIVARS AND VARIETIES: 'Nana' Commonly called dwarf Japanese garden juniper, this relatively popular clone is low growing, seldom more than 4 inches tall, slow spreading, compact, and eventually reaching 6 to 8 feet or slightly more across. Color is blue-green and becomes slightly purplish in winter.

'Nana Californica' Like 'Nana,' but deeper blue-green, finer textured; to 8 inches tall and spreading to 4 feet across.

'Nana Greenmound' Much like 'Nana Californica' but more bluish in color.

'Variegata' With creamy-white variegation.

CULTURE: Same as *J. horizontalis.*

*   *   *

J. sabina (så-bēn´å or så-bīn´å) Old Latin name for the savin.

COMMON NAME: Savin juniper.

NATIVE HABITAT: Europe.

HARDINESS: Zones 4 to 7.

HABIT: Varying from low to moderately tall, broad spreading woody shrub. The species is not usually considered a ground cover itself; however, many selections are excellent for this purpose.

SIZE: The species ranges from 4 to 6 feet tall and spreads 5 to 10 feet across. The size of ground covering types is listed with their individual descriptions.

RATE: Selections usually spread at relatively slow to moderate rates. Most of them can be spaced at 3 1/2 to 4 1/2 feet apart from 1 or 2 gallon sized containers.

FOLIAGE: Evergreen, 4 ranked, scalelike leaves are in opposite, overlapping pairs, about 1/20 inch long, bluntly pointed, convex below; leaves on young plants and on the older branches of mature plants are to 1/5 inch long, awl shaped, spreading, and sharply pointed; dark dull green, pungent when crushed, with bitter taste.

OF SPECIAL INTEREST: At one time, tea made from the foliage of *J. Sabina* was believed to be a powerful aphrodisiac when administered to women. Before attempting to substantiate the validity of this myth, it would be wise to find out if such a concoction is poisonous or not.

FLOWERS: Monoecious or dioecious, not of ornamental merit.

CONES: Male cones insignificant, female cones 4 to 6 scaled, 1 to 3 seeded, globose to ovoid, about 1/5 inch across, ripening brownish or dark bluish during the fall of their first year or the following spring.

SELECTED CULTIVARS AND VARIETIES: 'Arcadia' Similar to 'Tamariscifolia,' a dense, horizontal growing dwarf selection with arching branch tips; reaching a height of 20 inches and spreading 4 to 7 feet across; initial growth relatively fast; leaves vibrant green and mostly scalelike; reported to be resistant to juniper blight; believed to have originated from seed at D. Hill Nursery of Dundee, Illinois.
'Blue Danube' Semiupright clone; maturing to a slightly lower height and with more horizontal habit than the species; foliage is a pleasing feathery bluish-green.
'Blue Forest' Reaching 18 inches tall, this unique blue-green selection gives the appearance of a miniature forest.
'Broadmoor' A dwarf, low spreading, mound forming, decumbent selection from the same seedlot as 'Arcadia'; reaching 2 to 3 feet tall and spreading to 8 or 10 feet across; foliage is an interesting grayish-green and has proved to be resistant to juniper blight. Tested hardy to Zone 5. Creating a very dramatic wavelike effect when mass planted.
'Buffalo' Similar to 'Tamariscifolia,' but foliage is

bright green and mature height is only about 12 inches; spreading to 8 or more feet across.
'Calgary Carpet' Selected as a sport of 'Arcadia' and is similar in habit but lower; foliage soft green.
'Scandia' From the same seedlot as 'Broadmoor' and 'Arcadia,' this selection is broad spreading to 10 feet across and reaches 12 to 18 inches high. Foliage is mostly needlelike and grayish-blue green. Like the others, it has proved resistant to juniper blight.
'Tamariscifolia' Excellent, attractive, low spreading, mounding clone. Branches are layered horizontally; foliage awl shaped, short and bluish-green with feathery texture; nonfruiting; spreading to 15 feet across and reaching a height of 1 1/2 feet tall. Some claim this to be a form or variety originating in mountains of southern Europe.
'Tamariscifolia New Blue' Similar to 'Tamariscifolia,' but the foliage is more blue in color.

CULTURE: Soil: Adaptable to most well-drained soils. Tolerant to slight alkalinity.
Moisture: Very tolerant to drought and drying winds. Occasional deep watering in summer is beneficial.
Light: Full sun.

*   *   *

J. squamata (skwȧ-mā´tȧ) From Latin, *squama*, a scale, in reference to the scalelike nature of the foliage (Figure 3–113).

COMMON NAME: Singleseed juniper.

NATIVE HABITAT: China.

HARDINESS: Zones 4 to 8.

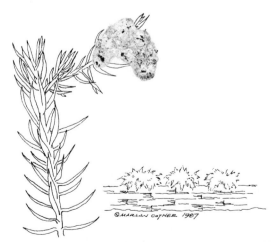

Figure 3–113 *Juniperus squamata* 'Blue Star' (life-size)

J. squamata is a variable species with respect to size and habit. For this reason it is seldom cultivated. Certain selections, however, make excellent ground covers. A general description of the species follows, and specifics of the selections are discussed later.

FOLIAGE: Evergreen, all awl shaped, ternate, linear-lanceolate, 1/8 to 5/16 inch long, curved, concave above and gray-green with two grayish-white bands, convex below and green, persisting on branches as brown scales after senescence.

FLOWERS: Monoecious, male flowers small and yellow; female flowers small, greenish, neither sex of ornamental worth in regard to fruit display.

CONES: Female cones are 1 seeded, elliptic, 1/4 to 1/3 inch diameter, 3 to 6 scaled, reddish-brown turning purplish upon ripening the second year.

SELECTED CULTIVARS AND VARIETIES: 'Blue Star' Increasingly popular, this beautiful, deep silvery-blue selection is low (usually not over 12 inches tall), dense, and spreading (have seen plants 4 to 5 feet across). Unfortunately, it has a high susceptibility to juniper blight. Even so, small-scale plantings, where cultural practices are adhered to, are entirely worth the effort to keep them disease free. Originated around 1950 as a branch sport of *J. squamata* 'Meyeri.'
'Prostrata' Low, slow growing, prostrate clone with short erect tips of leading shoots being nodding. Leaves are bluish-white above and green below.
'Variegata' Low growing, prostrate selection that reaches about 10 inches tall and spreads to 4 or 5 feet across. New growth is creamy colored.

CULTURE: Soil: Adaptable to most well-drained soils with pH neutral or acid.
Moisture: Relatively well adapted to dry conditions. Planting on elevated berm in a wind exposed site will help keep foliage dry and reduce the incidence of juniper blight. As with others, an occasional deep watering in summer may be beneficial.
Light: Full sun.

* * *

J. virginiana (vĕr-jin-i-ā´nȧ) Of Virginia.

COMMON NAME: Eastern red cedar.

NATIVE HABITAT: United States.

HARDINESS: Zones 2 to 9.

J. virginiana is itself an upright growing pyramidal tree; however, at least one selection is useful as a ground cover.

SELECTED CULTIVARS AND VARIETIES: 'Silver Spreader' A low, wide spreading selection with silvery-gray foliage. Introduced by Monrovia Nursery around 1955. It grows at a moderate rate and may have some tolerance to alkaline soils.

Generally Applicable to All Species

PATHOLOGY: Diseases: Juniper twig blight, cedar apple rust, root knot nematode. Some selections have resistance to these diseases. For a full account, many sources will have to be consulted, but good starting points are *Diseases and Pests of Ornamental Plants* and *Manual of Woody Landscape Plants*, both listed in the bibliography.

MAINTENANCE: Maintenance requirements are limited, consisting merely of cutting off branch tips as they outgrow their bounds. Seasonal shearing of taller types helps keep them dense and compact, while lower growing types stay that way naturally.

PROPAGATION: Cuttings: There are many special procedures for propagating junipers from cuttings. As it is a very economically important genus, considerable research has been expended to increase rooting percentages of juniper species. Generally, cuttings of junipers root well if they are taken after a couple of heavy frosts in fall (ensuring that they are dormant), stem wounded lightly to the cambial layer, base cut prior to dipping in rooting hormone preparation on the order of 3000 to 8000 ppm IBA/talc, apex not excised, and stuck in a very porous, sterile medium. Bottom heat of 65° to 75°F and a light watering daily are wise investments once cuttings are stuck.
Grafting: Grafting junipers is usually a simple matter; however, it generally is not necessary to graft ground covering types, as they tend to root well.
Seed: Seed is generally used to propagate species for the purposes of understocks or to look for new cultivars. Generally, seed should be gathered in the fall as berries ripen, the seed coat removed, then soaked in sulfuric acid for 30 minutes, and stratification begun. Stratification consists of 4 months at 40°F or, as an alternative, sow outdoors in a seedbed in fall for natural stratification.

✦✦✦✦✦✦✦✦✦✦✦✦✦✦✦✦

Lamiastrum (lă-mē-as´trum or lā-) From *Lamium*, a related genus, and *astrum*, meaning closely resembling, that is, like *Lamium*. The genus *Lamiastrum* is monospecific and previously was considered a species of *Lamium*. Today, too often it is listed erroneously as such.

FAMILY: LAMIACEAE (LABIATAE): Mint family.

L. galeobdolon (gŭ-lē-ob´dō-lon) Old generic name

meaning weasel and a bad smell. Just what this refers to is in question (Figure 3–114).

COMMON NAME: Yellow archangel, golden dead nettle.

NATIVE HABITAT: Woodlands from western Europe to Iran.

HARDINESS: Zones 4 to 9.

HABIT: Low, trailing, herbaceous ground cover.

SIZE: 8 to 18 inches tall, spreading indefinitely.

RATE: Relatively fast to invasive; space plants from 2 1/2 to 3 inch diameter containers 12 to 18 inches apart.

LANDSCAPE VALUE: Variegated selections are excellent in shade under trees where their silvery color helps to brighten up an area. Often used in large open spaces and on banks as a soil binder. In any event, the trailing types need room to spread and should not be planted in small areas where they will quickly outgrow their bounds. On a terrace or in a planter where they can cascade over, they are outstanding. The clump forming 'Herman's Pride' is versatile and can be used in small- and moderate-sized areas. No foot traffic.

FOLIAGE: Simple, evergreen, opposite, ovate to ovate-orbicular, to 3 inches long, ciliate, base truncate or somewhat cordate, margin coarsely toothed, occasionally crenate, aromatic when crushed, medium green.

INFLORESCENCE: Axillary or terminal 5 to 15 flowered whorls on erect shoots; flowers with bracts like leaves, calyx to 3/8 inch long; corolla hooded, double lipped, yellow with brown markings, 1/2 to 3/4 inch long, effective late spring.

FRUIT: Nutlets, not of ornamental significance.

SELECTED CULTIVARS AND VARIETIES: 'Compacta' Similar to 'Variegatum,' but slower growing and with smaller leaves.

'Herman's Pride' Leaves smaller, ovate, tapering somewhat toward apex, margins serrate, variegated intensely silver and green; very attractive throughout the year if sited in cool, moderately shaded location; fading excessively in summer if in a location that receives too much light and heat; clump forming, not trailing.

'Variegatum' Foliage silvery and green, developing blood-red to purplish center splotch in autumn; more common in use than the species; with aggressive trailing habit.

CULTURE: Soil: Fastest growth is in rich fertile loam, but tolerant to almost any soil. Many gardeners prefer to grow lamiastrum in infertile sandy soils as a means of slowing growth.
Moisture: Not greatly tolerant to drought; occasional watering in summer to maintain soil in a slightly moist condition is recommended.
Light: Full sun to moderate shade. Best in light to moderate shade, but fine in sun provided ample water is given.

PATHOLOGY: Diseases: Relatively free from disease and pest problems.

MAINTENANCE: Plantings can be kept looking healthy and compact if annually mowed in early summer after flowering.

PROPAGATION: Cuttings: Cuttings root easily and rapidly in summer or fall. Roots are produced at the nodes and cuttings can be made with a 1 inch stem section and a node containing two leaves. The stem is inserted into the soil to the point where the node is about even with the soil line. Treatment with a root inducing preparation of about 2000 ppm IBA/talc is helpful but not required.
Division: Simple division of clumps or rooted stems can best be accomplished in spring or fall.

OF SPECIAL INTEREST: The genus *Lamiastrum* differs from *Lamium* in that its anthers are without hairs, and the lower lip of the corolla has 3 lobes of nearly equal size, while the lower corolla lip of a typical species of *Lamium* has a central lobe that is much larger than the two lateral lobes, which are dentate (toothlike).

✦✦✦✦✦✦✦✦✦✦✦✦✦✦✦✦

Lamium (lā′mi-um) From Greek, *laimos*, throat, alluding to the throatlike appearance of the blossoms. This genus is composed of about 4 species of annual and perennial herbs that are usually

Figure 3–114 *Lamiastrum galeobdolon* 'Herman's Pride' (life-size)

decumbent in habit. They are native primarily to Asia, Africa, and Europe.

FAMILY: LAMIACEAE (LABIATAE): Mint family.

L. maculatum (mak-ū-lā´tum) Spotted, the foliage with a central silvery splotch (Figure 3–115).

COMMON NAME: Spotted dead nettle.

NATIVE HABITAT: Europe.

HARDINESS: Zones 3 to 8 or 9.

HABIT: Clump forming, horizontal spreading, herbaceous ground cover.

SIZE: 6 to 8 inches tall, usually spreading 12 to 16 inches across.

RATE: Moderate; space plants from 2 1/4 to 3 inch diameter containers 8 to 12 inches apart.

LANDSCAPE VALUE: Commonly used as a general cover in restricted areas where it cannot outgrow its bounds. Variegated foliage contrasts nicely with darker leaved plants. This makes it useful as a facing for hedges of *Taxus* or grouped plantings of shrubs such as rhododendron and others with dark foliage. No foot traffic.

FOLIAGE: Simple, evergreen, opposite, ovate, crinkled, crenate-dentate margined, basal leaves tending toward cordate, generally about 1 inch long by 1/2 inch wide, colored green with silvery-gray splash in center along midrib, unpleasantly pungent when crushed.

STEMS: Decumbent, squarish, lightly hairy, light green or brownish red.

INFLORESCENCE: Verticillasters in upper leaf axils; flowers with tubular calyx to 5/8 inch long, teeth unequal and divergent; corolla two-lipped, upper lip hooded, lower 3 lobed, middle lobe notched, prominent, lavender, effective spring and summer.

FRUIT: Nutlets, not ornamentally significant.

SELECTED CULTIVARS AND VARIETIES: 'Album' With flowers creamy white.

'Aureum' Leaves yellowish with whitish blotch along midrib, less aggressive, performing better in shade than sun.

'Beacon Silver' Excellent appearing (when not infected with leaf spot) selection. Leaves are radiant silver surrounded with a light border of green. Needs excellent drainage, relatively low humidity, light to moderate shade, and cool temperatures.

'Chequers' A commonly grown cultivar; flowers are reportedly pinkish, while foliage seems to be somewhat larger than the species, and habit compact.

'White Nancy' Similar to 'Beacon Silver' in respect to the foliage, but flowers white.

CULTURE: Soil: Adaptive to most well-drained soils, best in rich loam.

Moisture: Not highly tolerant of heat or drought, the leaf margins sometimes scorch in midsummer. Soil should be kept moist with regular watering.

Light: The species tolerates a range from full sun to moderate shade, while the cultivars need more shade and will be at their best in light to moderate shade.

PATHOLOGY: Diseases: Necrosis of foliage, damping off, leaf spots. The cultivars all seem more susceptible to disease than the species. Unfortunately, the species has much less ornamental value than the cultivars.

Pests: Slugs, aphids.

MAINTENANCE: Plantings can be kept dense and vibrant by mowing in summer after flowering.

PROPAGATION: Propagation is the same as listed for *Lamiastrum galeobdolon*, but more attention should be given to having a medium that drains very well, and mist should be discontinued immediately upon root formation.

OF SPECIAL INTEREST: See *Lamiastrum* for an explanation of the difference between the two genera.

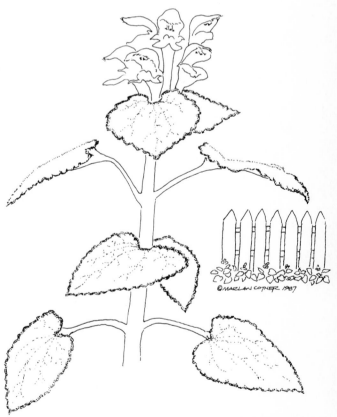

Figure 3–115 *Lamium maculatum* 'Beacon Silver' (life-size)

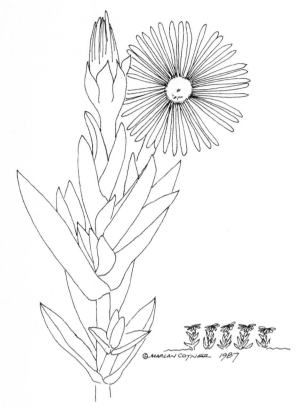

Figure 3–116 *Lampranthus aurantiacus* (life-size)

L**ampranthus** (lăm-prăn´thus) From Greek, *lampros*, brilliant, and *anthos*, flower, the flowers being very showy. Another ice plant, *Lampranthus* is composed of about 160 species of creeping, decumbent to erect, succulent perennials. In very generalized terms, the ground covering species can be described as follows:

FAMILY: AIZOACEAE: Carpetweed family.

HABIT: Fleshy, low, subshrubby or trailing, carpeting ground covers.

LANDSCAPE VALUE: General covers for use around shrubs and foundations. Colorful and functional as soil stabilizers and as edging plants for use along steps or walking paths. When planted on a terrace or next to a retaining wall, the branches create a spectacle of graceful color. No foot traffic.

FOLIAGE: Succulent, evergreen, opposite, numerous, cylindrical to 3 angled, more or less curved, glabrous, glandular dotted, colors ranging from gray to green.

INFLORESCENCE: Cymes or solitary flowers; flowers with 5 lobed calyx, many stamens, corolla red, pink, white, purple or yellow, attractive to bees.

FRUIT: 5 valved capsules, not ornamentally significant.

DISCUSSION: Many species could be utilized as ground covers; what follows is a list of some of the more common.

*   *   *

L. aurantiacus (â-ran-ti-ā´kus) Orange. (Figure 3–116).

COMMON NAME: Bush ice plant.

NATIVE HABITAT: South Africa.

HARDINESS: Zones 9 to 10.

SIZE: 10 to 15 inches tall, spreading 2 + feet across.

RATE: Moderate; space plants from pint- to quart-sized containers 12 to 18 inches apart.

DISCUSSION: *L. aurantiacus* has foliage that is gray-green, about 1 inch long, 3 sided; flowers 1 1/2 to 2 inches across, effective primarily in spring. Cultivars include 'Glaucus' With clear, bright yellow flowers.
'Gold Nugget' With bright orange flowers.
'Sunman' With golden-yellow flowers.

*   *   *

L. filicaulis (fil-i-câl´is) Meaning with threadlike stem.

COMMON NAME: Redondo creeper.

NATIVE HABITAT: South Africa.

HARDINESS: Zones 9 to 10.

SIZE: To 3 inches tall, spreading indefinitely.

RATE: Relatively slow; space plants from 3 to 4 inch diameter containers 10 to 12 inches apart.

DISCUSSION: Foliage is finely textured on thin creeping stems; pink flowers are small and borne in early spring.

*   *   *

L. productus (prō-duk´tus) Likely in reference to the many flowers.

COMMON NAME: Purple ice plant.

NATIVE HABITAT: South Africa.

HARDINESS: Zones 9 to 10.

SIZE: To about 15 inches tall, spreading 1 1/2 to 2 feet across.

RATE: Moderate; space plants from pint- or quart-sized containers 12 to 18 inches apart.

DISCUSSION: Gray-green leaves with bronzy tips; purple flowers are showy in late winter and spring.

* * *

L. spectabilis (spek-tab´i-lis) Showy, remarkable.

COMMON NAME: Showy ice plant, trailing ice plant.

NATIVE HABITAT: South Africa.

HARDINESS: Zones 9 to 10.

SIZE: To 15 inches tall, spreading 1 1/2 to 2 feet across.

RATE: Moderate; space plants from pint- or quart-sized containers 12 to 18 inches apart.

DISCUSSION: With foliage that is 2 to 3 inches long and about 1/4 inch wide, 3 angled to cylindrical, grayish with reddish tips; flowers solitary, 2 to 3 inches across on 3 to 6 inch long stems, very showy, bright purple in spring.

Generally Applicable to All Species

CULTURE: Soil: Adaptable to a wide range of soils, excellent drainage a requirement. Tolerant to, but much slower growing in infertile soils.
Moisture: Able to withstand drought very well once established. Occasional thorough watering in summer is beneficial.
Light: Full sun.

PATHOLOGY: No serious diseases.
Pests: Aphids, mealybugs.

MAINTENANCE: Little or no special maintenance required. Shearing after bloom will stimulate new growth and get rid of fruit capsules.

PROPAGATION: Cuttings: Easily rooted throughout the year.
Division: Simply divide plants and keep well moistened until roots are reestablished.

✝✝✝✝✝✝✝✝✝✝✝✝✝✝✝✝✝

Lantana (lan-tä´na or lan-tä´na) An ancient name for *Viburnum*, the foliage of the two genera sometimes being similar. This genus is composed of around 155 species of thorned and unarmed shrubs and herbaceous perennials. They are primarily native to the subtropics and tropical North and South America, with a few originating in Europe.

FAMILY: VERBENACEAE: Vervain family.

OF SPECIAL INTEREST: Fruit of these plants are reported to be poisonous.

HABIT: Ground covering species are trailing, woody, vinelike shrubs.

SIZE: 2 to 3 feet tall, spreading 6 to 8 feet across.

RATE: Relatively fast; space plants from gallon-sized containers 3 to 4 feet apart.

LANDSCAPE VALUE: Sturdy covers for use in larger areas, both on the flat or on slopes where they function well to control erosion. Often useful as accent plants in borders or as edging. No foot traffic.

* * *

L. montevidensis (mon-te-vid-en´sis) of Montevideo (Figure 3–117). Sometimes still referred to as *L. sellowiana*.

COMMON NAME: Polecat geranium, trailing lantana, weeping lantana.

NATIVE HABITAT: South America.

HARDINESS: Zones 9 to 10.

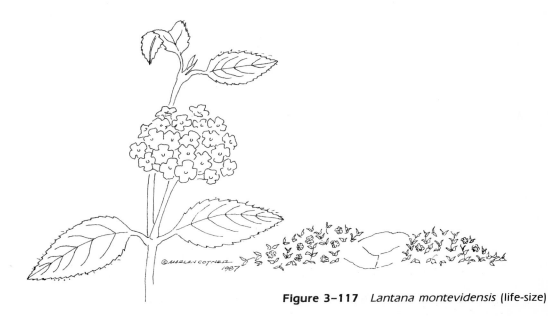

Figure 3–117 *Lantana montevidensis* (life-size)

FOLIAGE: Evergreen, simple, opposite or whorled, frequently rugose, ovate, 1/2 to 1 inch long by about 1/2 inch wide, coarsely toothed, ciliate, glabrous above, hairy below, dark green becoming darker with cold weather, pungent when crushed.

STEMS: Weak, trailing to 3 feet, hairy.

INFLORESCENCE: Terminal, compact heads to 1 inch or more in diameter; flowers 1/4 inch across, irregular, corolla 4 lobed, rosy pink to lilac, profusely born in spring, sparsely in other seasons.

FRUIT: Cluster of small, black drupes.

SELECTED CULTIVARS AND VARIETIES: 'Velutina White' White flowered selection.

*   *   *

L × hybrida (hīb ′ri-dȧ) A hybrid species.

COMMON NAME: Hybrid lantana.

DISCUSSION: Hybrids make up the bulk of cultivated lantanas that are suitable for ground covers. Parentage is not always well documented, but it is likely that the following cultivars are hybrids. Their habit is generally low spreading or trailing to a width of 6 to 8 feet. Hardy in Zones 9 and 10.

SELECTED CULTIVARS AND VARIETIES: 'Carnival' To 2 feet tall and 4 to 5 feet across; flowers mixed pink, yellow, lavender, and crimson.
'Confetti' To 3 feet tall by 8 feet wide; flowers yellow, pink, or purple.
'Cream Carpet' Flowers creamy white with bright yellow throat; to 3 feet tall and 8 feet across.
'Dwarf Pink' Pink flowered dwarf selection.
'Dwarf White' White flowered dwarf selection.
'Dwarf Yellow' Yellow flowered dwarf selection.
'Gold Mound' 2 feet tall by 6 feet wide; flowers orange-yellow.
'Golden Glow' Flowers golden-yellow, less than 3 feet tall.
'Kathleen' Flowers rose-pink with gold centers; to 2 feet tall and 6 feet across.
'Spreading Sunset' Flowers orange-red; to 3 feet tall and 8 feet across.
'Sunburst' Flowers bright yellow; to 3 feet tall and 8 feet across.
'Tangerine' Flowers orange-red; 2 to 3 feet tall, spreading 6 to 8 feet across.

Generally Applicable to the Species and Hybrids

CULTURE: Soil: Adaptable to almost any soil with good drainage.
Moisture: Very tolerant to drought; an occasional deep watering in summer is recommended.

Light: Full sun to light shade; flowering is best in full sun.

MAINTENANCE: May develop open dead centers in old plantings. This can be prevented, and plants kept juvenile and compact, by annual shearing (on moderate to heavy side) in early spring.

PATHOLOGY: Diseases: Black mildew, leaf spot, rust, wilt, less prevalent in full sun.
Pests: Caterpillars, greenhouse whitefly, mites, lantana aphid, southern root knot nematode, mealy bugs.

PROPAGATION: Cuttings: Softwood and hardwood cuttings are the standard means of propagation of the cultivars.
Seed: Seed is reported to be used in the propagation of the species and, of course, in the selection of new cultivars. I am not familiar with the particular methods involved.

Lathyrus (lath ′i-rus) Ancient Greek name for these leguminous plants. This genus is composed of about 100 or more species of herbs, some climbing with tendrils. Primarily, they are native to northern temperate regions and the mountains of Africa and South America.

FAMILY: FABACEAE (LEGUMINOSAE): Pea family.

L. latifolius (lat-i-fō ′li-us) Broad-leaved (Figure 3–118).

COMMON NAME: Perennial pea vine.

NATIVE HABITAT: Southern Europe.

HARDINESS: Zones 3 to 10.

HABIT: Herbaceous, trailing and climbing vine; if unsupported, a sprawling ground cover.

SIZE: 4 to 8 inches tall without support, spreading indefinitely.

RATE: Relatively fast; space plants from pint-sized containers 10 to 12 inches apart.

LANDSCAPE VALUE: Useful as a bank cover and soil binder in sunny locations. Especially good in gravelly, rugged terrain. No foot traffic.

FOLIAGE: Compound alternating; leaflets evenpinnate in one pair, ovate-lanceolate, to 4 inches long, acute and mucronate, 3 to 5 veined, medium green.

STEMS: Light green, glabrous, winged, tendrils stiff and stout, climbing if supported.

INFLORESCENCE: Elongated   many   flowered

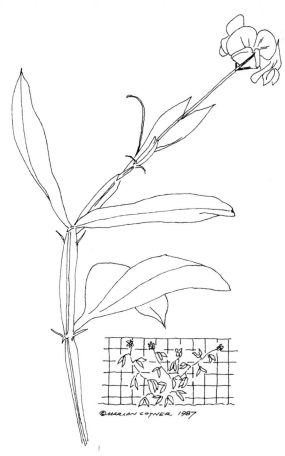

Figure 3–118 *Lathyrus latifolius* (life-size)

peduncles, about 1 inch across; flowers rather large, rose colored or white, effective from early summer to early fall.

FRUIT: 3 to 5 inch long legumes.

SELECTED CULTIVARS AND VARIETIES: 'Albus' Flowers white.

'Splendens' Flowers dark purple and red.

CULTURE: Soil: Adaptable to most soils that have good drainage and pH ranging from mildly acidic to mildly alkaline. Withstands rocky, infertile soils. Moisture: Well adapted to dry conditions; thus only infrequent deep watering during summer is needed. Light: Full sun to light shade.

PATHOLOGY: Diseases: Anthracnose, mosaic, black root rot, downy mildew.

Pests: Aphids, pea moth, mites, corn earworm, southern root knot nematode.

MAINTENANCE: Little or no maintenance needed.

PROPAGATION: Division: Simple division of clumps is the standard means of propagating the cultivars.

SEED: Seeds germinate in 2 to 3 weeks at temperatures from 68° to 80°F.

Laurentia (lâ-ren´she-à) Memorial. This genus is composed of about 25 species of annual and perennial herbs that are native to Africa, Australia, the Americas, and the Mediterranean region.

FAMILY: LOBELIACEAE: Lobelia family.

L. fluviatilis (floo-vē-at´i-lis) Growing in a river or flowing water (Figure 3–119).

COMMON NAME: Blue-star creeper.

NATIVE HABITAT: Southern Australia, Tasmania, and New Zealand.

HARDINESS: Zones 8 to 10.

HABIT: Low, herbaceous, mat forming ground cover.

SIZE: 2 to 5 inches tall, spreading indefinitely.

RATE: Moderate to relatively fast; space plants from 2 to 3 inch diameter containers 8 to 12 inches apart.

LANDSCAPE VALUE: Use as a general cover for smaller areas or as a filler between cracks and stepping stones. Tolerates light foot traffic.

FOLIAGE: Evergreen, alternate, simple, entire or slightly lobed, ciliate, to 1/2 inch long, ovate or orbicular, more linear toward end, glandular dotted and slightly pubescent on both sides, purplish below, medium dull green above.

Figure 3–119 *Laurentia fluviatilis* (life-size)

INFLORESCENCE: Racemes or flowers solitary; flowers to 3/8 inch across, irregular stellate; corolla 5 lobed, pale blue or purplish, effective in spring and then intermittently through summer.

FRUIT: Capsule, not ornamentally significant.

CULTURE: Soil: Adaptable to most fast draining soils. Best in sandy loam.
Moisture: Not notably tolerant of drought; soil should be kept moist with regular watering.
Light: Full sun to light shade; best in light shade and needing frequent watering when in full sun.

PATHOLOGY: Diseases: No serious diseases reported.
Pests: Aphids.

MAINTENANCE: Little or no maintenance required. The flowers can be removed after bloom to neaten appearance.

PROPAGATION: Cuttings: Cuttings can be rooted anytime under light mist.
Division: Divide in winter or early spring prior to growth.
Seed: Seed reportedly germinates readily.

*L*avandula (là-van´dū-là) Said to be derived from Latin, *lavo*, to wash, the Romans and Greeks having used lavender in their baths for fragrance. This genus, commonly referred to as lavender, and popular among herbalists, is composed of approximately 20 species of aromatic subshrubs, shrubs, and herbs. They are native to dry, hilly, often infertile areas of the Mediterranean region to Somalia, the Canary Islands, and India.

FAMILY: LAMIACEAE (LABIATAE): Mint family.

L. angustifolia ssp. *angustifolia* (an-gus-ti-fō´li-à) From Latin *angustiae*, narrowness, that is, narrow-leaved (Figure 3–120). Previously named *L. officinalis*, *L. spica*, and *L. vera*.

COMMON NAME: True lavender.

NATIVE HABITAT: Southern Europe and North Africa.

HARDINESS: Zones 5 or 6 to 9.

HABIT: Low growing, semiwoody subshrub.

SIZE: 1 to 3 feet tall, spreading from 12 to 24 inches across. The cultivars listed are generally low and spread broadly, and are more commonly used for ground cover than the subspecies.

RATE: Moderate; space plants from pint- to quart-sized containers 14 to 18 inches apart.

LANDSCAPE VALUE: Valuable as an edging for walks and borders where their pleasant fragrance can be most appreciated. Often planted as a specimen in a rock or herb garden, sometimes as a dwarf hedge. No foot traffic.

Figure 3–120 *Lavandula angustifolia* ssp. *angustifolia* (life-size)

FOLIAGE: Semievergreen, opposite or whorled, simple, oblong-linear or lanceolate, to 2 inches long by 1/4 inch wide, entire, margins rolled back, young leaves often clustered in axils, white tomentose in youth, becoming green with age, aromatic.

STEMS: Tomentose, glossy brown, squarish, 1 to 3 feet long.

INFLORESCENCE: Verticillasters on upright spikes, with 6 to 10 flowers; flowers are fragrant, calyx about 1/4 inch long, 13 nerved, densely pubescent, ovoid-tubular, slightly 5 toothed; corolla about 1/2 inch long, irregular, 5 lobed, usually colored purple, effective mid to late summer.

FRUIT: Of 4 glabrous nutlets, not ornamentally significant.

OF SPECIAL INTEREST: The flowers of lavender attract both bees and butterflies. Other uses include cut flowers and sachets made from dried flowers and leaves.

SELECTED CULTIVARS AND VARIETIES: 'Alba' With white flowers.
   'Carrol Gardens' With pale purple flowers.
   'Compacta' Compact habit, reaching only about 8 to 10 inches tall.
   'Dutch' With deep blue flowers.
   'Fragrance' Similar to the subspecies, but flowers heavily scented.
   'Hidcote' With rich purple flowers; vigorous, 15 to 20 inches tall; foliage silvery-gray.
   'Munstead Dwarf' Reaching only about 12 to 18 inches tall; flowers dark lavender-blue.
   'Nana' Dwarf selection to about 12 inches tall.
   'Rosea' Flowers rose pink.
   'Twinkle Purple' To 18 inches tall; flowers dark blue on short spikes; foliage silvery.
   'Waltham' Flowers deep purple.

CULTURE: Grows best in light loamy soils that have excellent drainage. Best with slightly acidic soil, but tolerates mild alkalinity.
   Moisture: Relatively well adapted to tolerate drought, but can benefit from occasional deep watering in hot, dry weather.
   Light: Full sun.

PATHOLOGY: Diseases: Leaf spot, root rot.
   Pests: Caterpillars (orange tortrix and yellow woolybear), four-lined plant bug, northern root knot nematode.

MAINTENANCE: Shear plants annually following bloom to keep them dense and compact.

PROPAGATION: Cuttings: Cuttings should be taken from side shoots in spring through early fall. Treat with a mild rooting compound that contains a fungicide to control damping off.

Division: Divide clumps in early spring.
Seed: Commercial supplies are available but generally not useful as cultivars are usually desired. At any rate, germination takes about 2 to 3 weeks at temperatures from 70° to 80°F.

Liriope (lă-rī´ă-pē) From Greek, *lirion*, a lily, or in Greek mythology, after Liriope, a fountain nymph and mother of Narcissus. This genus is comprised of about five species of evergreen, basal-leaved, herbaceous, lilylike, monocotyledonous perennials. They are tufted or rhizomatous in habit; originating in Japan, China, and Vietnam.

FAMILY: LILIACEAE: Lily family.

LANDSCAPE VALUE: Exceptional for edging border or walks or along a driveway. Often used underneath trees and in various-sized areas for general cover. Fine for accent or as a specimen in a rock garden. No foot traffic.

*   *   *

L. muscari (mus-kā´ri) Meaning unclear.

COMMON NAME: Big blue lily-turf.

NATIVE HABITAT: Japan and China.

HARDINESS: Zones 6 to 10.

HABIT: Clumplike, tuberous, grassy, lilylike cover.

SIZE: 18 to 24 inches high, crowns spreading 8 to 12 inches across.

RATE: Relatively slow growing; space 12 to 16 inches apart from pint- to quart-sized containers.

FOLIAGE: Evergreen, basal, linear, straplike, somewhat revolute, entire, ciliate, prominently parallel veined, midrib protruding below, glabrous and glandular dotted on both sides, dark green, to 2 feet long by 3/4 inch across.

INFLORESCENCE: Dense racemes 6 to 8 inches long, elevated on scapes to 12 inches or more long, extending above the foliage on young plants, partially hidden in mature plants; flowers regular, dark purple, 6 petaled, segmented, numerous, effective in mid to late summer.

FRUIT: 1 or 2 seeded, dark black, shiny, globose capsule to about 1/3 inch across.

SELECTED CULTIVARS AND VARIETIES: 'Christmas Tree' Inflorescence of light flowers tapering to a point like a Christmas tree. Flowers not elevated above foliage.
   'Evergreen Giant' Stiffly erect, taller selection.

'Gold Band' Leaves shorter, broader, and edged in gold.

'John Burch' Very large leaved selection with cockscomb-shaped inflorescences.

'Lilac Beauty' Lower growing with flowers deep violet.

'Majestic' To 2 feet high with many deep violet flowers.

'Monroes #2' With green leaves and white flowers.

'Monroe White' Flowers pure white. Zone 8 is approximate *northern* limit.

'Royal Purple' Colored dark purple.

'Silver Dragon' Dark green leaves with silvery-white longitudinal variegation.

'Silver Midget' Low growing to 8 inches; foliage narrowly banded with white.

'Silvery Sunproof' Leaves pale green with white to yellow stripes; tolerates more sun than the species and blooms rather profusely.

'Variegata' New leaves with yellow edges that become green their second year; 1 to 1 1/2 feet tall, with violet flowers. Zone 8 is approximate *northern* limit.

*   *   *

L. spicata (spi-kā´ta) Bearing spikes (Figure 3–121).
COMMON NAME: Creeping lily-turf.
NATIVE HABITAT: China and Japan.
HARDINESS: Zones 4 to 10.

HABIT: Rhizomatous, spreading, lilylike cover.

SIZE: 8 to 12 inches tall, spreading 6 to 12 inches across.

RATE: Moderate rate of growth; space 8 to 16 inches apart from pint- to quart-sized containers.

FOLIAGE: Evergreen, straplike, lanceolate, serrulate with translucent teeth, to 18 inches long by 1/4 inch wide, dark green until fall, then turning green-brown.

INFLORESCENCE: Terminal racemes on 8 to 10 inch violet scapes; flowers are small, 1/4 inch wide; corolla pale violet to white, effective mid to late summer.

FRUIT: 1 to 2 seeded, fleshy, blue-black, berrylike, globose capsule.

OTHER SPECIES: *L. exiliflora* Rhizomatous, dark-green leaved species to 1 1/2 feet tall, with less showy flowers of violet-blue in spring. Seldom planted in the landscape.

Generally Applicable to All Species

CULTURE: Soil: Adaptable to almost any well-drained soil type; not tolerant of alkaline soils. Moisture: With massive root systems that contain water storage nodules, liriope is able to withstand a substantial amount of drought. Infrequent summer watering will help to keep them attractive and healthy.

Figure 3–121 *Liriope spicata* (0.5 life-size)

Light: Full sun to moderately dense shade.

PATHOLOGY: Relatively free from severe problems; however, snails and slugs are partial to the foliage. Scale insects and mealy bugs may also pose a problem.

MAINTENANCE: Mow back foliage in spring before new growth to maintain a healthy, youthful appearance and stimulate vigor.

PROPAGATION: Division: Simply divide clumps in spring or fall.

Seed: Depulp ripe seed and stratify warm (75 °F) for 6 weeks prior to sowing. Many times in excess of half of the seedlings from variegated liriope exhibit variegation. They should not, however, be represented as cultivars.

*L*ithodora (lith-ō-dōr´à) From Greek *lithos*, a stone, and *dorea*, gift, the meaning of which is unknown. This genus is composed of seven species of rather low, hairy, subshrubs or shrubs. They are native to western and southern Europe, North Africa, and Asia Minor.

FAMILY: BORAGINACEAE: Borage family.

L. diffusa (di-fū´sà) Spreading. (Figure 3–122). Formerly *Lithospermum diffusum* and *L. prostratum*.

COMMON NAME: Spreading lithodora.

NATIVE HABITAT: South and western Europe and Morocco.

HARDINESS: Zones 6 to 7, not performing well in subtropics.

HABIT: Prostrate, low, spreading, dwarf evergreen shrub.

SIZE: 6 to 12 inches tall, spreading 24 to 30 inches across.

RATE: Moderate; space plants from gallon-sized containers 1 1/2 to 2 1/2 feet apart.

LANDSCAPE VALUE: Excellent companion for azaleas and other ericaceous plants. Good in rock garden or border as specimen. General cover for moderate-sized areas. No foot traffic.

FOLIAGE: Evergreen, alternate, simple, entire, linear to lanceolate, hairy, revolute, midrib conspicuous, indented above and protruding below, about 1/2 inch long by 1/8 to 1/4 inch wide, pubescent and glandular dotted both sides, dark green above and grayish below.

STEMS: Hairy in youth, glabrous in maturity, purplish.

INFLORESCENCE: Few flowered cymes, calyx and corolla 5 lobed; corolla regular, brilliant blue, 1/2 inch wide, effective spring to summer.

FRUIT: 4 striated nutlets that are not ornamentally significant.

SELECTED CULTIVARS AND VARIETIES: 'Alba' Flowers white.
'Grace Ward' Flowers larger than the species.
'Heavenly Blue' Flowers sky blue.

CULTURE: Soil: Best in fertile, moist but well-drained acidic soil. Not tolerant to alkalinity.
Moisture: Not highly tolerant to drought; keep soil moist with periodic watering.
Light: In areas of moderate temperature and relatively high humidity it will tolerate full sun; otherwise, light to moderate shade is needed.

PATHOLOGY: Diseases: Damping off.
Pests: Aphids.

MAINTENANCE: Light shearing following bloom will keep plantings dense and neat appearing.

PROPAGATION: Cuttings: Take cuttings during growing season. They will be helped by a mild rooting hormone preparation, and of necessity the medium must have excellent drainage and misting must cease as soon as rooted.

Figure 3–122 *Lithodora diffusa* (life-size)

✝✝✝✝✝✝✝✝✝✝✝✝✝✝✝✝✝

*L*onicera (lo-nis´ẽr-à) Named after Adam Lonicer, a german naturalist of the 16th century. This relatively large genus contains more than 150 species of primarily deciduous, but also evergreen or semievergreen, erect or climbing shrubs or vines. Native to many areas of the northern hemisphere.

FAMILY: CAPRIFOLIACEAE: Honeysuckle family.

HABIT: Woody, twining or trailing vines and low, broad, spreading shrubs.

SIZE: Vines generally reach 4 to 6 inches tall without support (they may occasionally mound to a couple of feet) and spread indefinitely. *Shrubs* are addressed individually.

RATE: Relatively fast to extremely fast (invasive), especially in warmer climates. Space vinelike plants from pint- or quart-sized containers 2 1/2 to 3 1/2 feet apart. Shrubby types grow at a moderate rate, and spacing of 2 1/2 to 3 1/2 feet from gallon-sized containers is usually acceptable.

LANDSCAPE VALUE: *Lonicera japonica* is most often used as a general cover or soil stabilizer on large open areas or banks that are free of other plants. Other vinelike species are also suitable for this purpose and may be interesting planted at the top of a wall where they can cascade over. Shrubby types are excellent for low hedges, general cover, or inside elevated planters. No foot traffic.

*   *   *

L. alpigena 'Nana' (al-pi-jē´nà) Of the Alps.

COMMON NAME: Dwarf Alps honeysuckle.

NATIVE HABITAT: Central and Southern Europe.

HARDINESS: Zone 5.

SIZE: To 3 feet tall, spreading to 5 feet across.

HABIT: Woody, low, upright stemmed, shrubby ground cover.

FOLIAGE: Simple, opposite, to 4 inches long, slightly crinkled, elliptic to oblong, dark green.

STEMS: Many branched, gray bark slightly exfoliating.

FLOWERS: Paired, on pubescent axillary peduncles; flowers about 1/2 inch long, yellow or greenish-yellow tinged brownish-red, effective in late spring.

FRUIT: Cherrylike drooping berries, effective late summer to early fall.

*   *   *

L. × *brownii* (brown-i-ī) After Brown.

COMMON NAME: Brown's hybrid honeysuckle.

HYBRID ORIGIN: *L. sempervirens* × *L. hirsuta.*

HARDINESS: Zones 3 to 9.

HABIT: Vine.

FOLIAGE: Simple, opposite, ovate to oblong, to 3 inches long, glaucous below, green above, upper leaves perfoliate.

INFLORESCENCE: Terminal spikes; flowers 2 lipped, glandular pubescent outside, 1 to 1 3/4 inches long, orange scarlet. Effective late spring and again in late summer.

SELECTED CULTIVARS AND VARIETIES: 'Dropmore Scarlet' Flowering profusely early summer to early fall, flowers red; very hardy; developed by F. L. Skinner of Dropmore, Manitoba, Canada.

*   *   *

L. caprifolium (kăp-ri-fō´li-um) Meaning derived from Latin, *capra,* goat, in reference not to the foliage as the name would imply, but rather to the ability of this plant to climb a hill with the ability of a goat.

COMMON NAME: Sweet honeysuckle, Italian woodbine.

NATIVE HABITAT: Europe.

HARDINESS: Zone 5.

HABIT: Vine.

FOLIAGE: Opposite, simple, elliptic, to 4 inches long, green above, blue-green below, upper leaves joined at the base.

INFLORESCENCE: Whorls; flowers white or purplish, to 2 inches long, 2 lipped, fragrant, effective in early summer to fall.

FRUIT: Orange-red berry.

*   *   *

L. etrusca (e-troos´kà) Of Tuscany.

COMMON NAME: Cream honeysuckle.

NATIVE HABITAT: Southern Europe.

HARDINESS: Zone 7.

HABIT: Vine.

FOLIAGE: Semievergreen, simple, opposite, obovate to elliptic, to 3 inches long, green above, glaucous and pubescent below, upper leaves perfoliate.

INFLORESCENCE: Dense spikes; corolla of flowers 2 lipped, opening cream, flushing to red, then finally becoming yellowish white; very fragrant, effective early in midsummer.

FRUIT: Red, berrylike.

SELECTED CULTIVARS AND VARIETIES: 'Superba' More vigorous with larger flowers than the species.

* * *

L. flava (flā'vȧ) Yellow, the flowers.

COMMON NAME: Yellow flowered honeysuckle.

NATIVE HABITAT: Southeastern United States.

HARDINESS: Zones 5 to 9 or 10.

HABIT: Vine.

FOLIAGE: Opposite, simple, elliptic, to 3 inches long, green above, bluish-green below, upper leaves perfoliate.

INFLORESCENCE: Terminal whorls; corolla of flowers orange-yellow, 2 lipped 1 1/4 inches long, trumpet shaped, fragrant, effective late spring to early summer.

FRUIT: Red, 1/4 inch diameter, berrylike.

* * *

L. heckrottii (hek-rot'i-ī) After Heckrott (Figure 3–123).

COMMON NAME: Everblooming honeysuckle, goldflame honeysuckle.

HYBRID ORIGIN: Unknown, likely *L. americana* × *L. sempervirens*.

HARDINESS: Zones 5 to 9.

FOLIAGE: Simple, opposite, elliptic, to 2 inches long, green above, glaucous below, semievergreen, turning purplish-red in winter.

STEMS: Stout, hollow, attractive purplish-red.

INFLORESCENCE: Spikes; corolla of flowers carmine in bud, opening to expose yellow interior, 2 lipped, to 2 inches long, effective late spring to late summer, very fragrant.

Figure 3–123 *Lonicera* × *heckrottii* (life-size)

SELECTED CULTIVARS AND VARIETIES: Many nurseries erroneously list this hybrid as cultivar 'Goldflame,' with the expected increase in sales and confusion in nomenclature.

* * *

L. japonica (ja-pon´-i-ka) Of Japan.

COMMON NAME: Japanese honeysuckle, gold and silver flower.

NATIVE HABITAT: Japan.

HARDINESS: Zones 4 to 10.

HABIT: Vine.

FOLIAGE: Semievergreen, short petioled, simple, opposite, ovate to oblong-ovate, 1 1/4 to 3 inches long, acute or short acuminate, base rounded to subcordate, pubescent both sides in youth, becoming glabrous with age, turning yellow in autumn.

FLOWERS: Paired, corolla 2 lipped, white-tinged pink or purplish, then turning yellow, 1 1/2 inches long, fragrant, effective in early summer.

STEMS: Reddish brown to light brown, pubescent, twining, 15 to 30 feet long.

FRUIT: Black, berrylike, 1/4 inch across, ripening in early to midautumn.

SELECTED CULTIVARS AND VARIETIES: 'Aureo-reticulata' (yellow-net honeysuckle) Much like the species but leaves smaller, to 2 inches long, colored bright green with striking yellow or golden colored venation that is viral induced; somewhat less cold hardy than the species (zones 5 to 9) and less vigorous; best in full sun; producing fewer flowers than the species. Introduced from China in 1862 by Robert Fortune.

'Halliana' Extremely popular selection with flowers white changing to yellow and very fragrant; frequently becoming naturalized and weedy, especially in the South.

'Purpurea' Foliage attractively colored bluish-purple on undersides of leaves, and greenish-purple above. Flowers purplish-red, white inside; zone 5.

Var. *repens* With veins that are often purplish; flowers white, sometimes with a hint of purple.

* * *

L. pileata (pi-le-a´ta) Having a cap, the berry being topped by a curious outgrowth of the calyx (Figure 3–124).

COMMON NAME: Royal carpet honeysuckle, privet honeysuckle.

NATIVE HABITAT: China.

HARDINESS: Zones 5 to 10; a protected site is needed in northern zones, as leaves tend to dessicate easily if exposed to winter sun and wind.

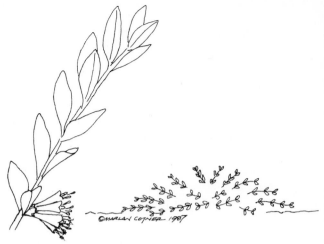

Figure 3–124 *Lonicera pileata* (life-size)

HABIT: Low, spreading, woody shrub.

SIZE: Generally around 12 inches tall, but frequently taller to about 2 feet in height, and spreading to 3 to 4 feet across.

FOLIAGE: Semievergreen, opposite, simple, ovate to oblong lanceolate, 1/2 to 1 1/4 inches long, 1/5 to 1/2 inch wide, entire, lustrous medium green above, pale green below, and sparsely pubescent on midrib (on occasion), margin sparsely ciliate.

FLOWERS: Paired, yellowish white, 5/16 inch long, fragrant, not very showy as they blend with the leaves, borne midspring.

FRUIT: Small, violet-purple, translucent, berrylike, as with flowers not very showy, but interesting and noticeable.

* * *

OTHER SPECIES: *L. prostrata* is a low, spreading or prostrate shrub with semievergreen foliage and relatively nonshowy flowers. It is not widely cultivated, but may be useful along coastal areas as it is reported to be tolerant of high salinity.

L. henryi is a semievergreen twining vine or ground cover (without support) that has a moderate rate of growth; simple foliage, opposite, ovate, 2 to 3 inches long with hairy margins and dark green coloration. Blooms produced early to late summer are colored yellowish to reddish-purple, and are followed by black, berrylike fruit that often persists through winter. Hardy to zone 4.

L. sempervirens is a semievergreen twining vine that makes a good ground cover when unsupported. Foliage is opposite, simple, from 1 to 3 1/4 inches long, colored reddish-purple in youth and per-foliate, becoming blue-green and opposite. Flowers borne in terminal clusters early to late summer are colored orange to red on the outside, yellow or orange on the inside. Hardy to zone 3.

L. xylosteum 'Nana' (emerald mound honeysuckle) A fine, low-mounded dwarf selection. Foliage is bluish-green and very attractive. Height at maturity is about 3 feet and spread to 6 feet. Flowers are rather inconspicuous. Hardy to zone 4 and in some cases zone 3.

Generally Applicable to All Species

CULTURE: Soil: Adaptable to most soils with good drainage and a pH of 4.0 to 8.0.

Moisture: Generally quite tolerant or moderately tolerant to drought. Allowing the soil to become somewhat on the dry side is often preferred as it will slow their growth. In most cases, only an in-frequent deep watering in hot, dry weather is needed.

Light: Full sun to light shade.

PATHOLOGY: Diseases: Leaf blight, leaf spots, powdery mildew, bacterial crown gall, twig blight, thread blight, hairy root.

Pests: Wooly honeysuckle aphid, leaf rollers, honeysuckle sawfly, four-lined plant bug, plant hopper, greenhouse whitefly.

MAINTENANCE: Prune heavily each year in early spring. An elevated lawn mower can be used on vines and will help remove dead undergrowth. Cut back leading stems as they outgrow their bounds or begin to climb trees or shrubs.

PROPAGATION: Cuttings: Softwood cuttings root easily and quickly in midsummer. Hardwood cuttings can also be rooted. In either case, the use of a mild rooting compound is beneficial, but not an absolute necessity; 3000 ppm IBA/talc is adequate.

Layering: Most honeysuckles root readily as branches touch the ground. Divide plantlets from rooted nodal segments in spring.

Seed: Propagation by seed is used primarily by the hybridizer, as great variability among seed of a species is frequently encountered.

✛✛✛✛✛✛✛✛✛✛✛✛✛✛✛✛✛

L*otus* (lō´tus) Old name adopted by Greek naturalists for a trefoillike plant. The genus is made up of about 100 species of subshrubs and herbs that are native to temperate regions throughout the world.

FAMILY: FABACEAE (LEGUMINOSAE): Pea family.

HABIT: Matlike, trailing or decumbent, semiwoody ground covers.

RATE: Moderate to relatively fast; space plants from pint- to quart-sized containers 18 to 24 inches apart.

LANDSCAPE VALUE: Usually used as general ground covers for moderate-sized areas. *L. corniculatus* is most often used for erosion control on banks and slopes along highways.

* * *

L. berthelotii (birth-e-lot´i-ī) Meaning unknown, likely a memorial.

COMMON NAME: Coral gem, parrot's beak, pelican's beak, winged pea.

NATIVE HABITAT: Tenerife, largest of the Canary Islands.

HARDINESS: Zone 10.

SIZE: 1 1/2 to 2 feet, spreading 4 to 6 feet across.

FOLIAGE: Evergreen, alternate, odd-pinnately com-pound; leaflets arranged in whorls or fascicles of 3 to 7 per group, linear (needlelike), entire, 1/2 inch long, 1/16 inch wide, pubescent, grayish.

STEMS: Pubescent, 2 to 3 feet long, becoming woody.

INFLORESCENCE: Flowers in clusters in leaf axils, irregular, to 1 inch long, pealike, corolla 5 lobed and colored scarlet, blooming profusely in early to midsummer.

FRUIT: Narrow, 2-valved legume.

* * *

L. corniculatus (kôr-nik-ū-lā´tus) A little horn, in reference to the shape of the flowers (Figure 3–125).

COMMON NAME: Bird's-foot trefoil (seed pods spreading like a bird's foot).

NATIVE HABITAT: Europe and Asia.

HARDINESS: Zones 5 to 10.

SIZE: 1 1/2 to 2 feet tall, spreading to 2 feet across.

FOLIAGE: Alternate odd-pinnately compound, leaf-lets in threes, obovate to oblanceolate, cloverlike, dark green.

INFLORESCENCE: 3 to 6 flowered umbels on long peduncles, corolla yellow or tinged reddish, effec-tive in summer.

FRUIT: Legume.

* * *

L. pinnatus (pin-nā´tus) Leaf morphology compoundly pinnate.

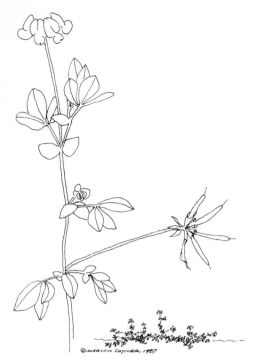

Figure 3–125 *Lotus corniculatus* (life-size)

COMMON NAME: Meadow deervetch.

NATIVE HABITAT: From central California to Washington state.

HARDINESS: Zone 8.

SIZE: 2 to 4 inches tall.

FLOWERS: Bearing yellow flowers throughout the summer.

*   *   *

L. tenuis (ten´ū-is) Slender or thin.

NATIVE HABITAT: Europe.

HARDINESS: Zone 4.

SIZE: 2 to 4 inches tall.

FLOWERS: Blooming yellow in late spring to early summer.

Generally Applicable to All Species

CULTURE: Soil: Adaptive to most well-drained soils. Moisture: Relatively tolerant to drought, but can be killed by extended hot, dry weather. It is a good practice to keep the soil moist with regular deep watering. Light: Full sun.

PATHOLOGY: Diseases: No serious diseases reported. Pets: Mealybug, black scale.

MAINTENANCE: Mow or shear periodically throughout the growing season to keep low and compact.

PROPAGATION: Cuttings: Cuttings root readily in early summer. Division: Divide in spring or fall. Seed: Commercial supplies are available.

*L*ysimachia (lī-si-mā´ki-a̠, mak´i-a̠, or lis-i-) From Greek *luo*, to loose, and *mache*, strife, hence the common name. Some would say that the name is for Lysimachus, King of Thrace (306 B.C.), who is reported to have discovered the loosestrife's supposed soothing properites. Regardless of where the name is derived from, this genus is composed of about 165 species of annual and perennial herbs, and rarely shrubs. They are native to many temperate and subtropical regions throughout the world.

FAMILY: PRIMULACEAE: Primrose family.

L. nummularia (num-ew-lā´rē-a̠) From Latin, *nummus*, a coin, in reference to the shape of the leaves (Figure 3–126).

COMMON NAME: Creeping Charley, creeping Jenny, moneywort, loosestrife.

NATIVE HABITAT: Europe.

HARDINESS: Zones 3 to 9.

HABIT: Herbaceous, matlike, trailing ground cover.

SIZE: 1 to 2 inches tall, spreading into the neighbor's yard.

RATE: Moderate to relatively fast; space plants from 3 to 4 inch diameter containers 10 to 16 inches apart.

LANDSCAPE VALUE: Well suited for a general cover in smaller, confined, shady or sunny locations. The contrast between the foliage and the flowers with the species is remarkable. The golden form is so bright yellow that it seems gaudy at times, surely capable of lightening the most dreary corner. Often used in planters or terrace where stems trail over and make a nice show. Useful, too, as a general cover next to pond or stream, or for use between stepping stones or in patio cracks. Tolerating infrequent foot traffic.

FOLIAGE: Evergreen, simple, opposite, nearly orbicular, to 1 inch long and wide, on 1/2 to 1 inch petioles, glabrous, glandular dotted both sides, bright shiny green.

STEMS: Trailing, rooting at nodes, glabrous, somewhat angular.

FLOWERS: Solitary, born in axils, regular, calyx 5 to 6 lobed, sepals cordate or lanceolate; corolla 5 lobed, about 3/4 inch across, bright yellow, effective early summer and then sometimes sparsely thereafter until fall.

SELECTED CULTIVARS AND VARIETIES: 'Aurea' Foliage exceedingly bright yellow in sun or shade, less vigorous than the species (but plenty so).

OF SPECIAL INTEREST: Was once considered to have potent healing properties. Supposedly used somewhere to dye the hair yellow; and, when the plant was burned, the smoke could drive away serpents.

Obviously, a very versatile plant that should be in every garden and incense burner.

CULTURE: Soil: Adaptable to most soils, excluding gravelly soils.

Moisture: Remarkably intolerant to drought; best when soil is kept moist to relatively wet.

Light: Full sun (if soil can be kept sufficiently moist) to moderate shade.

PATHOLOGY: Diseases: Fungal leaf blight.

Pests: Wooly aphids.

MAINTENANCE: Clip off shoots as they outgrow their bounds. Should it escape into a turf area, control is simple using broad-leaf herbicides.

PROPAGATION: Cuttings: Roots fast and reliably. The easiest means that I've found is to make 1 1/2 to 2 inch long cuttings from the leafy stems and simply place them (lower side down) on the soil surface; mist frequently enough to keep them wet. Roots usually form within 2 weeks.

Figure 3–126 *Lysimachia nummularia* (life-size)

+++++++++++++++++++++

*M*ahonia (ma-hō´ni-à) After Bernard M'Mahon of North America. The genus is composed of about 100 species of evergreen, broadleaf shrubs that are native to Asia and North and Central America.

FAMILY: BERBERIDACEAE: Barberry family.

HABIT: Low, stoloniferous, woody shrubs.

RATE: Moderate; space plants from quart- or gallon-sized containers 14 to 18 inches apart.

LANDSCAPE VALUE: Best when used for specimen planting in rock gardens. Also nice as walkway or border edging. Fruit, flowers, and foliage are all attractive and interesting, so plant them where they are visible and can be appreciated. No foot traffic.

*   *   *

M. nervosa (nẽr-vō´sà) Leaves veined, that is, nerved.

COMMON NAME: Cascades mahonia, longleaf mahonia, Oregon grape.

NATIVE HABITAT: Mountains of British Columbia southward to California.

HARDINESS: Zone 5.

SIZE: Reaching 12 to 18 inches tall, spreading 4 to 5 feet across.

FOLIAGE: Evergreen, alternate, compound, odd pinnate with overall length of 7 to 16 inches; leaflets numbering 11 to 19, primarily spiny toothed, shiny dark green above, to 3 inches long, 8 to 15 spined, turning reddish in winter.

INFLORESCENCE: Racemes to 8 inches long; flowers yellow, effective late spring.

FRUIT: Blue-black, globose berry.

*   *   *

M. repens (rē´penz) Creeping (Figure 3–127).

COMMON NAME: Creeping mahonia.

NATIVE HABITAT: Western coast of North America from British Columbia to California.

HARDINESS: Zones 5 to 9.

SIZE: 1 to 2 feet tall, spreading 4 to 5 feet across.

FOLIAGE: Semievergreen, alternate, odd pinnately compound; with 3, 5, or 7 leaflets that are 1 to 3 inches long, ovate, rounded at their apices, undulate margin with 6 to 14 bristle-tipped teeth per side, dull blue-green above, gray-green below and papillose.

INFLORESCENCE: Racemes; flowers small, fragrant bracts, 6 petaled, deep yellow, effective late spring and early summer.

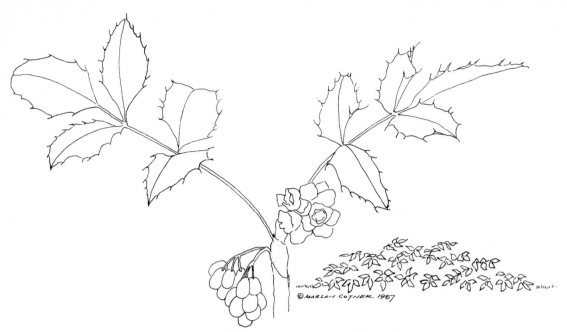

Figure 3-127 *Mahonia repens* (life-size)

FRUIT: Globose, blackish berries that are covered with blue bloom, effective early to late summer.

Generally Applicable to Both Species

CULTURE: Soil: Adaptable to various soils, but better if amended with liberal amounts of organic matter. Preference is for well-drained, slightly acid, rich loam. *M. nervosa* is more strict about requiring acidic soil.
Moisture: Tolerant to moderate drought, but better growth and appearance are realized if soil is maintained in a slightly moistened state.
Light: Full sun to light shade. In areas of hot, dry summers, plant in an area where they will be shaded in afternoon. In northern areas, protection from winter wind and sun is necessary if snow cover is inadequate to cover them.

PATHOLOGY: Diseases: Rusts; several types occur; however, *M. nervosa* and *M. repens* are both resistant to the wheat rust fungus that so many other plants of the barberry family are intermediate hosts for; leaf spots.
Pests: Barberry aphid, greedy scale, caterpillars.
Physiological: Leaf dessication from too much wind and sun in cold winter areas.

MAINTENANCE: Occasional light pruning in spring will keep plantings dense and compact. Covering with pine or spruce boughs is a good early winter practice in northern areas where snow cover is unreliable.

PROPAGATION: Cuttings: Leafy, semihardwood cuttings can be rooted readily when taken in late summer. Treatment with 3000 ppm IBA/talc is helpful. Late fall cuttings using 8000 ppm IBA/talc are reported to root in even higher percentages. Seed: Harvest when ripe, but do not allow to dry out. Stratify at 40 °F for 3 months prior to sowing the following spring.

✦✦✦✦✦✦✦✦✦✦✦✦✦✦✦

*M*azus (mā´zus) From Greek, *mazos*, a teat, in reference to the tubercles at the mouth of the flowers. This genus is made up of about 30 species of low growing, dense, mat forming herbs that are native to Asia, Australia, and New Zealand.

FAMILY: SCROPHULARIACEAE: Figwort family.

RATE: Relatively fast; space plants from 2 to 3 inch diameter containers 8 to 12 inches apart.

LANDSCAPE VALUE: Most commonly used for general, low ground cover in small areas; frequently under trees. They also make an interesting filler for between stepping stones and tolerate a little foot traffic.

*   *   *

M. pumilio (pew´mil-ē-ō) Dwarf.
COMMON NAME: Dwarf mazus.

NATIVE HABITAT: Australia and New Zealand.

HARDINESS: Zone 7.

HABIT: Rhizomatous, creeping, mat forming, low ground cover.

SIZE: 1 to 2 inches tall, spreading indefinitely.

FOLIAGE: Lower leaves opposite or in a rosette, becoming alternate, obovate, to 3 inches long, entire or coarsely toothed.

INFLORESCENCE: Terminal one-sided raceme; flowers to 5/16 inch long, calyx 5 lobed; corolla 2 lipped, upper lip 2 lobed, lower lip 3 lobed, white or bluish with yellow center, effective early to mid-summer.

FRUIT: Capsule, not ornamentally significant.

* * *

M. reptans (rep´tanz) Creeping and rooting (Figure 3–128).

COMMON NAME: Creeping mazus.

NATIVE HABITAT: Himalayas.

HARDINESS: Zones 3 to 9.

HABIT: Herbaceous, rooting while trailing, low, mat forming ground cover.

SIZE: 1 to 2 inches tall, spreading indefinitely.

FOLIAGE: Semievergreen, lower leaves opposite or in a rosette, becoming alternate, obovate to narrowly obovate, to 2 inches long, mostly entire, obtuse, very pilose.

FLOWERS: Solitary to groups of 3, to 3/4 inch long, on terminal peduncles, corolla 2 lipped, with upper lip 2 lobed and lower lip 3 lobed, purplish-blue, lower lip spotted white, yellow, and purple, effective late spring.

OTHER SPECIES: The selection 'Albiflorus' is a white flowered cultivar of *M. japonicus* that is similar in appearance to *M. reptans*, but somewhat larger in size.

Generally Applicable to All Species

CULTURE: Soil: Preference is for rich well-drained loam, but adaptable to almost any soil with good drainage.

Moisture: Not markedly tolerant to drought. Best when soil moisture is maintained with regular watering.

Light: Full sun to moderate shade.

PATHOLOGY: No serious diseases or pests.

MAINTENANCE: Little or none.

PROPAGATION: Division: Simple division remains the standard means of propagation. It can be accomplished best in early spring.

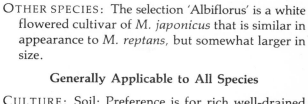

Mentha (men´thả) Greek name of a nymph. This genus, commonly recognized as the mints, is composed of about 25+ species of aromatic, square stemmed, erect or procumbent, herbaceous perennials. They are native to temperate regions of Europe.

FAMILY: LAMIACEAE (LABIATAE): Mint family.

FRUIT: Of 4 nutlets, not ornamentally significant.

OF SPECIAL INTEREST: Flowers of mints are particularly attractive to bees.

* * *

M. × piperita (pī-pẽr-ī´tả) From Latin *piper*, pepperlike.

COMMON NAME: Peppermint, black peppermint.

NATIVE HABITAT: Europe.

HARDINESS: Zone 3.

HYBRID ORIGIN: *M. spicata × M. aquatica*.

Figure 3–128 *Mazus reptans* (life-size)

HABIT: Herbaceous, rhizomatous/stoloniferous, spreading ground cover.

SIZE: 12 to 24 inches tall, spreading indefinitely.

RATE: Relatively fast; space plants from 2 1/2 to 3 1/2 inch diameter containers 12 to 16 inches apart.

LANDSCAPE VALUE: Useful as a soil binder along streams, pools, ditches, and ponds. No foot traffic.

FOLIAGE: Opposite, simple, lanceolate, to about 2 1/2 inches long, apex acute, margin serrate, petioled, pungent with strong peppermint odor, medium green.

STEMS: Usually glabrous, pungent, green, but frequently tinged purple, squarish in cross section.

INFLORESCENCE: Oblong spikes, 1 to 3 feet long, verticellasters congested; flowers sterile, calyx tubular, glabrous, and teeth ciliate, corolla lilac-pink to purple, effective in late summer.

*   *   *

M. pulegium (pū-li´ji-um) From Latin pulex, a flea, which the plant was thought to eradicate, presumably when applied topically.

COMMON NAME: Pennyroyal.

NATIVE HABITAT: Europe and Asia.

HARDINESS: Zone 7.

HABIT: Herbaceous, prostrate, mat forming ground cover.

SIZE: 1 to 3 inches tall, spreading indefinitely.

RATE: Relatively fast; space plants from 2 1/2 to 3 1/2 inch diameter containers 12 to 16 inches apart.

LANDSCAPE VALUE: Useful as a filler between stepping stones or as a general cover in small areas where it is contained. Very limited foot traffic.

FOLIAGE: Opposite, simple, ovate to orbicular, to 1 inch long, apex acute or obtuse, margin entire to serrate, pungent.

STEMS: Glabrous or tomentose, to 1 foot, decumbent, squarish.

INFLORESCENCE: Elongate verticillasters; flowers dense, small, corolla lilac-blue, effective in summer.

OF SPECIAL INTEREST: M. pulegium is (or was) used in medicine to relieve spasms.

*   *   *

M. requienii (rek-wē-en´ē-i) After Esprit Requien (1788–1851), a student of the flora of Southern France and Corsica.

COMMON NAME: Creeping mint, Corsican mint.

NATIVE HABITAT: Corsica.

HARDINESS: Zones 6 to 10.

HABIT: Herbaceous, rhizomatous, low, mosslike, creeping ground cover.

SIZE: 1/2 to 3 inches tall, spreading indefinitely.

RATE: Relatively fast; space plants from 2 1/2 to 3 1/2 inch diameter containers 12 to 16 inches apart.

LANDSCAPE VALUE: Same as M. pulegium.

FOLIAGE: Opposite, simple, entire, ovate to cordate to nearly orbicular, small—usually about 1/8 inch or slightly more long, 1/8 inch wide, mostly glabrous, some sparsely hairy, glandular dotted both sides, dark green, strongly aromatic.

INFLORESCENCE: Few flowered verticillasters; flowers small, irregular, corolla 4 lobed, lavender, effective in summer.

*   *   *

M. spicata (spi-kā´tȧ) Bearing flowers in spikes (Figure 3–129).

COMMON NAME: Spearmint, garden mint.

NATIVE HABITAT: Europe.

HARDINESS: Zone 3.

HABIT: Stoloniferous, herbaceous, spreading ground cover.

SIZE: 12 to 24 inches tall, spreading indefinitely.

RATE: Relatively fast, space plants from 2 1/2 to

Figure 3–129 *Mentha spicata* (0.5 life-size)

3 1/2 inch diameter containers 12 to 16 inches apart.

LANDSCAPE VALUE: Same as *M.* x *piperita.*

FOLIAGE: Lanceolate, opposite, to 2 inches long, apex acute, sessile, margins serrate, aromatic.

INFLORESCENCE: Verticillasters arranged in terminal spikes; flowers small, calyx campanulate, glabrous or hairy, teeth nearly equal; corolla 4 lobed, lilac, pink or white, effective in summer.

Generally Applicable to All Species

CULTURE: Soil: Adaptable to most any soil. Tolerating relatively wet soils.

Moisture: Generally not able to withstand a great deal of drought.

Watering should be frequent enough to maintain moderate to high levels of soil moisture.

Light: Full sun to light shade; more compact with increasing levels of sun.

PATHOLOGY: No serious diseases or pests reported.

MAINTENANCE: Occasional mowing will keep plantings looking neat.

PROPAGATION: Cuttings: Cuttings root easily when taken in early summer to early fall. Treatment with 1000 ppm IBA/talc is beneficial but not necessary.

Division: Simply divide anytime (prune back growth first if during growing season) and keep moist.

Seeds: Commercial supplies are readily available. Germination takes 1 to 2 weeks at 70° to 80°F. Sowing at a density of 125 to 175 seeds per square foot is ample.

+++++++++++++++++++

*M*esembryanthemum (mes-em-bri-an´the-mum, or me-zem-) From Greek *mesembria*, midday, and *anthemon*, flower—flowers open in the sun.

COMMON NAME: Ice plants, because of their glistening icylike appearance, caused by water-soaked cells. Sometimes also called fig marigold, pebble plant, or icicle plant. Years ago, the genus *Mesembryanthemum* was rather large, composed of approximately 1000 species of usually low growing, succulent, desert plants. The genus now is thought to contain about 74 species, while many of the other species have been allotted to different genera.

Characterized by brilliant, almost artificial appearing flowers, these succulent plants are annual or biennial and originated in Southwest Asia, Southern Europe, Africa, the Canary Islands,

Madeira, Chile, and Western North America. Perennial ice plants that have ground covering merits may be found under the genus headings *Carpobrotus*, *Cephalophyllum*, *Delosperma*, *Drosanthemum*, and *Lampranthus*.

*M*icrobiota (mi-kro-bi´o-ta) From Greek, *micros*, small, and *biota*, thuja, the foliage like thuja (white cedar), habit smaller. This relatively new genus has only one species. It is juniperlike, but differs in the morphology of the cones. Originating in the Valley of the Suchan to the east of Vladivostok, East Siberia.

FAMILY: CUPRESSACEAE: Cypress family.

M. decussata (de-ku-sa´ta) With leaves in pairs (Figure 3–130).

COMMON NAME: Siberian cypress, Russian cypress, microbiota.

NATIVE HABITAT: As above, discovered 1921, named 1923. Later received by Trompenburg Arboretum, Netherlands, in 1968, which was responsible for its mass distribution.

Figure 3-130 *Microbiota decussata* (life-size)

HARDINESS: Zone 2, being tested in zone 1.

HABIT: Low spreading, dwarf, woody shrub.

SIZE: To 8 inches tall, spreading 9 to 12 feet across.

RATE: Moderate, space plants from 1 to 2 gallon sized containers 3 1/2 to 4 1/2 feet apart.

LANDSCAPE VALUE: General cover for moderate to large areas or for facing building foundations. Interesting soft textured, feathery foliage is pleasing to the eye. No foot traffic.

FOLIAGE: Evergreen, opposite, scalelike or awl shaped, small, feathery, dull green in summer turning bronzy in autumn.

FLOWERS: Dioecious, small insignificant ornamentally.

CONES (FRUIT): Small, berrylike, about 1/4 inch across, on female plants. Most plants in cultivation are male clones and produce inconspicuous cones.

CULTURE: Soil: Tested little, likely adaptable to a wide range of soils provided drainage is good.
Moisture: The plantings I've seen have shown resistance to moderate drought, and likely will take a considerable amount. An occasional deep watering in summer is usually all that is needed to keep them looking lush.
Light: Full sun to light shade.

MAINTENANCE: Little or no maintenance is needed; a light shearing on an annual basis will help promote a compact habit.

PATHOLOGY: No serious diseases or pests reported.

PROPAGATION: Cuttings: Take cuttings in late fall after a couple of hard frosts. Treatment with a root inducing preparation of 8000 ppm IBA/talc is helpful, as is wounding to the cambium. Cuttings are best rooted in a porous medium such as 100% vermiculite and then transferred once roots are established.

Mitchella (mi-chel´á) After Dr. John Mitchell, a botanist of Virginia. The genus is composed of only 2 species of evergreen, herbaceous perennials. They are native to eastern North America and eastern Asia.

FAMILY: RUBIACEAE: Madder family (Figure 3–131).

M. repens (rē´penz) Creeping squaw vine.

COMMON NAME: Partridge berry, squawberry, twinberry, two-eyed Mary, running box.

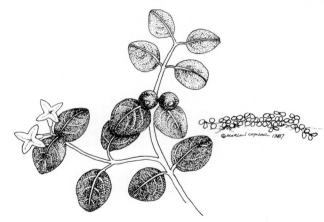

Figure 3–131   *Mitchella repens* (life-size)

NATIVE HABITAT: Woodlands of eastern North America.

HARDINESS: Zones 3 to 9.

HABIT: Semiwoody, low, trailing ground cover.

SIZE: 1 to 2 inches tall, spreading indefinitely.

RATE: Slow; space plants from 3 to 4 inch diameter containers 6 to 8 inches apart.

LANDSCAPE VALUE: Best use is in a naturalized setting as a general cover. Tolerates limited foot traffic.

FOLIAGE: Evergreen, opposite, simple, orbicular-ovate, to 3/4 inch long, slightly less in width, glossy dark green above with whitish venation, lighter below.

STEMS: Trailing, rooting, somewhat woody.

FLOWERS: Axillary or terminal pairs, 4 parted, to 1/2 inch long; corolla funnelform, tube elongated, short lobed, bearded inside, effective in summer.

FRUIT: 8 seeded, brilliant scarlet berry, about 1/4 inch in diameter, insipid but reported to be edible, effective in fall.

CULTURE: Soil: Requires organically rich, moist, acidic loam.
Moisture: Not greatly tolerant to drought; soil should be moderately moist at all times.
Light: Light to moderate shade.

PATHOLOGY: Diseases: Leaf spot, stem rot. No pests reported.

MAINTENANCE: No maintenance is usually necessary.

PROPAGATION: Cuttings: Cuttings root easily following treatment with 1000 ppm IBA/talc.
Division: Division is the most frequently used type of propagation. It should be done in early spring before new growth begins.

Seed: I've read that seed is an effective means of propagation but have no experience with it nor can I find any specific recommendations.

✦✦✦✦✦✦✦✦✦✦✦✦✦✦✦✦✦

Mitella (mi-tel´à) The diminutive of Latin, *mitra*, a miter, headdress of a bishop of the Western Church (a little miter), in reference to the two-cleft seed pod. Commonly known as bishop's cap or miterwort, this genus is composed of about 12 species of rhizomatous, herbaceous perennials. They are native to North America and northeast Asia.

FAMILY: SAXIFRAGACEAE: Saxifrage family.

M. diphylla (dĭ-fil´là) Two leaved in reference to paired leaves halfway up the scapes (Figure 3–132).

COMMON NAME: Common miterwort, double-leaved bishop's cap, coolwort.

NATIVE HABITAT: Woodlands of the United States from New England to South Carolina, westward to Minnesota.

HARDINESS: Zone 3.

HABIT: Herbaceous, low growing ground cover.

SIZE: 12 to 18 inches tall, spreading to 12 inches across.

RATE: Moderate; space plants from pint-sized containers 12 to 14 inches apart.

LANDSCAPE VALUE: Valuable as a general cover in naturalized woodland setting.

FOLIAGE: Small, primarily basal, with a pair of sessile, opposite, 3-lobed leaves about halfway up the stem; basal leaves are long-petioled, cordate, with irregularly toothed margin, glossy dark green.

INFLORESCENCE: Racemes, 6 to 8 inches long, on 1 1/2 foot tall scapes; scapes with one pair of opposite sessile leaves midway up; flowers tiny, 5-parted, petals pinnately cut, white, effective in spring.

FRUIT: Dehiscent capsule, not of ornamental significance.

OTHER SPECIES: *M. breweri*; similar to *M. diphylla*, but leaves rounded, blooming late spring to early summer.

CULTURE: Soil: Preference is for highly organic, acidic soil.
Moisture: Not notably tolerant of drought; ground should be kept moist and plants located so as to be sheltered from drying winds.

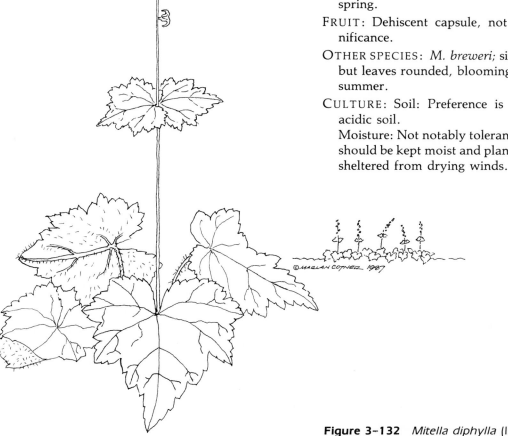

Figure 3–132 *Mitella diphylla* (life-size)

Light: Light to moderate shade.

PATHOLOGY: No serious diseases or pests reported.

MAINTENANCE: Periodic addition of peat or composted leaf mold is beneficial if grown in amended soils.

PROPAGATION: Division: Divide in early spring. Seed: Seed should be sown when ripe.

Muehlenbeckia (mū-len-bek´i-à) After Dr. Henry G. Muehlenbeck (1789–1845), French physician and botanist. This genus, commonly called wire plant, is composed of about 20 species of vining, climbing, or prostrate growing woody plants with wire-like stems. They are native to New Zealand, Australia, and South America.

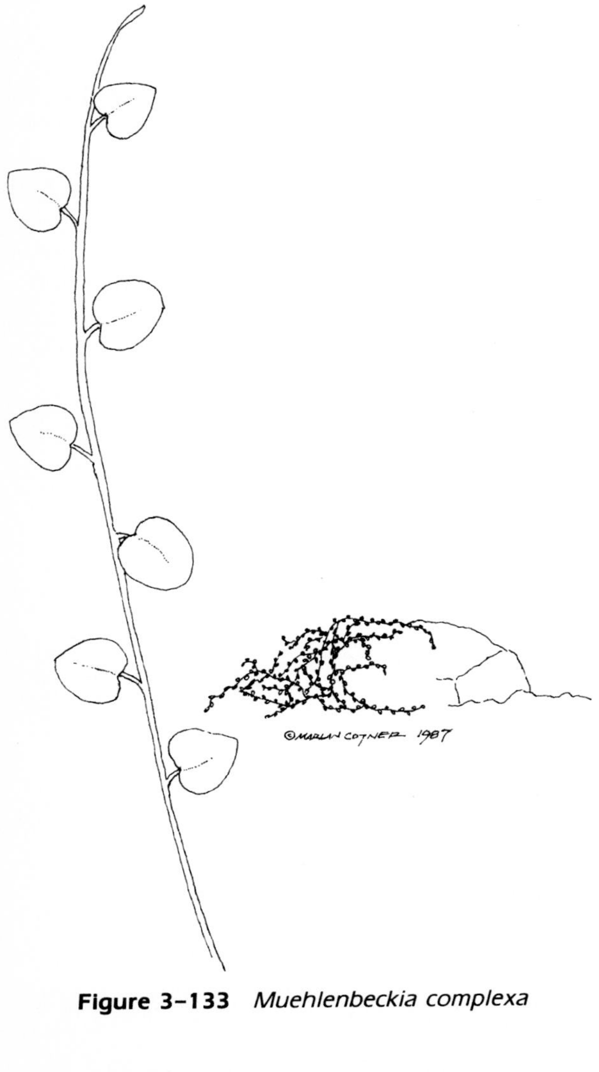

FAMILY: POLYGONACEAE: Buckwheat family.

M. axillaris (aks-i-lā´ris) (or ăks-i-lah´ris) Flowers borne in leaf axils.

COMMON NAME: Creeping wirevine.

NATIVE HABITAT: Australia, New Zealand, Tasmania.

HARDINESS: Zones 6 to 10.

HABIT: Small, prostrate, sprawling, mounding, semiwoody, matlike vine.

SIZE: 2 to 12 inches tall, spreading 4 to 5 feet across.

RATE: Relatively slow to moderate; space plants from gallon-sized containers 2 to 2 1/2 feet apart.

LANDSCAPE VALUE: Commonly used as a general cover for moderate- to larger-sized contained areas. Often it is combined with small, spring flowering bulbs for color, or is planted by itself as a specimen in the rock garden.

FOLIAGE: Evergreen, alternate, simple, oblong to nearly orbicular, 3/8 inch long or less by about the same in width, margin entire, glabrous, dark bronzy-green with reddish margins.

STEMS: Many branched, prostrate, sprawling, reddish, glandular dotted, very thin or wirelike.

FLOWERS: Solitary or paired, axillary, dioecious or polygamous (bearing both bisexual and unisexual flowers on the same plant), regular, 5-petaled, yellowish-green, 3/16 inch across, not ornamentally significant.

FRUIT: Black, shiny, ovoid, 3-angled achene, enclosed in a white, fleshy, berrylike calyx.

* * *

M. complexa (kom-pleks´à) Interwoven, entangled; reference is to the branches or possibly to the perianth, which swells to enclose the fruit (Figure 3–133).

COMMON NAME: Wirevine, mattress vine, maidenhair vine, necklace vine.

NATIVE HABITAT: New Zealand.

HARDINESS: Zones 6 to 10.

HABIT: Low growing, trailing vine.

SIZE: 2 to 18 inches tall, spreading several feet across.

RATE: Moderate to relatively fast; space plants from gallon-sized containers 3 to 4 feet apart.

LANDSCAPE VALUE: Good as a general cover in rather large open areas. Often planted on slopes along the seashore as salt tolerance is good. No foot traffic.

Figure 3–133 *Muehlenbeckia complexa*

FOLIAGE: Evergreen, small, alternate, simple, elliptic to fiddleform, 1/2 to 3/4 inch long, dark green.

STEMS: Many sprawling branchlets, mostly pubescent.

INFLORESCENCE: Dioecious, simple or dense panicle, 1/2 inch long, sometimes somewhat longer; minute greenish-yellow flowers that are for the most part insignificant, borne in fall.

FRUIT: Black, shiny, 3-angled achene that is partly fused with a white fleshy calyx.

Generally Applicable to Both Species

CULTURE: Soil: Best in light-textured, well-drained soils and tolerant to gravelly infertile soils. Seemingly adaptable as to pH, at least in the range of slight alkalinity to moderate acidity.
Moisture: Showing good drought tolerance, but an occasional watering is beneficial during the summer months. Don't hesitate to use these plants in exposed, windswept sites.
Light: Full sun to light shade.

PATHOLOGY: No serious diseases or pests reported.

MAINTENANCE: Mulch with pine or spruce branches in the fall if growing in northern locations where snow cover is minimal or unpredictable.

PROPAGATION: Cuttings: Softwood cuttings root well throughout summer.
Division: Rooted branches can be dug up and transplanted in spring, fall, or winter.

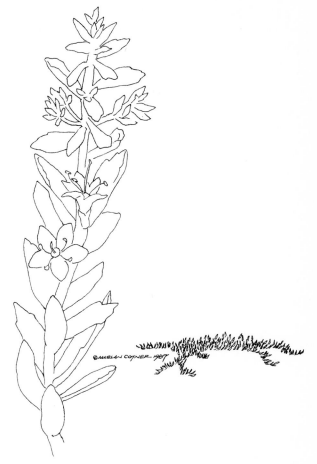

Figure 3–134 *Myoporum parvifolium* (life-size)

*M*yoporum (my-op´ôr-um or my-ō´pôr-um) Meaning unknown. This genus is composed of about 30 species of evergreen subshrubs, shrubs, and mostly trees. Primarily, they originate in Australia.

FAMILY: MYOPORACEAE: Myoporum family.

M. parvifolium (pär-wȧ-fō´lē-ûm or pär-vi-) Small leaved, from Latin *parva*, small, and *folium*, leaf (Figure 3–134).

COMMON NAME: Prostrate myoporum.

NATIVE HABITAT: New Zealand.

HARDINESS: Zones 9 to 10.

HABIT: Woody, prostrate, low, shrubby ground cover.

SIZE: 3 to 6 inches tall, spreading to 9 feet across.

RATE: Moderate to relatively fast; space plants from gallon-sized containers 3 to 5 feet apart.

LANDSCAPE VALUE: Considered to be quite fire resistant, and thus frequently used as a general cover in public locations such as parks, rest areas, and near entrances of buildings. Also effective for controlling erosion on slopes and along highway embankments. In either case, it is suited for moderate- to large-sized areas. No foot traffic.

FOLIAGE: Evergreen, alternate, simple, oblanceolate, to 1/2 inch long by 1/4 inch wide, irregularly toothed margin that is nearly entire, glabrous, glandular dotted on both sides, bright green, sometimes purplish tinted in winter.

STEMS: Prostrate, glabrous, glandular dotted, warty.

INFLORESCENCE: Clusters or solitary flowers, axillary; flowers to 3/8 inch across, regular, corolla 5 lobed and colored white, effective in summer.

FRUIT: Small, succulent, purplish drupe, about 1/8 inch in diameter.

SELECTED CULTIVARS AND VARIETIES: 'Putah Creek' Reported to be more symmetrical in habit and reaching 12 inches high and 4+ feet across.

OTHER SPECIES: *M. debile* is similar in habit to *M.*

parvifolium, but differs in having foliage that is narrower, almost to the point of appearing needlelike. It may be a couple of degrees less hardy, and reaches about 12 inches in height.

CULTURE: Soil: Quite adaptable, but preference is for well-drained, sandy loam with slightly acidic reaction.

Moisture: Relatively tolerant to drought. Growth and appearance are best when the humidity is relatively high and temperatures relatively cool. For this reason, plants are best grown along the coast rather than inland.

Light: Full sun.

PATHOLOGY: No serious diseases or pests reported.

MAINTENANCE: Periodic edging or trimming back of growth is needed when stems outgrow their bounds. Occasional mowing will rejuvenate and keep neat.

PROPAGATION: Cuttings: Cuttings root easily throughout the growing season. Light misting is beneficial but not always a necessity. No rooting hormones are needed, but root mass may be greater with 3000 ppm IBA/talc or solution.

*M*yosotis (mī-ō-sō′tis) From Greek, *mus*, a mouse, and *otes*, an ear, the foliage resembling the ear of a mouse. Sometimes referred to as forget-me-not, this genus is composed of about 50 species of annual, biennial, and perennial herbs. They are primarily native to temperate regions throughout the world.

FAMILY: BORAGINACEAE: Borage family.

M. scorpioides (skor-pē-oi′dez) Like a scorpion's tail, the inflorescence (Figure 3–135).

COMMON NAME: True forget-me-not.

NATIVE HABITAT: Europe.

HARDINESS: Zones 3 to 9.

HABIT: Herbaceous, decumbent, spreading ground cover.

SIZE: 10 to 18 inches tall, spreading to 12 inches across.

RATE: Moderate; space plants from 3 to 4 inch diameter containers 8 to 12 inches apart.

LANDSCAPE VALUE: Usually used as a general cover for small- to moderate-sized locations. Delicate, soft and lacy, the appearance of this plant helps to soften the harsh textures of trees and shrubs when used as a facing. Excellent in bog gardens, along streams and ponds, or poolside. No foot traffic.

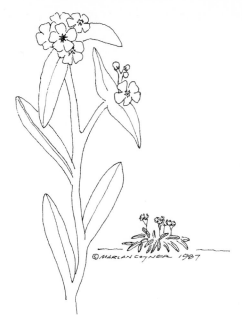

Figure 3–135 *Myosotis scorpoides* (life-size)

FOLIAGE: Simple, basal leaves petioled, stem leaves sessile, oblong-lanceolate or oblanceolate, entire, to 2 inches long, bright shiny green, deciduous.

INFLORESCENCE: Slender racemes; flowers small, 5-parted; corolla salverform, throat scaly, bright blue or occasionally pinkish with yellow, pink, or white eye. Effective late spring through summer.

FRUIT: Of 4 smooth, shiny nutlets, insignificant ornamentally.

SELECTED CULTIVARS AND VARIETIES: Var. *rosea* With pink flowers in summer.

Var. *semperflorens* Dwarf to 8 inches tall; blooming from early summer to early fall.

CULTURE: Soil: Adaptable to most soils.

Moisture: Not notably tolerant to drought; soil should be kept moist and plants located in a cool location sheltered from strong winds; thriving in high moisture situations.

Light: Full sun to moderate shade.

PATHOLOGY: Diseases: Blight, downy mildew, wilt, rust, aster yellows.

Pests: No serious pests reported.

MAINTENANCE: Dividing may be performed every fourth or fifth year to improve bloom.

PROPAGATION: Cuttings: Take cuttings in summer and root in a very porous medium. Treating with 1000 ppm IBA/talc may be to some benefit.

Division: Simply divide plants in spring before new growth begins.

***N**andina* (nan-di´nȧ) From *nandin,* the Japanese name for this shrub. The genus is monospecific.

FAMILY: BERBERIDACEAE: Barberry family.

N. domestica (dō-mes´ti-kȧ) Domestic, in reference to its various uses in Japanese households (Figure 3–136).

COMMON NAME: Heavenly bamboo, nandina, sacred bamboo.

NATIVE HABITAT: Central China and Japan.

HARDINESS: Zones 7 to 9.

HABIT: Woody based, broad to broad-spreading, rhizomatous shrub.

SIZE: The species may reach 6 to 8 feet tall and wide, but it is the dwarf selections that are most valuable for ground covers. Their sizes will be discussed later.

RATE: Relatively slow to moderate; space plants from gallon-sized containers 2 to 3 feet apart.

LANDSCAPE VALUE: Often used as edging along walkways and border plantings. Occasionally used as a dwarf hedge. In small groups, they make ex-cellent accent plants, and planted against a light-colored background they make excellent facers for foundation or garden ornaments. No foot traffic.

FOLIAGE: Evergreen, alternate, twice or thrice compound; leaflets 1/2 to 1 1/2 inches wide and 1 to 3 1/2 inches long, subsessile, elliptic-lanceolate, entire, apex acuminate, base cuneate, leathery, young leaves often tinged pink or bronzy as they open, becoming light green in summer, then bronzy with traces of red and occasionally purplish in fall.

STEMS: Erect, stout, unbranched, reddish-purple in youth becoming brown, yellow inside, as is true with many other members of Berberidaceae.

INFLORESCENCE: 8 to 15 inches long, erect, terminal panicles; flowers perfect, pinkish in bud, small, 1/4 to 1/2 inch wide, white, effective in early summer.

FRUIT: Rather exceptional, 2 seeded, bright red globose berries, about 1/3 inch across, profusely borne and causing branches to arch, effective in fall.

OF SPECIAL INTEREST: The fruit of nandina provides wintertime food for song birds.

SELECTED CULTIVARS AND VARIETIES: 'Atropurpurea Nana' Stiff upright habit to 2 feet tall; fall color reddish-purple.
'Harbor Dwarf' An excellent compact selection that develops into a 2 to 3 foot high, graceful mound. Introduced by Callaway Gardens.
'Nana' ('Pygmaea Compacta') Reaching 2 to 4 feet tall and forming a dense mounded shape; fruiting less than the species.

CULTURE: Soil: Adaptable to various soil types, but best in fertile loam with a pH of 5.0 to 6.5.
Moisture: Relatively tolerant of drought, with an occasional deep watering in summer being beneficial.
Light: Full sun to moderate shade, with best foliage color in full sun.

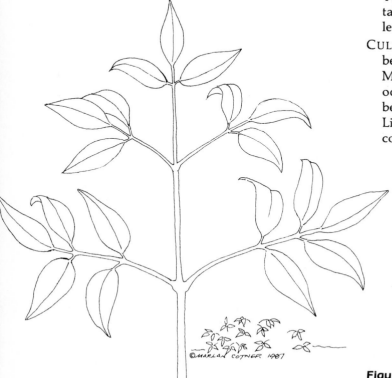

Figure 3–136 *Nandina domestica* 'Purpurea Nana' (life-size)

MAINTENANCE: Annual pruning should consist of removal of the oldest stems just above the crown and heading back of canes to promote compact, highly branched habit.

PATHOLOGY: Diseases: Red leaf spot, anthracnose, root rot.
Pests: Northern root knot nematode.

PROPAGATION: Cuttings: Softwood cuttings in spring or summer will root under mist following treatment with 1250 to 1500 ppm IBA/talc. As the cuttings become more woody, they become more difficult to root, but will respond to somewhat higher (1500 to 2000 ppm) concentrations of IBA. Root cuttings, too, are said to root, but are less practical.
Division: Rooted suckers around the base can be divided and transplanted in spring.
In vitro: Lately, many of the dwarf cultivars, primarily 'Harbor Dwarf,' are being propagated in a test tube.

*N**epeta* (nep′e-tȧ or ne-pē′tȧ) Possibly from Nepete, a city in Etruria (an ancient civilization roughly corresponding geographically to Tuscany in today's Italy). This rather large genus is composed of about 250 species of perennial and occasionally annual herbs. They are native to dry areas of temperate Europe, Asia, and northern Africa, along with the mountainous regions of tropical Africa.

FAMILY: LAMIACEAE (LABIATAE): Mint family.

N. × *faassenii* (fah-sen′i-ī) After J. H. Faassen, a Dutch nurseryman (Figure 3–137). Often erroneously listed as *Nepeta mussinii.*

COMMON NAME: Persian ground-ivy, Persian catmint, mauve catmint.

HYBRID ORIGIN: *N. mussinii* × *N. nepetella,* sometime before 1939 in Copenhagen Botanic Gardens, Denmark.

HARDINESS: Zone 3.

HABIT: Herbaceous, low, mounding, ground cover.

SIZE: 18 to 24 inches tall, spreading 2 to 3 feet across.

RATE: Moderate to rapid; space plants from 3 to 4 inch diameter containers 12 to 16 inches apart.

LANDSCAPE VALUE: Fine for edging a walk or perennial border. Makes a good specimen for use in a rock garden or general cover for use in small- to moderate-sized areas.

OF SPECIAL INTEREST: As with catnip (*N. cataria*), the foliage of this plant is attractive to cats. Additionally, the flowers supply nectar to hummingbirds.

FOLIAGE: Opposite, simple, lanceolate to oblong-ovate, wrinkled, to 1 1/4 inch long, coarsely crenate-dentate margin, base truncate, colored gray-green, evergreen.

INFLORESCENCE: Spikelike, elongate or interrupted raceme with axillary peduncled cymes; bracts linear-lanceolate, rather short, white pubescent; corolla to 1/2 inch long, violet blue, effective in summer.

FRUIT: Sterile, nonfruiting.

Figure 3–137 *Nepeta* × *faassenii* (life-size)

CULTURE: Soil: Adaptable to almost any well-drained soil.

Moisture: Showing good tolerance to drought once established; only periodic watering in the heat of summer is needed to keep this plant looking it's best.

Light: Full sun.

PATHOLOGY: Diseases: Leaf spots, both bacterial and fungal, stem rot, viral mosaic.

Pests: No serious pests reported.

MAINTENANCE: Shearing after bloom keeps plantings looking dense and may stimulate a second flowering period in fall.

Nierembergia (nē-rem-bēr´gi-à or -ji-à) After Juan E. Nieremberg, a Spanish Jesuit. This genus is composed of about 30 species of many branched, mostly glabrous, decumbent to erect, perennial herbs and subshrubs. They are native from Mexico to Chile and Argentina.

FAMILY: SOLANACEAE: Potato family.

N. repens (rē´penz) Creeping (Figure 3–138).

COMMON NAME: White cupflower.

NATIVE HABITAT: South America.

HARDINESS: Zones 7 or 8 to 10; said to survive in zones 5 and 6 when heavily mulched in fall.

HABIT: Herbaceous, trailing, mat forming ground cover.

SIZE: 4 to 6 inches tall, spreading to 12 inches across.

RATE: Moderate; space plants from pint-sized containers 12 to 14 inches apart.

LANDSCAPE VALUE: Use alone as a general cover for small areas or for the same purpose when combined with other low growing ground covers such as *Thymus* or *Cymbalaria*. No foot traffic.

FOLIAGE: Simple, alternate, entire, oblong or spatulate, to 1 1/4 inches long, petioled, colored bright green.

STEMS: Creeping, rooting at nodes.

FLOWERS: 5-parted, solitary, corolla 1 1/2 inches across, creamy white, streaked with purple, occasionally rose colored, effective in summer, similar to morning glories in appearance.

FRUIT: Many seeded, 2 valved capsules; not ornamentally significant.

OTHER SPECIES: *N. hippomanica* var. *violaceae* (dwarf cupflower) is a sprawling, evergreen, herbaceous perennial that reaches 6 to 12 inches tall and spreads about the same; blue to violet, bell-like,

Figure 3–138 *Nierembergia repens* (life-size)

solitary flowers are born terminally, reach 1 inch across and are very attractive throughout summer. A cultivar named 'Purple Robe' is commonly cultivated.

CULTURE: Soil: Adaptable to most well-drained soils.

Moisture: Not notably tolerant to drought; soil should be kept moist with additional watering as needed.

Light: Full sun if grown in a cool location with adequate soil moisture; otherwise, light shade.

PATHOLOGY: No serious diseases; aphids are reported to be a problem but are easily controlled.

MAINTENANCE: Mow plantings in fall or spring before new growth in order to keep plants dense and attractive.

PROPAGATION: Cuttings: Softwood cuttings taken in spring are easily rooted.

Division: Divide clumps in spring.

Seed: In northern locations these plants are treated as annuals. Commercial supplies of seed are available, and germination can be expected in 2 to 3 weeks at temperatures ranging from 68° to 75°F.

Omphalodes (om-fà-lō´dēz) From Greek, *omphalos*, a navel, and *eidos*, like, the shape of the seeds like a navel. This genus is composed of around

24 species of glabrous or sparsely hairy, annual or perennial herbs. They are native to Europe, Asia, and Mexico.

FAMILY: BORAGINACEAE: Borage family.

O. verna (vĕr´nȧ) Spring, the time of flowering (Figure 3–139).

COMMON NAME: Creeping navel-seed, blue-eyed mary, creeping forget-me-not.

NATIVE HABITAT: Europe.

HARDINESS: Zones 6 or 7 to 9.

HABIT: Herbaceous, stoloniferous, low, spreading ground cover.

SIZE: 8 inches tall, spreading 12 + inches across.

RATE: Relatively fast; space plants from 3 to 4 inch diameter containers 10 to 14 inches apart.

LANDSCAPE VALUE: Useful as a general cover in sunny locations where it has room to spread or can be contained. No foot traffic.

FOLIAGE: Semievergreen, alternate, simple, finely textured, ovate to ovate-lanceolate, 1 to 3 inches long, prominently veined.

INFLORESCENCE: Loose racemes; flowers 5-parted, corolla blue or sometimes whitish, 1/2 inch across, blooming mid to late spring.

FRUIT: Of 4 nutlets, not ornamentally significant.

OF SPECIAL INTEREST: Reported to have been a favorite plant of Queen Marie Antionette. Unfortunately, it is not widely cultivated and may be difficult to obtain.

SELECTED CULTIVARS AND VARIETIES: 'Alba' Flowers white.

CULTURE: Soil: Adaptable to wide range of soil types, provided drainage is good and organic content high enough to maintain some moisture. Moisture: Not notably tolerant to drought; soil should be kept moist with periodic watering. Light: Moderate shade.

PATHOLOGY: No diseases or pests reported.

MAINTENANCE: No special maintenance required.

PROPAGATION: Division: Divide plants in early spring.

*O*noclea (on-ō-klē´ȧ) From Greek, *onokleia*, a plant with leaves rolled up into the semblance of berries, referring to the capsulelike spore capsules. Monospecific, this genus is composed of a coarsely textured, rhizomatous, terrestrial fern.

FAMILY: POLYPODIACEAE: Polypody family.

O. sensibilis (sen-si´bil-is) Sensitive, in reference to its sensitivity to early frost in fall or, as some say, because its picked leaves tend to roll up at the edges. (Figure 3–140).

COMMON NAME: Sensitive fern, bead fern.

NATIVE HABITAT: Marshy soils of North America, Europe, and Asia.

HARDINESS: Zone 3.

HABIT: Medium size, rhizomatous, creeping fern.

SIZE: 12 to 24 inches tall, spreading indefinitely.

RATE: Moderate; space plants from pint- or quart-sized containers 14 to 20 inches apart.

LANDSCAPE VALUE: Dense network of roots that this plant forms makes it well suited to bind soil and act as a general cover in areas with gentle grade. Looks nice when naturalized along the perimeter of a wooded setting. No foot traffic.

FOLIAGE: Deciduous, dimorphic; sterile leaves solitary and scattered, to 4 1/2 feet long but usually about 2 feet long, pinnatifid, light yellowish-green, about 12 pairs of rounded lobes, wavy toothed margins, white hairs on lower surface; fer-

Figure 3–139 *Omphalodes verna* (0.5 life-size)

tile leaves about 12 inches long (sometimes to 2 1/2 feet), twice pinnately compound, pinnules rolled up into beadlike segments that open to discharge spores, colored dark brown or black in maturity.

PETIOLE: Usually somewhat longer than the leaf, yellow, few scaled, slightly furrowed, base brown and stout.

FRUIT: Spores are encased in hard, beadlike segments of fertile subleaflets.

CULTURE: Soil: Rich, acidic, highly organic loam is preferred, but this plant will also do well in light textured soils.

Moisture: Somewhat tolerant to drought, but best growth and appearance are exhibited when the soil is kept quite moist.

Light: Full sun to moderate shade.

PATHOLOGY: See **Ferns: Pathology.**

MAINTENANCE: Relatively maintenance free; old leaves from the previous fall can be cleaned up by mowing in early spring. Shearing back in midsummer (or mowing) induces new growth and increases the density of plantings. If grown in amended soils, an annual application of rotted leaf mold is a good practice.

PROPAGATION: See **Ferns: Propagation.**

***O**phiopogon* (ō-fi-ō-pō´gon) From Greek, *ophis*, a serpent, and *pogon*, a beard. This genus, commonly called lily-turf, is composed of about 10 species of evergreen, basal leaved, eventually sod forming, herbaceous, lilylike ground covers. They are native from India to Korea and Japan.

FAMILY: LILIACEAE: Lily family.

HABIT: Fountain shaped, grassy (or lilylike) ground covers.

LANDSCAPE VALUE: Use en masse for general cover in large, medium, or small areas. Often very attractive as a facing to trees or shrubs. Excellent as an edging along border, path, or drive. Widely used in Oriental garden settings. Limited foot traffic.

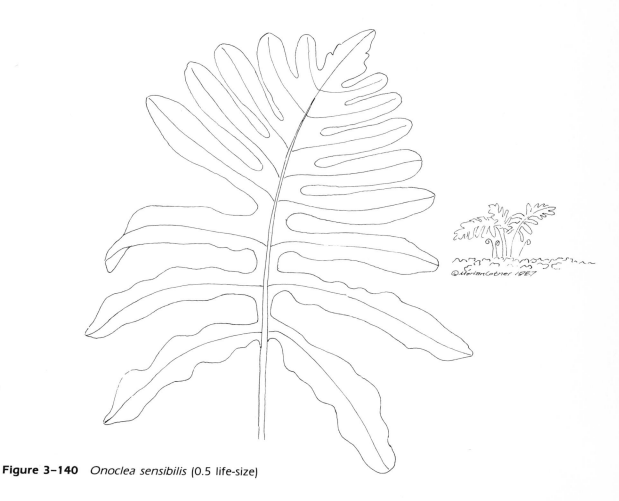

Figure 3–140 *Onoclea sensibilis* (0.5 life-size)

*   *   *

O. jaburan (jab-ū ´ran) Oriental name for the species (sometimes represented as *Liriope gigantea*).

COMMON NAME: White lily-turf, Jaburan lily-turf, snakebeard.

NATIVE HABITAT: Japan.

HARDINESS: Zone 7.

SIZE: 24 to 36 inches tall, tufted, nontuberous, spreading 6 to 8 inches across.

RATE: Moderate; space plants from pint- or quart-sized containers about 6 inches apart.

FOLIAGE: Evergreen, basal, grasslike, linear, to 2 feet long, sometimes somewhat more, 1/2 inch wide, dark green.

INFLORESCENCE: Short racemes with flowers in axillary fascicles; flowers 1/2 inch or more long, 6-parted, corolla white, effective in summer.

FRUIT: Berrylike, oblong, violet blue.

SELECTED CULTIVARS AND VARIETIES: 'Aureus' Leaves longitudinally striped with yellow. 'Variegatus' Leaves striped with white.

*   *   *

O. japonicus (jä-pon ´i-kus) Of Japan.

COMMON NAME: Dwarf lily-turf, Mondo grass.

NATIVE HABITAT: Japan and Korea.

HARDINESS: Zones 7 to 10.

SIZE: 6 to 15 inches tall, roots tuberlike, rhizomatous and stoloniferous, forming sods, spreading indefinitely.

RATE: Moderate; space plants from pint-sized containers 8 to 12 inches apart.

FOLIAGE: Evergreen, basal, linear, grasslike, to 15 inches or more in length, 1/8 inch wide or narrower, often curved, serrate, almost bristly margin, glabrous and scabrous, dark green.

INFLORESCENCE: Short, loose flowered raceme; flowers light lilac to white, 1/4 inch across, regular, 6 parted, effective early summer.

FRUIT: Globose capsule to about 1/4 inch across, colored blue.

SELECTED CULTIVARS AND VARIETIES: Var. *compacta* reaches 2 inches tall and 6 inches across. 'Nippon' Dark green foliage; reaching only 2 inches tall and spreading 3 to 4 inches across.

*   *   *

O. planiscapus 'Nigrescens' (plahn-i-skah ´pus) With flat (planelike) scape, (nī-gres ´enz) with blackish leaves (Figure 3–141).

COMMON NAME: Black lily-turf.

NATIVE HABITAT: Japan, the species native to there.

HARDINESS: Zones 6 to 10.

SIZE: 6 inches tall, rhizomatous, spreading indefinitely.

FOLIAGE: Evergreen, basal, grasslike, linear to about 12 or more inches long, about 1/4 inch across, many veined, dark, rather flat, black above (sometimes with a hint of purple), black and white (stomates) longitudinally striped below.

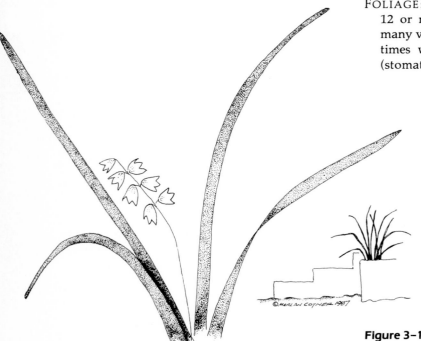

Figure 3–141 *Ophiopogon planiscapus nigrescens* (0.75 life-size)

INFLORESCENCE: Short racemes; flowers pinkish to purplish, about 1/4 inch long, borne in summer and attractive, but compared to the unusual foliage their ornamental value is little appreciated.

FRUIT: Blackish berries.

OF SPECIAL INTEREST: Exceptional and unique for its unusual foliage coloration. Outstanding contrast is attained when combined with light foliaged (especially those with yellow coloration) plants. Relatively difficult to obtain at present; however, in vitro culture techniques have been developed by Clarence Falstad at Walter's Gardens and could make this plant readily available.

Generally Applicable to All Species

CULTURE: Soil: Adaptable to most well-drained soils.

Moisture: Not notably tolerant to drought; soil should be maintained in a slightly moist condition with regular watering.

Light: Full sun to moderate shade; usually only good in full sun when grown along the coast; elsewhere, best in light to moderate shade. *O. jaburan* usually best with light to moderate shade.

PATHOLOGY: No serious diseases or pests reported.

MAINTENANCE: Little or no maintenance required.

PROPAGATION: Division: Simply divide offsets in spring or fall.

Seed: Ripe seed should be depulped, then stratified warm at 75 °F for 6 weeks prior to sowing.

✦✦✦✦✦✦✦✦✦✦✦✦✦✦✦✦✦

Origanum (ō-rig´á-num or or-i-gā´num) From Greek *oros*, a mountain, and *ganos*, beauty, the usual habitat and attributes of these plants. Composed of 15 to 20 species of annual, biennial, and perennial herbs and dwarf shrubs, *Origanum* is native to Europe and the Mediterranean region.

FAMILY: LAMIACEAE (LABIATAE): Mint family.

LANDSCAPE VALUE: Most commonly used as general covers in smaller locations. No foot traffic.

* * *

O. dictamnus (dik-tăm´nus) Old generic name, adopted from Virgil, plants of the genus once being common on Mt. Dicte (Figure 3–142).

COMMON NAME: Dittany of Crete, Crete dittany, hop marjoram.

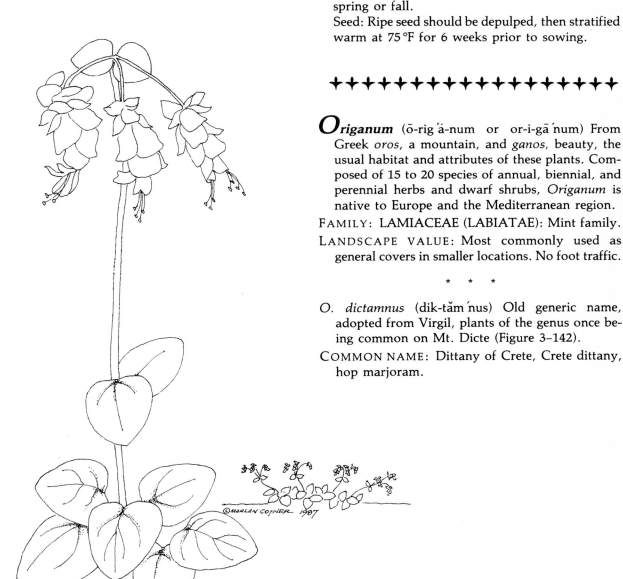

Figure 3–142 *Origanum dictamnus* (life-size)

NATIVE HABITAT: Greece and Crete.

HARDINESS: Zones 9 to 10.

HABIT: Herbaceous, procumbent ground cover.

SIZE: 12 inches tall, spreading 10 to 12 inches across.

RATE: Moderate; space plants from pint-sized containers 8 to 10 inches apart.

FOLIAGE: Opposite, simple, ovate to orbicular, to 1 inch long by 1 inch across, wooly, whitish gray-green, aromatic.

STEMS: To 12 inches, arching, squarish.

INFLORESCENCE: Terminal, loose panicles, spikelike, bracts to 3/8 inch long, showy purple, calyx shorter than bracts; corolla tubular, to 1/2 inch long, pink, effective late summer until fall, bracts persisting after corolla fades.

FRUIT: Of 4 nutlets, not ornamentally significant.

*   *   *

O. vulgare 'Compacta Nana' (vul-gā´rē) Common.

COMMON NAME: Creeping oregano, dwarf marjoram, dwarf oregano.

NATIVE HABITAT: The species is native to Europe.

HARDINESS: Zone 5.

HABIT: Low, creeping, herbaceous ground cover.

SIZE: 4 inches tall, spreading indefinitely.

RATE: Relatively slow to moderate; space plants from pint-sized containers 12 to 16 inches apart.

FOLIAGE: Opposite, simple, ovate to lanceolate-ovate, to about 3/4 inch long, often waxy margined, entire or somewhat toothed, glabrous or hairy, aromatic.

INFLORESCENCE: Corymbose or paniculate spikelets; bracts usually purple, corolla white to purplish, effective late summer.

FRUIT: Of 4 nutlets, not ornamentally significant.

Generally Applicable to Both Species

CULTURE: Soil: Adaptable to almost any well-drained soil, tolerant to infertility.
Moisture: Well adapted to withstand drought; only occasional watering in summer may be needed.
Light: Full sun.

PATHOLOGY: No serious diseases or pests reported.

MAINTENANCE: Mow in early spring to encourage vibrant new growth.

PROPAGATION: Cuttings: Take cuttings from nonflowering wood in mid to late summer. Treatment with 1000 ppm IBA/talc may be to some benefit.
Division: Simply divide plants in spring.

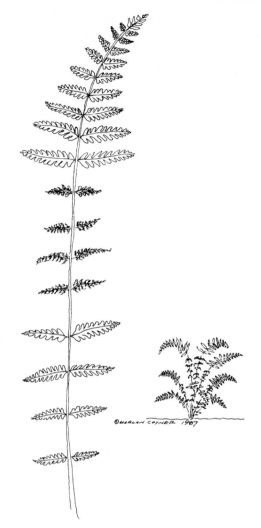

Figure 3–143 *Osmunda claytoniana* (0.25 life-size)

Osmunda (oz-mun´dȧ or os-) Various derivations (all legendary) have been proposed to explain this generic name. The most likely is the Saxon word signifying strength, *O. regalis* being tall and robust. Another explanation is that it is derived from Osmunder, one of the names of Thor, the Scandinavian god. This genus is composed of about 10 species of coarse textured ferns that are native to temperate and tropical areas of eastern Asia and North and South America.

FAMILY: OSMUNDACEAE: Osmunda family.

OF SPECIAL INTEREST: There are many myths associated with this genus. Biting the crosiers of osmunda was at one time believed to ensure an entire year without toothaches. The royal fern was supposedly good for healing injuries, especially those envolving the bones.

O. claytoniana (klā-tō-nē-ā´na) After John Clayton, an American botanist (1686–1773) (Figure 3–143).

COMMON NAME: Interrupted fern.

NATIVE HABITAT: North America and Asia.

HARDINESS: Zone 3.

HABIT: Spreading, medium-sized fern.

SIZE: 2 to 4 feet tall, spreading to 2 feet across.

RATE: Moderate; space plants from quart-sized containers 2 to 2 1/2 feet apart.

LANDSCAPE VALUE: Commonly used for accent with woodland plants in a naturalized setting. Good for use along the bank of a stream or pond or en masse as a general cover in a naturalized area.

FOLIAGE: Deciduous, dark green; sterile leaves to 4 feet long but usually shorter, twice pinnately compound, lanceolate in outline, arching, leaflets lanceolate and tapering to a blunt tip, ascending, with 18 or more pairs; fertile leaves similar, but more erect and taller with 4 or more pairs of dark green leaflets in the middle that bear spore cases in dense cylindrical clusters that become dark brown at maturity, leaflets below and above the fertile leaflets are similar to those of sterile leaves; leaflets in basal portion spaced farther apart than those above fertile leaflets.

PETIOLE: Round, somewhat furrowed, green, stout.

SPORES: Bright green spores that are short lived with chlorophyll, encased; spore cases short stalked, relatively large, clustered, green turning brown.

CULTURE: Soil: Adaptable to nearly any soil; uncommonly quite tolerant of dry, stony soils. A pH that is mildly acidic is preferred.
Moisture: Relatively tolerant to drought, but periodic watering throughout the summer months is a good practice as appearance is best if soil is kept slightly moist.
Light: Full sun to moderate shade.

PATHOLOGY: See **Ferns: Pathology.**

MAINTENANCE: Annual topdressing with compost will keep soil supplied with nutrients.

PROPAGATION: See **Ferns: Propagation.**

*O*steospermum (ost-ē-ō-sperm´um) Derived from the Greek word *osteon*, a bone, and *sperma*, seed, likely in reference to the hardness of the seed. This genus of mostly South African plants consists of about 70 species of herbs (both annual and perennial), subshrubs, and shrubs.

FAMILY: ASTERACEAE (COMPOSITAE): Sunflower family.

O. fruticosum (frō-ti-kō´sum) Shrubby, the habit (Figure 3–144).

COMMON NAME: Trailing African daisy, freeway daisy.

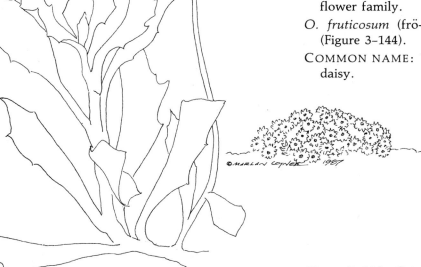

Figure 3–144 *Osteospermum fruticosum* (life-size)

NATIVE HABITAT: South Africa.

HARDINESS: Zones 9 to 10.

HABIT: Thick, mat forming, trailing, prostrate to decumbent, herbaceous ground cover.

SIZE: 6 to 14 inches tall, spreading to 4 or more feet across.

RATE: Moderate to relatively fast; space plants from 2 1/4 inch diameter containers 12 to 20 inches apart.

LANDSCAPE VALUE: Excellent as a general cover or soil stabilizer on slopes of moderate to large size. Useful along the shore as well as inland. Often seen along highways in the Los Angeles area. No foot traffic. Considered relatively fire retardent and salt tolerant.

FOLIAGE: Evergreen, simple, alternate, obovate to spatulate, to about 2 inches long and 3/4 inch across, sessile or nearly so, thick, almost succulent appearing, gray-green, glandular hairy, becoming glabrous with age.

STEMS: Trailing and rooting.

INFLORESCENCE: Heads to 2 inches across, elevated on peduncles that may reach 4 inches tall; disc flowers lilac-purple, ray flowers lilac on upper surface, which soon fades to white, deeper lilac on lower surface, effective and very showy late fall to early spring, but also sporadically during other seasons.

FRUIT: Smooth, triangular achene, not ornamentally significant.

SELECTED CULTIVARS AND VARIETIES: 'African Queen' Leaves deeper green; flowers deep purple, habit similar but somewhat taller than the species. 'Burgundy Mound' With deep purple flowers. 'Snow White' ('White Cloud') Said to be a hybrid; flowers showy white in late winter and early spring, somewhat taller and more upright in comparison to the species.

CULTURE: Soil: Adaptable to a wide range of soil types, provided drainage is good. Prefers moderately fertile loam.
Moisture: Very tolerant of drought once established; however, occasional thorough watering is needed to maintain the best appearance.
Light: Full sun.

PATHOLOGY: Diseases: Damping off can be a problem if soil drains poorly or plants are frequently overwatered.
Pests: No serious pests reported.

MAINTENANCE: Trim back stems as they outgrow their bounds.

PROPAGATION: Cuttings: Cuttings root quite well under mist. The addition of a mild rooting compound is to some benefit.
Division: Rooted stem sections are easily dug and transplanted, usually during winter or early spring.

✦✦✦✦✦✦✦✦✦✦✦✦✦✦✦✦

O*xalis* (ok´sȧ-lis or oks-ā´lis) From Greek, *oxis*, acid, alluding to the acidity of the leaves of many species. This huge genus is made up of about 850 species of small, annual, and perennial herbs and suffrutescent (woody base with herbaceous shoots) subshrubs, and rarely shrubs. They are native to various regions on all continents, yet primarily they come from South Africa and South America.

FAMILY: OXALIDACEAE: Wood sorrel family.

O. oregana (ôr-ē-gā´nȧ) From Oregon (Figure 3–145).

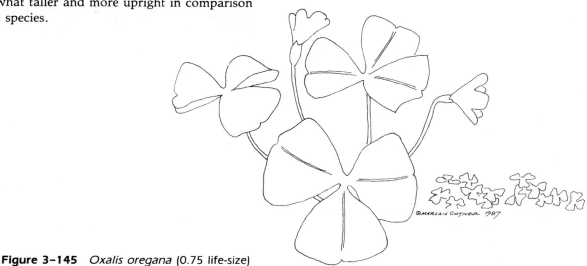

Figure 3–145 *Oxalis oregana* (0.75 life-size)

COMMON NAME: Oregon oxalis, redwood sorrel.

NATIVE HABITAT: Pacific northwest coastal areas of the United States.

HARDINESS: Zones 9 to 10.

HABIT: Low, herbaceous, rhizomatous, more or less stemless, creeping cloverlike ground cover.

SIZE: To 10 inches tall, spreading indefinitely.

RATE: Relatively slow; space plants from 2 to 4 inch diameter containers 10 to 16 inches apart.

LANDSCAPE VALUE: Fine for use as edging along shaded walkway, especially through naturalized setting. As a companion to ferns or rhododendrons as facing, there are few rivals. It may also be used as general cover in areas of all sizes. No foot traffic.

FOLIAGE: On long petioles of 6 to 8 inches, cloverlike, to 4 inches across with 3 leaflets, each broadly obcordate to 1 1/2 inches long and wide, colored medium green, midrib lighter green.

FLOWERS: Solitary, on peduncles that may be 6 inches long, about 1 inch across; corolla 5-petaled and colored pink, white, or rose, blooming in spring and sometimes again in fall.

FRUIT: Capsule, of no ornamental significance.

CULTURE: Soil: Needing organically rich, moderately acidic soil.
Moisture: Not notably tolerant to drought; soil should be kept moist with regular watering and plants should be located in a cool location that is sheltered from strong winds.
Light: Moderate to dense shade.

PATHOLOGY: Diseases: Beet curly top virus, seed smut, root rot, red rust and other rusts, leaf spot. Pests: No serious pests reported.

MAINTENANCE: Little or no special maintenance needed.

PROPAGATION: Division: Divide clumps in spring or fall and water regularly until roots are well established.

✛✛✛✛✛✛✛✛✛✛✛✛✛✛✛✛✛✛

*P*achysandra (pak-i-san´dra) From Greek *pakys*, thick, and *aner*, a man (alluding to the unusually thick stamens). This genus is composed of only five species of herbaceous perennials and subshrubs.

FAMILY: BUXACEAE: Box family.

LANDSCAPE VALUE: Both species of pachysandra that are discussed here are excellent, reliable covers for medium to large areas. They grow well on shaded banks and under trees, make nice edging along walks, and combine well with dwarf conifers, rhododendrons, pieris, and many other shrubs. Both tolerate very infrequent foot traffic. *P. terminalis* is likely the most versatile and widely used ground cover in the United States.

*   *   *

P. procumbens (prō-kum´benz) Procumbent, that is, trailing without rooting.

COMMON NAME: Allegheny pachysandra, Allegheny spurge.

NATIVE HABITAT: Southeastern United States.

HARDINESS: Zones 4 to 9.

HABIT: Herbaceous, clump forming, procumbent, dense cover.

SIZE: 6 to 12 inches tall, spreading 4 or more feet across.

RATE: Relatively slow growing; space 10 to 16 inches apart from pint- to quart-sized containers.

LEAVES: Semievergreen, simple, alternate, crowded toward end, broadly ovate to suborbicular, to 3 1/2 inches long, toothed apically, narrowing to long petiole, medium rather dull green, becoming reddish-green in autumn, very pleasant soft texture.

STEMS: Procumbent reddish-green, somewhat rough to the touch.

FLOWERS: In spikes to 5 inches long, male above female, apetalous; male flowers with four stamens, female with 4 to 6 sepals and 3 or 4 stigma, sometimes pungent smelling; greenish or purplish borne near the base of stems in spring, prior to leaves in areas where deciduous.

FRUIT: Drupaceous, mottled green and purple, 3-parted.

*   *   *

P. terminalis (tĕr-mi-nā´lis) Flowers terminally located (Figure 3–146).

COMMON NAME: Japanese spurge, Japanese pachysandra.

NATIVE HABITAT: Japan.

HARDINESS: Zones 5 to 9.

HABIT: Semiwoody, rhizomatous, spreading, dense cover.

SIZE: 6 to 12 inches, spreading indefinitely.

RATE: Relatively slow to moderate growing; space 6 to 10 inches apart from 2 to 3 inch diameter containers.

LEAVES: Evergreen, alternate, crowded at ends, obovate, 2 to 4 inches long, 1/2 to 1 1/2 inches wide, toothed above the middle, glabrous, three

Figure 3–146 *Pachysandra terminalis* (0.75 life-size)

nerved at base, dark shiny green above, dull green below, petioles 2/5 to 1 1/4 inches long.

STEMS: Greenish-yellow, erect, 6 to 12 inches long.

INFLORESCENCE: Spikelike, 1 to 2 inches long, apetalous, terminally borne; flowers separated, male above female, female with two stigmas, 4 to 6 sepals, white; male flowers of four stamens, pleasantly fragrant, borne in early spring. Interestingly, plants that bloom branch more than those without flowers, as apical dominance seems to be destroyed.

FRUIT: Seldom produced and is naturally self-sterile in monoclonal stands. Fruit is produced when various clones are interplanted and appears as a green, berrylike drupe, which turns white and reaches about 3/8 inch across.

SELECTED CULTIVARS AND VARIETIES: 'Emerald Carpet' See 'Green Carpet.'
'Green Carpet' More compact with leaves often broader, margin frequently entire or teeth fewer. Seems to retain its color better through winter and in sun.
'Kingwood' Selected by Logan Monroe of Kingwood Nursery in Ohio. This cultivar is medium green and glossy, easily distinguished by the deeply cut teeth on the apex of the leaves. Texture is rather soft and lacy; a fine selection.
'Variegata' Leaves smaller and variegated, creamy white. An excellent selection for a shady or sunny location; with slower growth than the species. Sometimes listed by the cultivar name 'Silver Edge.'

Generally Applicable to Both Species

CULTURE: Soil: Best growth is in well-drained organically rich, loamy soil, pH from 5.0 to 7.0. Moisture: Not notably drought tolerant. Soil should be kept moist with periodic watering, especially if planted in areas exposed to drying winds.
Light: Best in light to dense shade. Plants become chlorotic when grown in full sun. The cultivar *P. terminalis* 'Variegata' seems to perform well in sun provided it is grown in rich soil and supplied adequate water.

MAINTENANCE: Little or no maintenance is required. Occasionally, plants will come up underneath edging and require irradication from lawn areas.

PATHOLOGY: Diseases: Fungal leaf blight. *P. procumbens* is said to be resistant.

Pests: Two spotted mite and others, northern root knot nematode, occasionally slugs and aphids.

PROPAGATION: Cuttings: Make cuttings out of new season's terminal growth. No node is necessary and plants respond very well to 1000 to 3000 ppm IBA/talc or solution. Rooting takes about 4 to 6 weeks at 70°F.

Rhizomes: Rhizomes may be cut into 1 to 2 inch segments, covered with 1/2 inch of peat/vermiculite, 3 : 1, and kept moist. Roots and shoots form in 1 to 2 months. Treatment with auxin has been found inhibitory to shoot formation.

Division: Simply divide the plants during spring or fall. *P. procumbens* roots rather sparsely and is usually best propagated by division.

***P**arthenocissus* (pär-then-o-sis´us) Meaning unknown. This genus, commonly called woodbine, is composed of about 15 species of deciduous, woody vines that climb by tendrils that are frequently disclike at their ends. They are native to temperate regions of North America and Asia.

FAMILY: VITACEAE: Grape family.

P. quinquifolia (kwin-kwe-fō´li-à) Five-leaved, referring to 5 leaflets per leaf (Figure 3–147).

COMMON NAME: Virginia creeper, woodbine, American ivy, five-leaved ivy.

NATIVE HABITAT: Woodlands of eastern United States.

HARDINESS: Zones 3 to 9.

HABIT: Low, woody, creeping vine that makes a dense, rather coarse ground cover when unsupported.

SIZE: 3 to 8 inches tall (without support), spreading indefinitely.

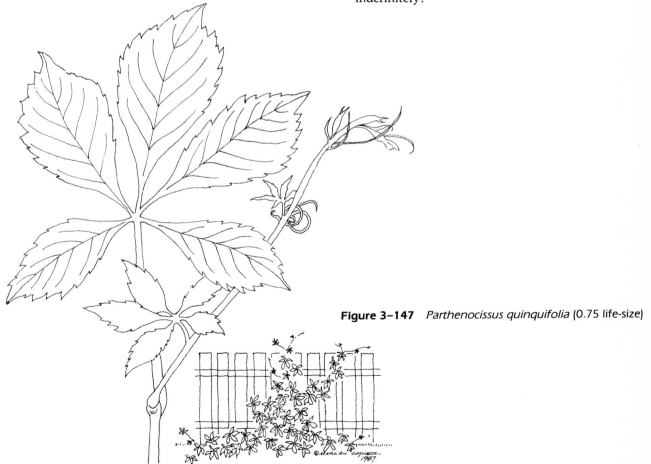

Figure 3–147 *Parthenocissus quinquifolia* (0.75 life-size)

RATE: Relatively fast; space plants from gallon-sized containers, 3 to 4 feet apart.

LANDSCAPE VALUE: Useful as general cover for large areas; especially good on naturalized slopes for erosion control. No foot traffic. Tolerates salt spray.

FOLIAGE: Deciduous, alternate, petioled about 1/3 inch long, palmately compound; 5 leaflets, each elliptic-ovate, to 6 inches long by up to 2 1/2 inches across, acuminate, coarsely toothed, shiny green above, paler green below, waxy bronze to reddish in youth, purplish-red to crimson in early autumn.

INFLORESCENCE: Cymes that usually form terminal panicles; flowers small, 5 parted, petals greenish-white, obscured by foliage, insignificant ornamentally, born early to midsummer.

FRUIT: 1/4 inch diameter, bluish-black, bloomy berry, effective in fall as it persists after the foliage has dropped.

OF SPECIAL INTEREST: Many species of birds favor the fruit of Virginia creeper. They include the eastern kingbird, pileated woodpecker, the great creasted flycatcher, brown thrasher, American robin, chickadees, scarlet tanagers, and others.

SELECTED CULTIVARS AND VARIETIES: 'Englemanii' With leaflets that are smaller than the species. 'Saint Paulii' With smaller leaflets that are hairy below, and often with arial rootlets in addition to tendrils, which accounts for its greater ability to cling than the species.

CULTURE: Soil: Adaptable to most well-drained soils; pH best in range from 5.0 to 7.0.
Moisture: Fairly tolerant to drought, but best when soil is kept moist with occasional watering in the heat of summer.
Light: Full sun to moderate shade.

PATHOLOGY: Diseases: Canker of stem, downy mildew, leaf spots, powdery mildew, wilt, anthracnose, blight, root rot.
Pests: Beetles, eight spotted forester, leaf hoppers, several types of scales, aphids, yellow woolybear, mites.

MAINTENANCE: Cut back stems as they outgrow bounds or as they attempt to climb upon shrubs. *Parthenocissus* generally does not harm trees, but rather enhances their appearance, especially in autumn when it turns red.

PROPAGATION: Cuttings: Softwood cuttings root easily in early summer.
Layering: Plants are easily started from leaf and node sections that have naturally rooted.
Seed: Stratify seed at 40°F for 3 months, then sow in spring. Commercial supplies are available.

✦✦✦✦✦✦✦✦✦✦✦✦✦✦✦✦✦

***P**axistima* (păks-i´sti-mȧ) From Greek, *pachys*, thick, and *stigma*, the female reproductive apparatus, alluding to thick stigmas. Containing two species of North American evergreen subshrubs, this genus was formerly called Pachystima and likely will be changed back to this name in the near future.

FAMILY: CELASTRACEAE: Staff-tree family.

HABIT: Low, dense spreading to decumbent, woody, shrubby ground covers.

RATE: Relatively slow to moderate; space plants from quart- to gallon-sized containers 12 to 24 inches apart.

LANDSCAPE VALUE: Useful as general covers for moderate-sized areas and as a facing to ericaceous plants such as rhododendron, pieris, and azalea. *P. myrsinites* is occasionally put to use as a low hedge. No foot traffic.

*   *   *

P. canbyi (kan´bē-ī) After William Marriot Canby (1831–1904), who discovered this species (Figure 3–148).

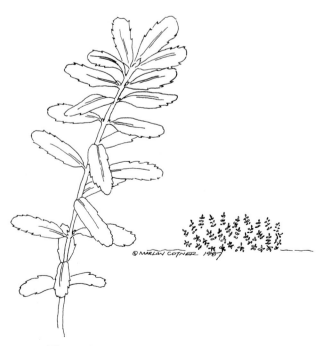

Figure 3–148 *Paxistima canbyi* (life-size)

COMMON NAME: Canby paxistima, ratstripper, cliff-green, mountain lover.

NATIVE HABITAT: Eastern North America, central Appalachian Mountains.

HARDINESS: Zones 4 or 5 to 6 or 7; zone 4 northward if snowcover is reliable, and zone 7 southward in cool microclimates.

SIZE: 12 inches tall, spreading 10 to 14 feet across.

FOLIAGE: Evergreen, opposite, simple, glabrous, linear-oblong, to 1 inch long and 3/16 or less across, apex obtuse, base revolute, margin minutely serrulate toward apex, short petioled, rather coarse in texture.

STEMS: Decumbent to ascending, to 16 inches long, slender, glabrous, brownish, spongy, somewhat 4-sided.

INFLORESCENCE: Few flowered cymes; flowers perfect, small, 4-parted, corolla greenish or reddish, 1/5 inch across, borne in late spring, but not very showy.

FRUIT: White, leathery capsule, about 1/16 inch long, not of ornamental significance.

* * *

P. myrsinites (mur-sin-it´ēz) Like *Myrtus*, the foliage.

COMMON NAME: Myrtle paxistima, Oregon paxistima, Oregon boxwood.

NATIVE HABITAT: West coast of North America from British Columbia to California and New Mexico.

HARDINESS: Zones 5 to 7.

SIZE: To 18 inches tall, spreading to 3 feet across.

FOLIAGE: Evergreen, opposite, simple, glabrous, ovate, oblong, or lanceolate, to 1 1/4 inch long, apex acute to obtuse, margin finely serrate along the upper half, dark shiny green, leathery.

STEMS: Prostrate, to 40 inches long.

INFLORESCENCE: Axillary cymes or flowers solitary; flowers perfect, 4-parted, purplish, born late spring, not particularly valuable ornamentally.

FRUIT: White, 2-valved capsules, not of ornamental worth.

Generally Applicable to Both Species

CULTURE: Soil: Adaptable to most well-drained soils. The pH is best when between 4.5 and 6.0. At a higher pH, the foliage quickly becomes chlorotic and growth is slowed. Tolerant to rocky soil.
Moisture: Very tolerant to moderate periods of drought, but best appearance is attained when soil is kept slightly moist with watering as needed.

Light: Full sun to light shade. Most attractive and dense growth attained in full sun.

PATHOLOGY: Diseases: Leaf spot; root rot can be counted on if planted in soil that does not drain adequately.
Pests: Scale insects are occasionally troublesome as are mites.

MAINTENANCE: Relatively little maintenance is required. A light shearing in spring will help keep plantings dense and compact.

PROPAGATION: Cuttings: Softwood cuttings root relatively easily, yet slowly under mist. Be sure medium is very well drained (50 percent peat plus perlite or sand) and turn mist off as soon as rooted. Hardwood cuttings can also be rooted. Make them in fall. Whether the cuttings are made from hard or soft wood, they root better if a rooting preparation of 3000 ppm IBA is applied and heat supplied to the bottom so as to keep medium temperature at about 70°F.

✚✚✚✚✚✚✚✚✚✚✚✚✚✚✚

*P*elargonium (pel-är-gō´ni-um) From Greek, *pelargos*, a stork, the ripe seed head supposedly resembling the head and beak of a stork. This genus is commonly called geranium, and, unlike the genus *Geranium*, is composed of about 280 species of those plants that most people recognize as geraniums. Commonly used throughout the country as house plants, these annual and perennial herbs are primarily native to South Africa.

FAMILY: GERANIACEAE: Geranium family.

P. peltatum (pel-tā´tum) Shieldlike, the leaf shape (Figure 3–149).

COMMON NAME: Ivy geranium, trailing geranium, hanging geranium.

NATIVE HABITAT: South Africa.

HARDINESS: Zones 9 to 10.

HABIT: Herbaceous, trailing, carpeting, ground covers.

SIZE: 12 to 24 inches tall by 3 feet or more across.

RATE: Relatively fast; space plants from pint- or quart-sized containers 16 to 24 inches apart.

LANDSCAPE VALUE: Very attractive, bright and lively general covers for almost any wide open area. Excellent when planted on gently sloping banks. No foot traffic. Best where temperature is usually not higher than 80°F.

FOLIAGE: Evergreen, alternate, peltate (shield shaped), broadly ovate to orbicular in outline, shallowly 5 lobed, lobes pointed, margin entire, 2 to 3 inches long and wide, leathery, venation conspicuous, succulent, light to medium, vibrant green.

STEMS: Trailing to 4 feet or more, squarish, sparsely hairy, rough, glandular dotted, color similar to foliage.

INFLORESCENCE: 5 to 7 flowered axillary umbels, 2 to 3 inches across, on long stems; flowers short pedicelled, regular, 5-petaled, rose-carmine varying to white, upper flowers dark veined, blooming continually from late winter to early fall if soil is not allowed to dry out.

FRUIT: 5 valved, dehiscent, shaped like a stork's bill, not of ornamental significance.

SELECTED CULTIVARS AND VARIETIES: 'Charles Monselet' Flowers red.
'Charles Turner' Flowers rose-pink.
'Comtesse Degrey' Flowers salmon pink.
'Giant Lavender' Flowers lavender with red markings.
'Jeanne d'Arc' Flowers lavender with darker striping.
'L'Elegante' Foliage variegated; flowers pink.
'Riga' Flowers rosy-pink and abundant.
'Tavira' Flowers profuse, bright red.

OTHER SPECIES: Many other species of geraniums exist and vary in their ability to form a dense cover. Of the most interesting plants on earth are those geraniums with scented foliage. Commonly grown as house plants and available in most garden centers, these herbs have foliage that is scented in the most peculiar fragrances. The rubbed leaf of scented geraniums may reveal odors ranging from rose to lemon to pineapple or even chocolate-mint. Such geraniums, although most often grown as pot plants or annuals in northern climatic zones, might also be considered for small-scale ground covers in southern states.

CULTURE: Soil: Adaptable to most well-drained soils and a wide pH range.
Moisture: Not notably tolerant to drought; soil should be kept constantly moist (but not saturated) with regular watering.
Light: Full sun; can tolerate more heat if located in an area of partial afternoon shade.

PATHOLOGY: Bacterial leaf spot, crown gall, bacterial fasciation, alternaria leaf spot, black stem rot, blossom blight, leaf spots caused by fungi, root rots, rust; viral leaf curl, mosaic, and wilt.
Pests: Aphids, caterpillars, geranium plume moth, mealybugs, four lined plant bug, black vine weevil, greenhouse whitefly, eastern subterranean termite, southern root knot nematode, northern root knot nematode, Japanese weevil, slugs, scales, mites.
Physiological: Oedema.

PROPAGATION: Cuttings: Cuttings from leafy stem sections are easily rooted in a porous medium under mist. Pathogen-free stock plants should be carefully maintained.
Seed: Commercial supplies are available. Seed germinates well at temperatures around 70 °F.

✦✦✦✦✦✦✦✦✦✦✦✦✦✦✦✦✦

*P*ernettya (pĕr-net´i-à) Named for Don Pernetty, author of a book about the Falkland Islands. This genus is made up of about 25 species of low growing evergreen shrubs that are native to New Zealand and Tasmania and from Mexico to southern South America.

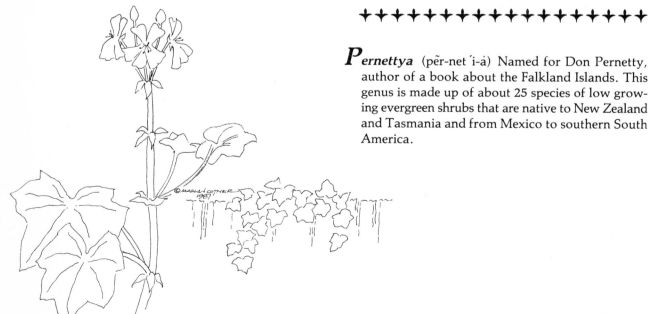

Figure 3–149 *Pelargonium peltatum* (0.5 life-size)

OF SPECIAL INTEREST: The leaves and nectar of *Pernettya* species are both reported to be poisonous. Ingestion has been associated with such symptoms as burning in the mouth, diarrhea, headache, weakness, and cardiac maladies. Severe cases have resulted in coma or death. The causitive agents reportedly are grayanotoxins.

*   *   *

P. mucronata (mū-krō-na´tȧ) Leaves ending in a point; or as its said in Latin, *mucro* (Figure 3–150).

COMMON NAME: Chilean pernettya.

NATIVE HABITAT: Chile and Argentina.

HARDINESS: Zones 6 or 7.

HABIT: Low, rhizomatous, spreading shrub.

SIZE: Ranging from 18 to 36 inches tall, spreading to 3 feet across.

RATE: Moderate; space plants from gallon-sized containers 2 to 2 1/2 feet apart.

LANDSCAPE VALUE: Good for edging along path or leading into building entrance. Sometimes used to good effect as an edging around the perimeter of a swimming pool, pond, or stream. Exceptional for colorful fruit display. No foot traffic.

FOLIAGE: Evergreen, alternate, simple, ovate-lanceolate to ovate, to 3/4 inch long by 3/8 inch across, apex sharply pointed, somewhat revolute serrate margin, glabrous and glandular dotted both sides, midrib conspicuous, dark glossy green becoming bronzy in cold weather.

STEMS: Rigid, many branched, to 3 feet, finely hairy, reddish on upper side.

FLOWERS: Solitary, regular, perfect, axillary, calyx 5 parted; corolla gamopetalous (margins of petals fused toward base forming a tube), 5 lobed, 1/4 inch long, white to pink, effective spring and summer.

FRUIT: Very showy, metallic, variously colored, persistent (from fall to spring), about 1/2 inch diameter globose berry. Different clones must be present to ensure fruit production as cross-pollination is needed. The exceptional fruit of this species is believed by many to be the most outstanding in existence for ornamental effect.

SELECTED CULTIVARS AND VARIETIES: Not surprisingly, there are many selections that differ in the color and size of their fruit.
'Alba' With white fruit that is slightly pink-tinged.
'Cherry Ripe' Berries large and bright cherry-red.
'Coccinea' With scarlet fruit.
Davis's hybrids Hybrid group with large and variously colored fruit.

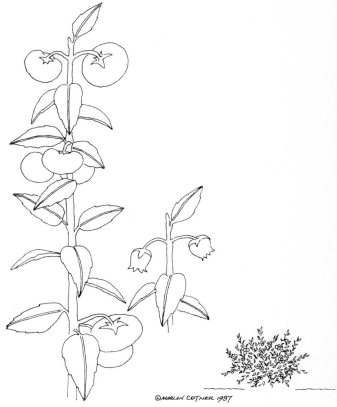

©MARLAN COTNER 1987

Figure 3–150 *Pernettya mucronata* (life-size)

'Lilacina' Fruit lilac-purple.
'Pink Pearl' Berries medium sized and lilac-pink.
'Purpurea' Fruit violet-purple.
'Rosie' Young stems red, foliage dark green, berries pinkish-rose and large.
'Sea Shell' Berries pinkish, ripening to rose.
'Thymifolia' Reportedly a sterile nonfruiting clone. It is noteworthy for its diminutive habit and many white flowers.
'White Pearl' Reported to be a sport of 'Alba' with medium to large, shiny white berries.

*   *   *

P. tasmanica (tas-man-i´kȧ) From Tasmania.

COMMON NAME: Tasmania pernettya.

NATIVE HABITAT: Tasmania.

HARDINESS: Zone 7.

HABIT: Low, creeping, dwarf, usually prostrate, woody shrub.

SIZE: 1 to 3 inches tall.

RATE: Relatively slow to moderate; space plants from quart-sized containers 10 to 12 inches apart.

LANDSCAPE VALUE: Use as a general cover for

small- or moderate-sized areas. Looks neat and attractive as an edging leading up to the entrance of a building.

FOLIAGE: Evergreen, alternate, simple, oblong, to 5/16 inch long, somewhat crenate margins, pointed apex, bright green, leathery.

STEMS: Slender, wiry, creeping.

FLOWERS: Solitary, axillary, to 1/4 inch long, 5 parted, corolla campanulate, white, effective in early summer.

FRUIT: Berry to about 3/8 inch across, colored bright red.

Generally Applicable to Both Species

CULTURE: Soil: Best in acidic, loamy soil with good drainage.
Moisture: Able to tolerate short periods of drought, but best when soil is kept moist with regular watering.
Light: In hot climates, plants should be placed so as to be shaded in the afternoon. In moderate climates, plant in full sun.

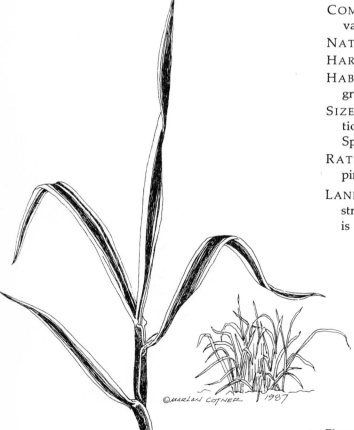

PATHOLOGY: No serious diseases or pests reported.

MAINTENANCE: Occasionally prune lightly to keep plants looking dense and compact.

PROPAGATION: Cuttings: Softwood cuttings root readily.
Division: Suckers with roots can be divided and transplanted in early spring or winter.
Layering/Seed: Both have been used in the past, but generally are not practiced on a large scale.

*P*halaris (fal´ă-ris or fa´lăr-is) Ancient Greek name, possibly from *phalaros*, shining, referring to the polished seeds. This genus is made up of about 15 species of annual and perennial grasses that originated in North America, Europe, and North Africa.

FAMILY: POACEAE (GRAMINEAE): Grass family.

P. arundinacea var. *picta* (a-run-di-nā´sē-ā) From Latin, *aurndo*, reedlike (Figure 3–151). Common to see listed under any of the following: *P. aurundinacea* 'Variegata,' 'Tricolor,' var. *elegantissima*.

COMMON NAME: Ribbon grass, gardener's garters, variegated reed canary grass.

NATIVE HABITAT: Europe and North America.

HARDINESS: Zone 3.

HABIT: Upright stemmed, rhizomatous, spreading grass.

SIZE: 2 to 3 feet tall, varying with cultural conditions; taller in shade and with moist, fertile soil. Spreading indefinitely.

RATE: Relatively fast to invasive; space plants from pint- or quart-sized containers 14 to 24 inches apart.

LANDSCAPE VALUE: Used for erosion control along stream banks, open slopes, and in ditches, this plant is both attractive and effective. Interesting as a

Figure 3–151 *Phalaris arundinacea* var. *picta* (life-size)

general cover, accent plant in border, or as facing to tall shrubs, *provided spread can be controlled.* Tolerates only infrequent foot traffic.

FOLIAGE: Deciduous, flat, 6 to 12 inches long by 1/2 to 3/4 inch across, rough to the touch, longitudinally striped white alternating with green, sometimes with hints of purplish-red or yellow. When grown in dry conditions, bottom leaves often turn brown.

INFLORESCENCE: Lanceolate panicle that is loose at the base and becomes denser distally; blooming in summer, not ornamentally significant.

CULM (STEM): Erect or bent at base, becoming erect, 4 to 6 swollen nodes.

FRUIT: Caryopsis, not significant ornamentally.

CULTURE: Soil: Adaptable to most soil types; growth will be slower and more dense (both desirable) if grown in heavy clay or sandy infertile soils. Moisture: Best when grown in either very wet or very dry conditions (at least from the standpoint of minimizing maintenance), as growth will be slower and more dense. If grown in moist, rich loam, the plantings will often open up and become extremely vigorous, and leaves may lose much of their variegation. Soil fertility, regardless of soil type, is best kept low.
Light: Full sun to moderate shade.

PATHOLOGY: No serious diseases or pests.

MAINTENANCE: Mow in spring to rid plantings of dead stems and leaves from the previous year. To do so periodically during the growing season will help to keep plantings dense, neat, and compact.

PROPAGATION: Cuttings: Cuttings can be prepared of rhizome shoots. Make them about 2 inches long; do not treat with a root inducing preparation as it is likely to repress shoot formation.
Division: Divide clumps any time.

Phlox (floks) From Greek, *phlego*, to burn, or *phlox*, a flame, in reference to the brightly colored flowers. Very common among herbaceous perennial enthusiasts, this genus is composed of about 60 species of annual and perennial herbs and subshrubs that are native to North America, with one species from Siberia. Many are noteworthy for their profuse, bright colored blossoms that commonly bloom so densely that they completely obscure the foliage.

FAMILY: POLEMONIACEAE: Phlox family.

LANDSCAPE VALUE: Generally useful in rock gardens as specimens or for edging along foreground of perennial border, path, or walkway. Quite striking as general covers in small- to moderate-sized areas. Often they work well in front of the sunny side of a low hedge as a facing. No foot traffic.

FRUIT: 3 valved capsule, not ornamentally significant.

OF SPECIAL INTEREST: The planting of phlox is common among gardeners who are interested in attracting hummingbirds and butterflies.

* * *

P. divaricata (di-văr´i-kā-tà) From Latin, *divarico*, spreading or to stretch out, referring to the growth habit.

COMMON NAME: Wild blue phlox, wild sweet william, blue phlox.

NATIVE HABITAT: Western United States.

HARDINESS: Zone 3.

HABIT: Semiwoody, low spreading, mat forming ground cover.

SIZE: 8 to 12 inches tall, spreading 18 to 36 inches across.

RATE: Moderate; space plants from pint- or quart-sized containers 14 to 20 inches apart.

ADDITIONAL LANDSCAPE VALUE: Makes a nice accent plant when mixed with lower covers such as *Vinca minor.*

SPECIAL CULTURE: Moisture: Watering should be frequent enough to maintain soil in a slightly moistened condition.
Light: Full sun to moderate shade.

FOLIAGE: Opposite with terminal leaves often becoming alternate, simple, entire, ovate to oblong, to 2 inches long, apex acute, more rounded on sterile shoots than reproductive shoots.

STEMS: Decumbent, rooting, to 1 1/2 feet long.

INFLORESCENCE: Terminal cymes or solitary; flowers showy, about 1 inch across, perfect, regular, 5-parted; corolla pale violet-blue to lavender, salverform, lobes flat and spreading, usually notched or erose (margin irregularly incised as if chewed by an insect), effective midspring to early summer, only mildly fragrant.

SELECTED CULTIVARS AND VARIETIES: 'Fullers White' Flowers white, reported to be more tolerant of shade than the species.
'Lamphami' With fragrant lavender-blue flowers that have rounded, obtuse, entire corolla lobes; ovate leaves. More vigorous with longer flowering period.

*   *   *

P. nivalis (ni-vā´lis) Noble, famous.

COMMON NAME: Trailing phlox.

NATIVE HABITAT: Eastern United States from Virginia to Florida.

HARDINESS: Zones 6 to 9.

HABIT: Semiwoody, prostrate spreading ground cover.

SIZE: 12 inches tall, spreading to 18 inches across.

RATE: Moderate; space plants from pint- to quart-sized containers 10 to 12 inches apart.

SPECIAL CULTURE: Moisture: Fairly tolerant to drought, but best when ground is kept slightly moist with periodic watering.
Light: Full sun to light shade.

FOLIAGE: Simple, opposite, sometimes becoming alternate terminally, subulate to linear-subulate (awl shaped), to 1/2 inch long.

STEMS: Decumbent, to 6 inches long.

INFLORESCENCE: Terminal cymes, more or less erect, occasionally flowers are solitary; flowers regular, about 1 inch across, corolla lobes entire to erose or shallowly notched, purple to pink or white, blooming early summer and often again in early fall.

SELECTED CULTIVARS AND VARIETIES: 'Camla' With salmon-pink flowers.
'Gladyn' White flowers with repeated flowering in fall.

*   *   *

P. × procumbens (prō-kum´benz) Meaning trailing without rooting.

COMMON NAME: Trailing phlox, hybrid phlox.

HYBRID ORIGIN: *P. stolonifera* × *P. subulata*.

HARDINESS: Zone 5.

HABIT: Procumbent, semiwoody, mat forming ground cover.

SIZE: 10 to 12 inches tall, spreading 12 to 18 inches across.

RATE: Moderate; space plants from pint- or quart-sized containers 10 to 12 inches apart.

SPECIAL CULTURE: Moisture: Best when soil is only slightly moist; good to let soil dry somewhat between infrequent deep waterings.
Light: Full sun to light shade.

FOLIAGE: Simple, opposite, elliptic or oblanceolate, to 1 inch long.

FLOWERS: Bright purple, about 3/4 inch across, effective mid to late spring.

*   *   *

P. stolonifera (stō-lon-if´ẽr-á) Having stolons.

COMMON NAME: Creeping phlox.

NATIVE HABITAT: United States from Pennsylvania to Georgia.

HARDINESS: Zone 3.

HABIT: Low, mat forming, trailing, semiwoody ground cover.

SIZE: 6 to 12 inches tall, spreading indefinitely.

RATE: Moderate; space plants from pint- or quart-sized containers 10 to 12 inches apart.

SPECIAL CULTURE: Soil: Best in organically rich loam with pH from 5.0 to 7.0.
Moisture: Not notably tolerant to drought; soil should be kept moist with regular watering.
Light: Light to moderate shade.

FOLIAGE: Evergreen, simple, opposite, spatulate-obovate or oblong to ovate, to 3 3/4 inches long.

INFLORESCENCE: Glandular, pubescent, loose cymes on erect stems; flowers about 1 inch across, regular; corolla salverform, purple or violet, effective mid to late spring. Pink, white, and rose flowered selections have also been made.

*   *   *

P. subulata (sub-ū-lā´tá) Awl shaped, the leaves (Figure 3–152).

COMMON NAME: Moss-pink, ground-pink, mountain phlox, moss phlox.

NATIVE HABITAT: Sandy woodlands from New York to North Carolina.

HARDINESS: Zones 2 or 3 to 9.

HABIT: Low, mosslike, mat forming, semiwoody ground cover.

SIZE: 4 to 6 inches tall, sometimes lower; spreading to 24 inches across.

RATE: Moderate; space plants from 3 to 4 inch diameter containers 8 to 12 inches apart.

SPECIAL CULTURE: Soil: Adaptable to most well-drained soils with a pH of 6.0 to 8.0; tolerating gravelly soils.
Moisture: Fairly tolerant of drought, but usually best when soil is maintained in a slightly moistened condition with periodic watering.
Light: Full sun.

FOLIAGE: Simple, opposite, sometimes becoming alternate terminally, evergreen, linear to subulate, to 1 inch long, crowded, sharply pointed, glabrous and glandular dotted both sides, ciliate margined.

STEMS: Reddish or brownish, hairy, squarish, becoming woody at the base.

INFLORESCENCE: Terminal panicles; flowers numerous, perfect, 5 parted; corolla salverform, red-purple to violet-purple, pink or white, 3/4 inch across, lobes shallowly notched, effective mid to late spring.

SELECTED CULTIVARS AND VARIETIES: There are numerous cultivars, only a few of which are described here.

'Alba' Flowers white.

'Alexander's Surprise' Flowers pink.

'Atropurpurea' Flowers rose-purple with a crimson ring.

'Blue Hills' Showy sky-blue flowers.

'Brilliant' Flowers magenta.

'Chuckles' Flowers pink.

'Crimson Beauty' Flowers bright red.

'Emerald Cushion' Smaller plant with pink flowers.

'Millstream Daphne' Vigorous compact selection; flowers pink with yellow eye.

'Red Wings' With red flowers.

'Rosea' Flowers rosy-pink.

'Scarlet Flame' Flowers scarlet; vigorous grower.

'White Delight' Flowers large, profuse, pure white.

Generally Applicable to All Species

CULTURE: Soil: Best in sandy, well-drained loam.

PATHOLOGY: Diseases: Leaf spots, powdery mildew, rusts, crown rot, stem blight, bacterial crown gall.

Figure 3–152 *Phlox subulata* (life-size)

Pests: Beetles, scales, wireworms, two spotted mite, bulb and stem nematode, stalk borer, aster leafhopper.

Physiological: Leaf blight, occuring in spring, was reported by P. P. Pirone (the researcher and author who discovered the cause of this disease) to be due to the inability of the stems to conduct enough water to supply all the foliage while new growth is taking place in spring. As a result the basal leaves desiccate and eventually die. The disease usually affects older, established plants.

MAINTENANCE: Shear or mow plantings after bloom to help keep them compact and neat. Division every 3 years is often performed to induce rejuvenation and increased bloom. Only slow release fertilizers are recommended as roots are shallow and tend to be damaged by rapid changes in salinity levels.

PROPAGATION: Cuttings: Softwood cuttings taken from new shoots in spring or summer are quickly rooted. They should be stuck in a very porous medium and treated with a fungicide that controls damping off. Root cuttings have become more popular as of late, since severe problems with damping off often occur with softwood cuttings. Clumps should be dug in the fall, and all large roots removed within a few inches of the crown (replant crown). Roots then should be cut into 2 inch segments, placed in flats, lightly covered with medium, and given mild bottom heat. Transplant the following spring.

Division: Simply divide nonwoody clumps in the fall or early spring. Cutting back shoots to one-half their length is helpful as the roots are sparse.

In vitro: In vitro methods have been developed using shoot explants.

Seed: Commercial supplies are available but variability must be expected. Seeds generally germinate in 3 to 4 weeks at 70°F.

✦✦✦✦✦✦✦✦✦✦✦✦✦✦✦✦✦

***P**hyla* (fy´la) From Greek, *phyla* (a tribe, in reference to the compound flower heads). This small genus is composed of about 15 species of procumbent or trailing herbaceous perennials. They are native to tropical and subtropical regions.

FAMILY: VERBENACEAE: Vervain family.

P. nodiflora (nō-di-flō´ra) With flowers borne from the nodes (Figure 3–153).

COMMON NAME: Creeping lippia, frogfruit, garden lippia, mat grass, turkey-tangle, capeweed.

Figure 3–153 *Phyla nodiflora* (life-size)

NATIVE HABITAT: South America.

HARDINESS: Zones 9 to 10.

HABIT: Herbaceous, matlike, stoloniferous ground cover.

SIZE: 1 to 6 inches tall, spreading indefinitely.

RATE: Relatively fast; space stolons or divisions about 12 to 14 inches apart.

LANDSCAPE VALUE: Commonly used as a lawn substitute, especially in informal landscapes. Also effective as a general ground cover for use around a pond, pool, or in the rock garden. Tolerates considerable foot traffic, along with salt spray.

FOLIAGE: Evergreen, opposite, simple, spatulate to cuneate-obovate, to 1 3/4 inch long by 1/4 inch wide, toothed toward apex, coarsely hairy and glandular dotted on both sides, grayish green to medium green.

INFLORESCENCE: Dense heads on axillary peduncles, to 1/2 inch wide; flowers irregular; corolla with two equal lips, small, white or lilac, effective spring to fall.

FRUIT: Of 2 nutlets, not ornamentally significant.

SELECTED CULTIVARS AND VARIETIES: Var. *canescens* Flowers lilac with yellow throat. 'Rosea' Flowers rose colored.

CULTURE: Soil: Adaptable to most soils.
Moisture: Once established, highly tolerant to drought. Water deeply but only infrequently during the heat of summer. If plants should wilt, they will recover if watered soon after.
Light: Full sun to moderate shade. Plants reach about 6 inches tall in shade and only 1 to 2 inches in full sun.

MAINTENANCE: Mow as one would a turf lawn in order to maintain desired height and compactness. Flowers are often mowed off while still in their infancy as they attract many bees.

PATHOLOGY: Reported to be susceptible to nematode infestations.

PROPAGATION: Division: Divide in late fall through early spring.

Picea (pī′sē-à or pis′ē-à) Ancient Latin name for the spruce or fir, probably derived from *pix* (pitch). The genus *Picea*, or spruce as most recognize it, is composed of about 45 species of monoecious, coniferous, usually short needled, evergreen trees of pyramidal outline. They are usually tall and symmetrical and originate from cool areas of the northern hemisphere. A few low growing selections are useful and rather interesting evergreen ground covers.

FAMILY: PINACEAE: Spruce, pine, fir family.

* * *

P. abies (ā′bez, or ab′i-ēz) Latin for the silver fir (Figure 3–154).

COMMON NAME: Norway spruce.

NATIVE HABITAT: Northern and Central Europe.

HARDINESS: Zones 2 to 7.

HABIT: Ground cover types are generally low, flat topped, wide spreading, woody shrubs.

SIZE: 12 to 24 inches tall, spreading 3 to 5 feet across.

RATE: Relatively slow; spacing of 3 to 4 feet for plants from 1 to 2 gallon sized containers is usually acceptable.

LANDSCAPE VALUE: Often used as specimens in rock gardens or bed of dwarf conifers. When planted en masse, they make a dense general cover for large- or moderate-sized areas. No foot traffic.

DISCUSSION: Commonly called Norway spruce, this species is a widely planted, vigorous tree. It may reach 100 feet tall (or more) when fully mature.

Foliage is light to dark green, short, stout, and needlelike. Many sports have arisen with habits that make them useful as ground covers. They are briefly described as follows:

'Nidiformis' (Bird's-nest spruce) A low, dense, spreading shrub with medium green, 1/2 to 1 inch long, needlelike foliage. Sometimes the center opens up with age and thus the common name bird's-nest spruce. May reach 3 to 6 feet tall, but rarely over 1 1/2 to 2 feet in the landscape. Spread may be 4 or more feet across. Relatively slow grower.

'Procumbens' A rather vigorous, flat topped selection with compact habit and thin, stiff, horizontal branches. To about 2 or 3 feet tall.

'Pseudoprostrata' Seldom cultivated, medium green, prostrate selection. A specimen at the National Arboretum is about 6 or 8 feet wide by little more than 8 to 12 inches tall and somewhat mounded in the center.

'Pumila Nigra' With branches held at 30 to 45 degree angles, this selection is not truly procumbent although it spreads very wide and seldom exceeds 3 feet tall.

'Repens' A fine, compact, wide spreading, slightly mounded selection. Foliage is medium green; mature height around 2 feet tall. Worthy of greater attention than presently given.

'Tabuliformis' Prostrate growth with horizontal branch pattern.

* * *

P. pungens (pun´jenz) Pointed, the leaves, or sharp, the taste.

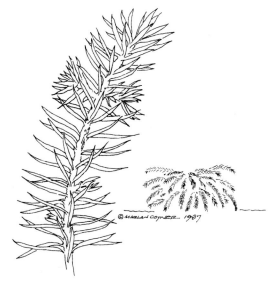

Figure 3-154 *Picea abies* 'Nidiformis' (life-size)

COMMON NAME: Colorado spruce, Colorado blue spruce.

NATIVE HABITAT: Southwestern United States, Rocky Mountains from Colorado to Utah and New Mexico.

HARDINESS: Zones 2 to 7.

DISCUSSION: The species (Colorado spruce) is not itself a ground cover. Rather, it is a popular, conical tree that may reach over 100 feet tall at maturity. Foliage is needlelike, rigid, ranging from dull green to bluish or silvery-white. The following cultivar is an interesting ground cover and is well worth the effort needed to find it.
P. pungens 'Glauca Pendula' is a very striking, bright silvery blue-green, prostrate growing selection. Branches are held horizontally from 12 to 24 inches above the ground and reach out to give it a maximum spread of over 10 feet across. It spreads at a slow rate and should be spaced 4 to 5 feet apart from 2 gallon sized containers if mass planted.

LANDSCAPE VALUE: Can be massed for general cover, but the best use is as a specimen. The effect when a single plant is displayed next to the top of a stone retaining wall is exceptional. Branches will cascade over and command attention.

MAINTENANCE: Occasionally upright branches arise and should be trimmed off as they become noticeable.

PROPAGATION: When taking a scion or cutting from this plant, it should be from horizontal, rather than slightly erect, shoots, as topophysis is displayed.

OF SPECIAL INTEREST: This plant may be identical to the cultivar that is sometimes sold under the name 'Kosters' or Koster's weeping blue spruce.

Generally Applicable to Both Species

CULTURE: Soil: Adaptable to most well-drained soils with a pH in the range from 5.5 to 6.5. Intolerant of alkaline soils.
Moisture: Once established, spruces are relatively tolerant to drought; however, best performance is realized when the soil is kept moderately moist with deep irrigation as needed.
Light: Full sun.

PATHOLOGY: Diseases: Canker, needle casts, rusts, wood decay.
Pests: Aphids, budworms, spruce bud scale, spruce needle miner, pine needle scale, spruce epizeuxis, sawflies, white pine weevil, spruce spider mite, bagworm.

MAINTENANCE: All varieties discussed can be kept

neat and more compact by lightly pruning in spring when new growth is about one-half developed.

PROPAGATION: Cuttings: To achieve the highest percentages take cuttings in November or December. Wounding to the cambium, treatment with an 8000 ppm IBA/talc root inducing compund, bottom heat to 65°F, and high light intensity are all helpful.

Grafting: In most cases, grafting onto a 2 or 3 year seedling of the species is successful. A side cleft graft is the most common technique.

✛✛✛✛✛✛✛✛✛✛✛✛✛✛✛✛

***P**inus* (pī´nus) Ancient classical name for a pine tree. The pines, as they are commonly called, number about 90 species of coniferous, evergreen, monoecious, needle-leaved trees from the northern hemisphere. Their foliage is of two types; scalelike needles are deciduous, while long, needlelike leaves are arranged in fascicles of 2 to 5 (sometimes one) and are primarily evergreen. Selections of various species are in use as ground covers and in most cases are derived from witches'-brooms. (See OF SPECIAL INTEREST section for a discussion of witches'-brooms.)

FAMILY: PINACEAE: Spruce, pine, fir family.

LANDSCAPE VALUE: Those cultivars that form button or cushionlike mounds are suited for specimen planting in rock garden or dwarf conifer border. Spaced close together, they make an interesting general cover on sloping terrain. The sprawling selections are about impossible to beat as specimens when planted atop a rocky ledge or terrace and allowed to cascade over. Next to stairs or on a bank, where the grade change places them at eye level, allows one to appreciate them at their best. No foot traffic.

*   *   *

P. banksiana (bank-sē-ā´nȧ) Meaning unknown, likely a commemorative.

COMMON NAME: Jack pine, scrub pine, gray pine.

NATIVE HABITAT: North America from the Arctic Circle south to northern New York and Minnesota.

HARDINESS: Zones 2 to 6 or 7.

DISCUSSION: *P. banksiana* as a species develops into a tree to 70 feet tall, with 2 needle fascicles of olive-green, 3/4 to 2 inch long needles. 'Uncle Foggy' is a weeping form with prostrate habit that may reach 2 feet tall and spread to 15 feet across.

RATE: Moderate; use plants from 2 gallon or larger sized containers.

CULTURE: Soil: Best in well-drained sandy soils. Adaptable to pH range from slightly alkaline to relatively acidic.

*   *   *

P. densiflora (den-si-flō´rȧ) Dense flowering.

COMMON NAME: Japanese red pine.

NATIVE HABITAT: Japan, Korea, China (the species).

HARDINESS: Zone 4.

DISCUSSION: *P. densiflora* as a species develops into a tree that exceeds 100 feet. Foliage is 2 needle fascicled, 3 to 5 inches long, lustrous bright green. The selection 'Prostrata' has horizontal habit and hugs the ground.

SIZE: 12 to 24 inches tall, spreading 8+ feet across.

RATE: Relatively slow; select material from 2 gallon or larger sized container.

CULTURE: Soil: Best in well-drained sandy soils with a pH between 5.0 and 6.5.

*   *   *

P. flexilis (fleks´i-lis) Flexible, in reference to the trunk.

COMMON NAME: Limber pine.

NATIVE HABITAT: Western North America.

HARDINESS: Zones 4 to 7.

DISCUSSION: *P. flexilis* is a 5 needle fascicled tree. Needles are 2 1/2 to 3 1/2 inches long, dark to glaucous green. The tree may reach a mature height of 50 or more feet with a pyramidal outline. The selection 'Glauca Pendula' has prostrate habit with bluish-green foliage.

CULTURE: Soil: Best in well-drained sandy loam, with pH from 7.5 to more acidic.

*   *   *

P. mugo (mū´gō) Italian vernacular.

COMMON NAME: Swiss mountain pine, mugo pine.

NATIVE HABITAT: Europe.

HARDINESS: Zones 2 to 7.

DISCUSSION: This commonly cultivated species has 2 needle fascicles with 1 to 2 inch long, rigid, medium to dark green foliage that often turns yellowish in winter. A brief description of the species is given, but it is the lower growing clones and varieties that are primarily useful as ground covers. Descriptions of their unique features are addressed later.

HABIT: Usually low, broad spreading, sometimes prostrate, woody, upright branched shrub.

SIZE: In general, the height will range from 1 to 6 feet tall. Spread, too, is quite variable, to more than 10 feet across. Although mugo pines root relatively well, they are often propagated by seed, which produces variable plants.

RATE: Relatively slow; space plants from 1 to 3 gallon sized containers at distances in accord with the expected mature spread of the cultivar.

SELECTED CULTIVARS AND VARIETIES: 'Gnome' Low, compact selection with dark green color. Cushionlike in outline; reaching 15 inches tall and about 3 feet across.

Var. *mugo* Low growing variety, which usually is less than 4 to 5 feet tall by about twice as wide.

Var. *pumilo* A low spreading variety that may reach 5 to 6 feet tall but is usually much lower; spreading to 10 or more feet across.

CULTURE: Soil: Best in sandy, well-drained loam with a pH of 7.0 to more acidic.

* * *

P. nigra (nīˊgrȧ) Dark, the leaves.

COMMON NAME: Austrian pine.

NATIVE HABITAT: Europe.

HARDINESS: Zones 4 to 7.

DISCUSSION: The species *P. nigra* is a rapid growing, large tree. It may reach 90 feet tall and has foliage that consists of stiff, straight or curved, 4 to 6 inch long, 1/16 to 1/12 inch wide needles. Dark green and sharply pointed, the needles are fascicled in twos. The cultivar 'Hornibrookiana,' which originated as a witches'-broom, is a low growing, dwarf selection. It is characterized by dense habit and low height. Discovered by B. H. Slavin in Seneca Park, Rochester, New York. The following description applies to this cultivar:

HABIT: Low, compact, horizontal spreading, woody shrub.

SIZE: 18 to 24 inches tall, spreading to 6 feet across.

RATE: Relatively slow; use plants from 2 to 3 gallon sized containers.

CULTURE: Soil: Best in well-drained sandy loam, with pH of 7.0 to more acidic.
Moisture: Well suited to tolerate drought; occasional deep watering in summer is to some benefit.
Light: Full sun.

* * *

P. parviflora 'Bergman' (par-wȧ-flōˊrȧ or par-vi-) Small leaved, from Latin.

COMMON NAME: Bergman's Japanese white pine.

NATIVE HABITAT: The species originated in Japan.

HARDINESS: Zones 4 or 5 to 7.

HABIT: Low, horizontal spreading, woody shrub (the selection).

SIZE: To about 16 inches tall, spreading to 3 feet across.

RATE: Relatively slow.

CULTURE: Soil: Adaptable to wide range of well-drained soils.

* * *

P. strobus (strōˊbus) Old generic name, referring to the strobili (Figure 3–155).

COMMON NAME: Eastern white pine, white pine.

NATIVE HABITAT: Eastern North America.

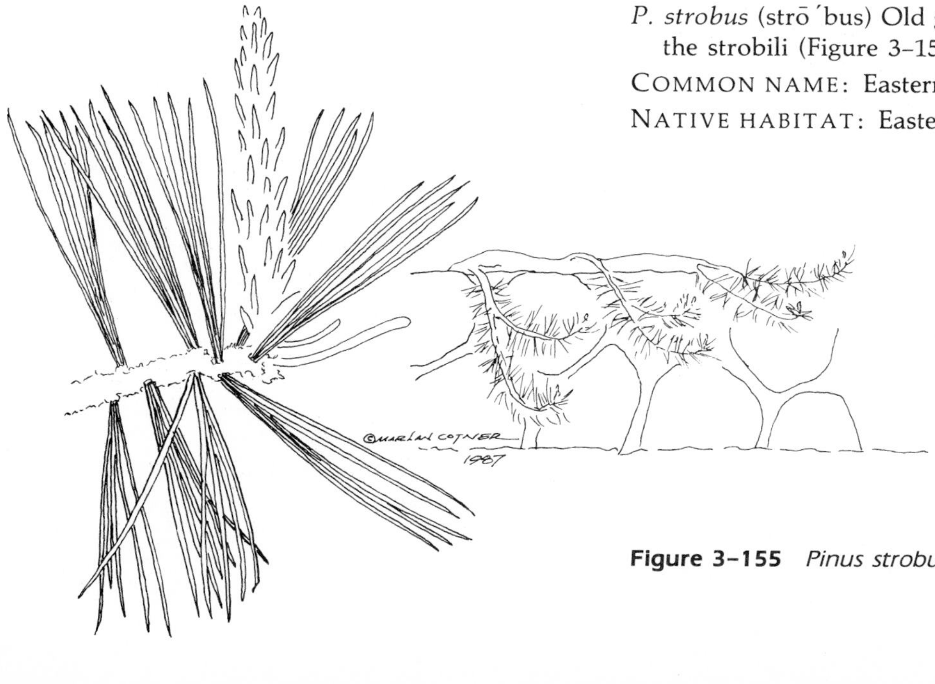

Figure 3–155 *Pinus strobus* 'Ottawa' (life-size)

HARDINESS: Zones 3 to 8.

DISCUSSION: The state tree of Michigan and Maine, *P. strobus* is fast growing and may reach over 150 feet tall. Generally, it reaches 60 to 80 feet. Foliage is comprised of 3 to 5 inch long, slender, bluish green, stiff pointed, soft, flexible needles that are bundled 5 to a fascicle. Cultivars with ground covering merit are described as follows:

HABIT: Low, horizontal spreading, woody, shrubby ground covers.

RATE: Relatively slow; use material from 2 to 3 gallon sized containers.

SELECTED CULTIVARS AND VARIETIES: 'Hairds Broom' Compact, dwarf, prostrate, cushionlike selection that reaches about 12 inches tall and spreads to over 3 feet across.
'Merrimack' Similar to 'Hairds Broom,' but foliage bluish green and height to 2 or 3 feet tall.
Forma *ottawa* A weeping tree form that I discovered in Muskegon, Michigan, in 1984. Foliage is typical of the species, but branches (with repeating serpentine pattern) trail along the ground horizontally. Height is to about 12 inches and spread is unknown, likely to exceed 20 feet; named to commemorate the Ottawa Indians.

CULTURE: Soil: Adaptable to a wide range of well-drained soils with pH of 7.0 to more acidic.

* * *

P. sylvestris (syl-ves´tris) Of the woods.

COMMON NAME: Scotch pine, Scot's pine, Scotch fir.

NATIVE HABITAT: Ranges from Norway and Scotland to Spain, western Asia, and northwestern Siberia.

HARDINESS: Zones 2 to 8.

DISCUSSION: The species *P. sylvestris* is a tree that may reach a height of 90 feet. Generally growing in the range of 30 to 65 feet tall, it has needles that are arranged in fascicles of two, are from 1 to 4 inches long, stiff and colored blue-green. It is one of the most common species for use as a Christmas tree. Ground covering types have arisen and are described as follows:

HABIT: Low, horizontal spreading, erect branched, woody shrubs.

RATE: Relatively slow; select plants from 2 gallon or larger sized containers.

SELECTED CULTIVARS AND VARIETIES: 'Albyn's Prostrate' A rather loose, prostrate clone, which I have not seen to exceed 10 or 12 inches tall and 4 feet across. Holds color well in winter.

'Hillside Creeper' Similar to 'Albyn's Prostrate,' but coarser textured, becoming yellow in winter.

CULTURE: Best in well-drained, sandy, acidic loam.

Generally Applicable to All Species

CULTURE: Soil: Addressed individually with species and cultivar descriptions.
Moisture: Overall, these species are quite tolerant to drought once established. Thereafter, occasional thorough watering in summer may prove to be beneficial, but is not essential in most years.
Light: Full sun preferred, light shade tolerated.

OF SPECIAL INTEREST: Witches'-brooms are branches on a tree or shrub that display extremely dense and busy growth. Frequently, this is seen on cherry, blueberry, hackberry, spruce, fir, and pine. Interestingly, these variations in morphology are caused by fungi. Not considered injurious to the plants, witches'-brooms can be propagated by cutting or grafting and develop into plants that are characteristically dwarf and often useful ground covers.

PATHOLOGY: Diseases: Diseases and pests that may affect the pines are extensive, but when proper cultural practices are followed, they are nothing to be overly concerned about. Diseases include late damping off, root rot, dieback, tip blight, stem blister, rust, comandra twig blight, leaf casts, needle blight, needle rust, white pine blister rust, wood decay.
Physiological: Tip mottling and necrosis from gaseous pollutants; leaf desiccation due to exposure to airborne salt.
Pests: Pine bark aphid, pine leaf chermid, white pine aphid, European pine shoot moth, Nantucket pine moth, sawfies, pine webworm, pine false webworm, pine needle scale, pine needle miner, pine spittlebug, pine tortoise scale, red pine scale, pales weevil, white pine shoot borer, tube moth, bark beetles, and Zimmerman pine moth.

MAINTENANCE: Plants are kept more dense and compact by shearing off one-half of the new shoots (candles) in late spring or early summer. This causes formation of lateral buds below and around the edge of the cut.

PROPAGATION: Cuttings: In general, pine cuttings are difficult to root. *P. mugo* roots readily when taken in midsummer. Treatment with a root inducing compound is helpful; 8000 ppm IBA/talc has been found to work well.
Grafting: Side cleft grafting is commonly used for propagation of select clones. Two or three year seedlings of the species are used as rootstalks. Scions should be taken from partly matured wood of the

new growth from the previous year in late winter when dormant; understalks should be showing active root growth after a period of dormancy.
Seed: Propagation by seed is usually avoided as variability is great.

*P*ittosporum (pi-tos´pō-rum, or pit-o-spō´rum) From Greek, *pitte*, tar, and *sporos*, seed, the seed being coated with a resinous substance. The genus is composed of about 100 species of evergreen shrubs and trees that are native to warm temperate, subtropical, and tropical regions of the Old World.

FAMILY: PITTOSPORACEAE: Tobira family.

P. tobira 'Wheeler's Dwarf' (tō-bī´rá) A Japanese name for the species (Figure 3–156).

COMMON NAME: Wheeler's dwarf Japanese pittosporum, dwarf Australian laurel, dwarf mock orange.

NATIVE HABITAT: The species is native to China and Japan.

HARDINESS: Zones 8 to 9, possibly zone 7 if sheltered and protected from the winter sun and wind.

DISCUSSION: The species itself is not useful as a ground cover; rather the low, compact selection Wheeler's Dwarf is useful for this purpose. It is described thusly:

HABIT: Compact, low, horizontal spreading, upright branched, woody shrub.

SIZE: 2 to 3 feet tall, spreading 3 to 4 feet across.

RATE: Relatively slow; space plants from 1 to 2 gallon sized containers 2 1/2 to 3 feet apart.

LANDSCAPE VALUE: Often used as an edging along walkways. Fine as a specimen, low foundation facer, or low hedge. En masse as a general cover for moderate to large areas, an interesting mounding effect is created. No foot traffic. Tolerates airborne salt quite well.

FOLIAGE: Evergreen, simple, alternate, densely set, obovate, about 2 to 4 inches long and 1/2 inch across, edges rolling under, apex rounded, base tapering to a short petiole, glabrous, leathery, dark shiny green.

INFLORESCENCE: Many flowered umbellate clusters; flowers fragrant, creamy white, perfect, 5 parted, about 3/8 inch across, effective mid to late spring, but relatively nonshowy in relation to their enticing fragrance, which is similar to orange blossoms.

FRUIT: 3 valved, pear-shaped capsule, green turning brown in fall.

CULTURE: Soil: Adaptable to most well-drained soils with a pH in the range from 5.0 to 7.5.
Moisture: Moderately tolerant of drought, but plants are best when soil is kept slightly moist by regular watering as needed.
Light: Full sun to light shade.

PATHOLOGY: Diseases: Leaf spots, stem rot, thread blight, root rot, viral mosaic.
Pests: Scales (many species), aphids, mealybugs, southern root knot nematode.

MAINTENANCE: Very little if any maintenance is required. A light pruning on occasion will keep plants at a desired height.

PROPAGATION: Cuttings: Cuttings root readily if taken when semihard. IBA treatments of 1000–3000 ppm have proved beneficial.

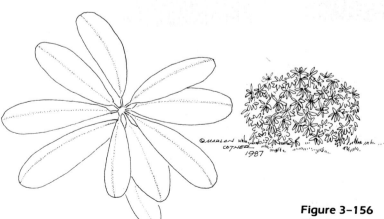

Figure 3–156 *Pittosporum tobira* 'Wheeler's Dwarf' (0.5 life-size)

✚✚✚✚✚✚✚✚✚✚✚✚✚✚✚✚✚

P*olygonum* (pō-lig´ō-num) Many joints, in reference to the stem formation. Some authorities feel that the reference is to the roots. This genus is made up of about 150 species of primarily herbaceous annuals and perennials, along with a few that are partly woody. Occasionally they inhabit ponds, but most are terrestrial. They are native to many places throughout the world.

FAMILY: POLYGONACEAE: Buckwheat family.

*   *   *

P. affine (à-fin´ē) Kindred, or allied, that is, to some other species (Figure 3–157).

COMMON NAME: Himalayan fleeceflower.

NATIVE HABITAT: Himalayas.

HARDINESS: Zone 3.

HABIT: Mat forming, erect leaved, decumbent stemmed, herbaceous, trailing ground cover.

SIZE: 4 to 6 inches tall, spreading indefinitely.

RATE: Moderate; space plants from 2 1/2 to 4 inch diameter containers 10 to 16 inches apart.

LANDSCAPE VALUE: General cover for small- to moderate-sized areas; color is attractive both in bloom and when not blooming. Good for facing the south or west sides of loose deciduous or open evergreen shrubs and small trees. Tolerates very infrequent foot traffic.

FOLIAGE: Semievergreen, primarily basal, narrowly oblong, to 4 inches long, an inch or so across, margin finely serrate, dark to medium green, turning rusty red in fall before becoming brown (in the north).

INFLORESCENCE: Dense, erect, 3 inch long terminal spikes on sparsely leafy stems that elevate them to 1 to 1 1/2 feet; flowers small, bright rose, from early to midautumn. Very dainty and attractive.

FRUIT: Achene, not ornamentally significant.

SELECTED CULTIVARS AND VARIETIES: 'Border Jewel' Similar to the species, but foliage larger and blooming spring to fall.
'Donald Lowndes' Flowers bright pink and quite showy.

*   *   *

P. capitatum (käp-i-tä´tum) Flowers in heads.

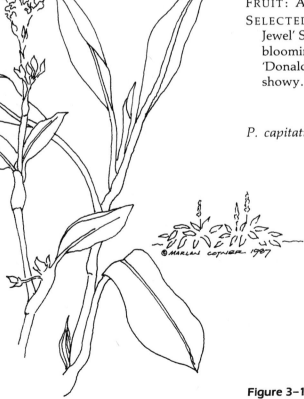

Figure 3–157 *Polygonum affine* 'Border Jewel' (life-size)

COMMON NAME: Pinkhead knotweed, magic carpet polygonum.

NATIVE HABITAT: Himalayas.

HARDINESS: Zones 7 to 10.

HABIT: Low, herbaceous, trailing, matlike ground cover.

SIZE: 3 to 6 inches tall, spreading indefinitely.

RATE: Relatively fast; space plants from 2 1/2 to 3 inch diameter containers 10 to 14 inches apart. Freely self-sowing.

LANDSCAPE VALUE: Use as a general cover in smaller-sized confined areas; especially attractive between large rocks on slopes. Tolerates very limited foot traffic.

FOLIAGE: Semievergreen, alternate, simple, entire, ciliate, elliptic, to 1 1/2 inches long, about 3/4 inch across, variegated bronzy and purplish-red, becoming crimson in fall, pubescent both sides.

INFLORESCENCE: Dense, racemose heads to about 3/4 inch across; flowers tiny, 6 parted, pink, effective in spring and occasionally during summer.

FRUIT: Small achenes, not ornamentally significant.

* * *

P. cuspidatum var. *compactum* (kus-pi-dā´tum) Cusped, the leaves.

COMMON NAME: Japanese fleeceflower, Mexican bamboo, Japanese knotweed.

NATIVE HABITAT: Japan.

HARDINESS: Zones 3 to 10.

HABIT: Sprawling, rhizomatous, deciduous, herbaceous ground cover.

SIZE: 18 to 35 inches tall, spreading indefinitely.

RATE: Relatively fast to invasive; space plants from pint- or quart-sized containers 16 to 24 inches apart.

LANDSCAPE VALUE: Commonly used as a low hedge or border where it can be contained. Exceptional in northern areas as a border around parking lots or alongside building entrances or walks where it withstands reflected summer heat and dies back in winter, thus saving it from damage due to mounded snow. No foot traffic. *P. cuspidatum* itself may reach 8 feet tall and usually is not considered worth cultivating. The variety *compactum* is most commonly seen and is described here.

FOLIAGE: Opposite, simple, deciduous, entire, broadly ovate to nearly orbicular, 3 to 6 inches long, apex bluntly pointed, margin wavy, petiole about 1 inch long, medium green when mature, reddish in youth and sometimes crimson in fall.

INFLORESCENCE: Dense panicled racemes, axillary;

flowers perfect, without petals, buds red, 5 sepals, white or pink, effective in early fall, overall appearing soft and lacy.

FRUIT: Showy pink to red, winged achenes, effective in fall until first hard frost.

OTHER SPECIES

* * *

P. reynoutria (rā-noo´trē-ȧ) Commemorative.

COMMON NAME: Reynoutria fleeceflower.

NATIVE HABITAT: Japan.

HARDINESS: Zone 4.

HABIT: Similar to *P. cuspidatum* var. *compactum* but lower. Some claim them to be identical.

SIZE: 4 to 6 inches tall, spreading indefinitely.

RATE: Relatively fast to invasive; space plants from pint- or quart-sized containers 12 to 16 inches apart.

LANDSCAPE VALUE: General cover in small- or medium-sized confined areas. No foot traffic.

* * *

P. bistorta (snakeweed) Increasingly cultivated, this herbaceous ground cover reaches about 1 1/2 to 2 feet tall. Foliage is oblong-ovate, green, and mostly basal. Flowers are pink or white, in dense terminal spikes, from late spring through late summer. It is clump forming and its cultivar 'Superbum' reaches 2 to 3 feet tall with more profuse flower formation. This species is hardy to Zone 4 and is useful in the same situations as *P. affine*.

Generally Applicable to All Species

CULTURE: Soil: Adaptable to most well-drained soils. Tolerant to infertility and gravelly soils.
Moisture: Most fleeceflowers show good drought tolerance, yet occasional watering in summer is useful for all species.
Light: Full sun to light shade.

MAINTENANCE: Plantings of taller types such as *P. cuspidatum* var. *compactum* may be sheared when taller, or frequently mowed to keep them compact and neat. All species may be mowed in fall or spring to clean up dead foliage in northern regions.

PATHOLOGY: Diseases: Leaf spot.
Pests: Japanese beetle, aphids, slugs, snails, spittlebugs.

PROPAGATION: Cuttings: Most types are easily rooted from cuttings in early summer. *P. cuspidatum* var. *compactum* roots best when cuttings are very soft (rubbery). Mowing will bring a new crop of such cuttings within a few weeks.

Treating with a mild rooting preparation may speed the process.

Division: Simple division in spring or fall is widely practiced.

Seed: Commercial supplies are available for some species. Usually, seed germinates in 3 to 4 weeks at 70 °F.

***P**olypodium* (pol-i-pō´di-um) From Greek *polys*, many, and *pous*, a foot (many little feet), in reference to the furry footlike divisions of the creeping stems. This large genus, commonly called polypody, is composed of mostly tropical, often epiphytic, rhizomatous ferns.

FAMILY: POLYPODIACEAE: Polypody family.

HABIT: Rhizomatous, spreading ferns.

RATE: Moderate growth rate; space plants from pint- to quart-sized containers 1 1/2 to 2 feet apart.

* * *

P. virginianum (vĕr-jin-i-ā´num) Of Virginia (Figure 3–158).

COMMON NAME: Rock polypody, American wall fern.

NATIVE HABITAT: Eastern North America.

HARDINESS: Zone 3.

SIZE: Usually grows to 10 inches, infrequently somewhat taller, spreading to 5 feet across.

LANDSCAPE VALUE: Good in border as a companion to ericaceous plants or in a shaded area of a rockery. Nice when planted with native woodland flowers in a shady location. No foot traffic.

FOLIAGE: Evergreen, oblong-lanceolate to triangular, deeply pinnatifid to 1 foot long and 3 inches wide; lobes broadly rounded, 10 to 20 pairs, entire or toothed wavy margin, alternate to subopposite, green on both sizes.

STEMS: About one-third the length of the leaf, thin, round, light dull green and smooth to the touch.

SORI: Large, red-brown, round, each borne on the end of a free veinlet, nearly marginal, usually more numerous on upper leaflets, no indusia.

OTHER SPECIES

* * *

P. aureum (â´rē-um) Golden (the sori or rhizome scales).

COMMON NAME: Golden polypody fern, hare's-foot fern, rabbit's-foot fern.

NATIVE HABITAT: Tropical Florida through the West Indies to Argentina.

HARDINESS: Zone 10.

SIZE: To 36 inches tall, spreading to 6 feet across.

LANDSCAPE VALUE: Interesting specimen in rock garden or in border grown near a large rock or contrasting shrubs.

GENERAL DESCRIPTION: With pinnatifid leaves to 4 feet long and 2 feet across, lobes to 1 foot long and 2 inches wide; sori are in two rows, often bright golden yellow.

Generally Applicable to Both Species

CULTURE: Soil: Richly organic, moist soil of relatively low pH.

Moisture: For best growth, soil should be kept moist and plants located where temperatures are relatively cool.

Light: Light to dense shade.

PATHOLOGY: See **Ferns: Pathology.**

MAINTENANCE: Little or no maintenance required.

PROPAGATION: See **Ferns: Propagation.**

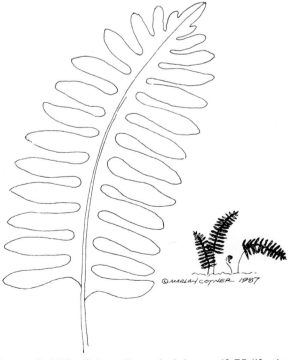

Figure 3–158 *Polypodium virginianum* (0.75 life-size)

✦✦✦✦✦✦✦✦✦✦✦✦✦✦✦✦

*P*olystichum (po-lis´ti-kum) From Greek, *polys*, many, and *stichos,* a row, in reference to the several rows of spore cases. This genus is composed of about 120 species of evergreen, woodland, coarsely toothed, rhizomatous ferns of worldwide distribution. Many species of this genus are suitable for ground cover use. The more common are briefly described here.

FAMILY: POLYPODACEAE: Polypody family.

HABIT: Coarse, rigid, rhizomatous ferns.

RATE: Relatively slow to moderate; space plants from pint- to quart-sized containers or bare root divisions 1 1/2 to 2 feet apart.

LANDSCAPE VALUE: Excellent covers when planted along a stream bank or around a pond or pool in filtered sun. Fine for accent or specimen in naturalized areas or a shaded section in the rock garden. No foot traffic.

*   *   *

P. acrostichoides (à-kros-ti-koi´dēz) Acrostichumlike (Figure 3–159).

COMMON NAME: Christmas fern, dagger fern, canker brake.

NATIVE HABITAT: Eastern North America.

HARDINESS: Zones 3 to 9.

SIZE: To 36 inches tall but usually much shorter, spreading to about 4 or more feet across.

FOLIAGE: Evergreen, pinnate, to 3 feet long (usually about 2 feet), 4 to 5 inches across; lance-shaped leaflets are eared and number 20 to 40 pairs, leathery, medium green.

SORI: Numerous, round, in 2 or more rows at the end of veins on terminal leaflets, often confluent.

OF SPECIAL INTEREST: The common name Christmas fern stems from the use of this plant for decoration at Christmastime by early New England settlers.

*   *   *

P. braunii (brän´i-ī) After Braun.

COMMON NAME: Braun's holly fern, shield fern.

NATIVE HABITAT: Northeastern North America.

HARDINESS: Zone 3.

SIZE: 18 to 24 inches tall, spreading to 2 feet across.

FOLIAGE: Semievergreen, twice pinnately compound, to 2 feet long, 6 to 8 inches wide, narrowly elliptical in outline; leaflets lanceolate, oblong or

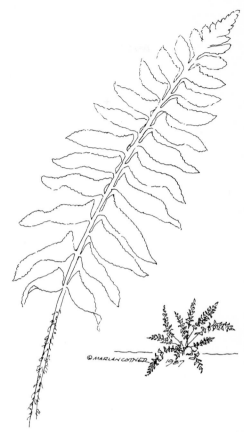

Figure 3–159 *Polystichum acrostichoides* (0.75 life-size)

ovate, slightly auricled, toothed, leathery, scaly, crowded, overlapping, 6 to 18 pairs, deep shiny green.

PETIOLE: Short, flat on front surface, brown scaly, pale hairy.

SORI: Small, round, near midvein in two rows, indusia circular and shieldlike.

*   *   *

P. lonchitis (lon-kī´tis) Spearlike, the leaves.

COMMON NAME: Mountain holly fern, holly fern.

NATIVE HABITAT: Europe and North America.

HARDINESS: Zone 3.

SIZE: 9 to 20 inches tall, spreading to 3 feet across.

FOLIAGE: Evergreen, to 2 feet long, pinnate; pinnae lanceolate, to 1 1/2 inches long, base eared, margin spiny toothed, reducing to small triangular lobes toward petiole, leathery, rich green.

Generally Applicable to All Species

CULTURE: Soil: Generally, species of this genus are

quite adaptable to varying soil types. Best growth, however, is usually attained in an organically rich loam with a pH close to the neutral point.

Moisture: Most luxuriant growth is obtained when planted in a cool location and soil is kept moderately moist.

Light: Light to dense shade.

MAINTENANCE: Mulch annually to maintain the organic content of the soil if growing in a location where this does not occur naturally.

PATHOLOGY: See **Ferns: Pathology.**

PROPAGATION: See **Ferns: Propagation.**

*P*otentilla (pō-ten-til ´à) From Latin *potens*, powerful, some species having active medicinal properties. This genus, commonly called cinquefoil, is composed of about 500 species of annual and perennial herbs and woody shrubs. They are native to northern temperate, boreal, and arctic regions, with a small number from the southern hemisphere.

FAMILY: ROSACEAE: Rose family.

LANDSCAPE VALUE: Excellent general covers for moderate-sized areas. Floral display is noteworthy, often contrasting strongly with foliage. Sometimes potentillas are used to good effect in a rock garden setting. Nice on gentle slopes. Tolerating only limited foot traffic, but useful as lawn substitutes where walking paths exist.

FRUIT: Achenes, not ornamentally significant.

*   *   *

P. cinerea (ki-ne ´rē-à or sin-ẽr ´ i-à) Gray-hairy; syn. *P. tommasiniana.*

COMMON NAME: Rusty cinquefoil.

NATIVE HABITAT: Alps Mountains.

HARDINESS: Zones 3 to 9.

HABIT: Low, herbaceous, creeping ground cover.

SIZE: 2 to 4 inches tall, spreading indefinitely.

RATE: Moderate to rapid growth; space 10 to 16 inches apart from pint- to quart-sized containers.

FOLIAGE: Semievergreen, palmately compound; five leaflets, oblong to obovate, margin dentate, gray-green.

FLOWERS: Perfect, 5 parted, corolla pale yellow: effective in late spring.

CULTURE: Soil: Adaptable to most well-drained soils.

Moisture: Growth is best when soil is maintained slightly moist.

Light: Full sun to light shade.

*   *   *

P. nepalensis (ne-pa-len ´sis) Himalayas of Nepal.

COMMON NAME: Nepal cinquefoil.

NATIVE HABITAT: Himalayas.

HARDINESS: Zones 5 to 9.

HABIT: Low, herbaceous, trailing cover.

SIZE: 12 inches tall, spreading indefinitely.

RATE: Moderate growing; space 10 to 14 inches apart from pint- to quart-sized containers.

FOLIAGE: Semievergreen, palmately compound; five leaflets, 1 to 2 1/2 inches long, obovate-oblong to oblanceolate, margin coarsely toothed, dark green.

INFLORESCENCE: Panicles on long stalks, flowers 1 inch across, rose red in summer.

SELECTED CULTIVARS AND VARIETIES: 'Willmottiae' Dwarf selection with many magenta-rose flowers. Plantings I have seen in the Midwest have had difficulty with mildew and rust.

CULTURE: Same as *P. cinera.*

*   *   *

P. tabernaemontani (ta-bẽr-nē-mon-tān ´ ī) From Latin *tabernae*, place, cottage, hovel, and *montana*, mountain, a description of its native habitat. Often this species is listed as *P. verna*, its former name. It is frequently confused with *P. crantzii*, which apparently grows more upright and taller.

COMMON NAME: Spring cinquefoil.

NATIVE HABITAT: Europe.

HARDINESS: Zones 5 to 9.

HABIT: Low, mat forming, prostrate, trailing, herbaceous ground cover.

SIZE: 3 to 6 inches tall, spreading indefinitely.

RATE: Moderate to relatively fast growing; space 10 to 16 inches apart from pint- to quart-sized containers.

FOLIAGE: Semievergreen, palmately compound, alternate; 5 leaflets (sometimes seven), narrowly ovate, 1 1/2 inches long by 1/2 inch wide, margins coarsely toothed, medium-dark green, sessile or short petioled, somewhat hairy.

STEMS: Spreading to 18 inches long, rooting, with short ascending branchlets.

INFLORESCENCE: Terminal cymes or solitary; flowers 5 parted, corolla bright golden-yellow, 1/2 to 1 inch across, effective late spring to early summer, often blooming again, but sparsely, during late summer to early fall.

SELECTED CULTIVARS AND VARIETIES: 'Nana' Very compact, reaching 2 to 3 inches high. Dark green foliage, not creeping but clumplike; spreading by crown multiplication; flowers profuse, yellow and contrasting well with foliage.
'Orange Flame' Reported to be similar to 'Nana,' but with orange flowers and slightly taller maximum height.

CULTURE: See *P. cinerea.*

*   *   *

P. × *tonguei* (ton-gū´i-ī) Meaning unclear, likely a commemorative (Figure 3–160).

HYBRID ORIGIN: *P. anglica* × *P. nepalensis.*

HARDINESS: Zones 4 to 9.

HABIT: Low, herbaceous, trailing ground cover.

SIZE: 3 to 8 inches tall, spreading to 18 + inches across.

RATE: Moderate rate of growth; space 10 to 14 inches apart from pint- to quart-sized containers.

FOLIAGE: Semievergreen, palmately compound; 3 to 5 leaflets, obovate, coarsely toothed, to about 1 1/2 inches long, dark green.

STEMS: Prostrate, hairy, to 12 inches long.

FLOWERS: To 1/2 inch across, petals obovate, colored yellow with red base, effective in midsummer to early fall.

*   *   *

P. tridentata (tri-den-tā´tà) Three-lobed, the leaves.

COMMON NAME: Wineleaf cinquefoil, three-toothed cinquefoil.

NATIVE HABITAT: Greenland and Manitoba, Wisconsin, extending south to Georgia and west to Iowa. Introduced in 1787.

HARDINESS: Zone 2.

HABIT: Low, mat forming, semiwoody, rhizomatous ground cover.

SIZE: 2 to 12 inches tall, spreading indefinitely.

RATE: Moderate rate of growth; space 10 to 16 inches apart from pint- to quart-sized containers.

FOLIAGE: Semievergreen, palmately compound; 3 leaflets, cuneate-oblong, 1/2 to 2 inches long, margin 3 to 5 toothed at apex, nearly glabrous, leathery, dark glossy green, oldest leaves turning yellow and exfoliating in fall.

INFLORESCENCE: Cymes, with a few or many flowers, 5 petaled, 1/4 inch across, white, effective throughout the summer.

CULTURE: Soil: Prefers sandy, well-drained loam. A pH from 4.0 to 5.0 is best.
Moisture: Relatively tolerant of drought, but benefits by periodic, thorough watering to maintain soil moisture.
Light: Best in light shade. Tolerates full sun.

*   *   *

OTHER SPECIES: *P. astrosanguinea* 'Gibson's Scarlet' A low, herbaceous perennial ground covering type. Reaching 1 1/2 feet high and spreading 2 feet or more across. This plant is remarkable for its large scarlet flowers throughout summer.

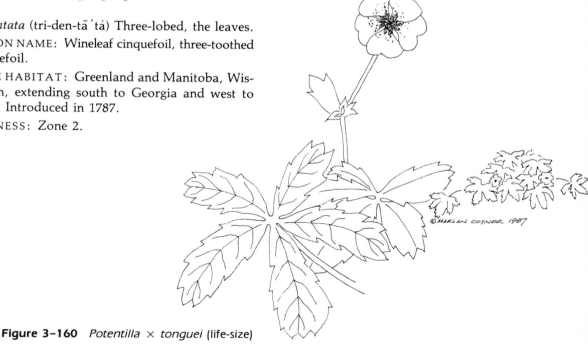

Figure 3–160 *Potentilla* × *tonguei* (life-size)

CULTURE: Soil: Adaptable to most well-drained soils. Tolerant of and generally best in soils of low fertility. A pH in the range of 7.0 to 8.0 is best, but they will tolerate acidic soils.
Moisture: Somewhat tolerant of drought, but plantings are best when soil moisture is maintained with periodic watering during the summer months.
Light: Best in light shade, but tolerant of full sun.

Generally Applicable to All Species

PATHOLOGY: Diseases: Downy mildew, leaf spots, powdery mildew, leaf rust.
Pests: Rose aphid, strawberry weevil, rabbits, slugs.

MAINTENANCE: Mowing after bloom will help keep plantings dense and full.

PROPAGATION: Division: Simply divide plantlets in spring or fall.

Figure 3–161   *Pratia angulata* (life-size)

*P**ratia** (prā´ti-à) After C. L. Prat-Bernon, a French Naval officer. The genus is composed of about 25 species of primarily prostrate herbs. They are native mostly to Australia, New Zealand, and tropical Asia. Others come from tropical Africa and North and South America.

FAMILY: LOBELIACEAE: Lobelia family.

P. angulata (an-gū-lā´tà) Angled, presumably the growth habit (Figure 3–161).

COMMON NAME: Pratia, panakenake.

NATIVE HABITAT: New Zealand.

HARDINESS: Zones 7 to 10.

HABIT: Herbaceous, low, trailing, mat forming ground cover.

SIZE: 2 to 6 inches tall, spreading indefinitely.

RATE: Relatively fast; space plants from 3 to 4 inch diameter containers 8 to 12 inches apart.

LANDSCAPE VALUE: General cover for smaller confined areas or in a rock garden where it easily establishes itself between the stones. Best growth occurs in the Pacific Northwest where summers are cool and moist, and winters are rather mild.

FOLIAGE: Evergreen, simple, alternate, nearly orbicular, 1/2 inch long by 1/2 inch wide, margin dentate, glabrous, grooved longitudinally, medium green.

FLOWERS: Numerous, solitary, axillary, 3/4 inch across, irregular; corolla 2 lipped, 5 lobed, 3/8 to 3/4 inch wide, white or bluish-white, purple veined, effective spring and summer.

FRUIT: 1/4 inch diameter reddish-purple berry, effective in winter.

SELECTED CULTIVARS AND VARIETIES: Var. *treadwellii* With larger flowers than the species.

CULTURE: Soil: Adaptive to a range from organically rich loamy soil to sandy soils (where flowering is more profuse). Presumably adaptive to pH.
Moisture: Intolerant to extended hot, dry weather; moisture should be maintained with regular watering.
Light: Full sun where climate is moderate; light shade in warmer areas.

PATHOLOGY: No serious diseases reported.
Pests: Aphids, mites.

MAINTENANCE: Mow in spring or fall to keep plantings dense and compact.

PROPAGATION: Division: Fall, winter, or early spring, when not rapidly growing, are best times. Seed: Sow seed in cold frame in early fall. The following spring plants will be ready to transplant into small containers, and should be kept in a shaded area until fall.

*P**rimula** (prim´ū-là or pri-mū´là) From Latin *primus*, first, referring to the early flowering of many of the species. This genus is composed of about 400 species of primarily herbaceous perennials. They are native to the temperate zone of the northern hemisphere, with a few from the southern hemisphere.

FAMILY: PRIMULACEAE: Primrose family.

HABIT: Low, carpeting herbaceous ground covers.

RATE: Moderate rate of growth; space 10 to 14 inches apart from pint- or quart-sized containers.

LANDSCAPE VALUE: Very attractive as edging along a path or border in shaded or natural area. Excellent and well suited to growing along stream or pond bank. Often used to good effect interplanted with hostas as the leaves of primroses benefit from extra shading later in the season. Generally, these primroses are not successful in areas of hot summers and cold winters. No foot traffic.

*   *   *

P. × polyantha (pol-i-an´the) Many flowered (Figure 3–162).

COMMON NAME: Polyantha primrose, polyanthus.

HYBRID ORIGIN: *P. veris*, *P. elatior*, and *P. vulgaris*, in what order or combination I am not sure.

HARDINESS: Zones 3 to 8.

SIZE: To 12 inches tall, spreading 6 to 10 inches across.

FOLIAGE: Evergreen, arranged in a rosette, simple, obovate tapering to a winged petiole, to 10 inches long and 3 to 4 inches across, wrinkled, margin entire and wavy, light green.

FRUIT: Capsules, not ornamentally significant.

INFLORESCENCE: Many flowered erect umbels, on scapes 8 to 15 inches high; flowers to 1 inch across, single or double, regular, perfect, corolla funnelform or salverform, 5 lobed, effective in spring, many colored in combinations of white, gray, yellow, blue, copper, brown, purple, orange; bracts are involucral or leaflike.

SELECTED CULTIVARS AND VARIETIES: There are many unnamed types of polyanthus primrose. Named cultivars are also numerous and a complete description should be obtained prior to purchase.

OTHER SPECIES: *P. auriculata* (â-rik-ū-lā´tà) From Latin *auricula*, an ear, in reference to the earlike leaves.

COMMON NAME: Auricula primrose, auriculata, eared primrose.

NATIVE HABITAT: Alps of Europe.

HARDINESS: Zone 2 or 3.

SIZE: To 8 inches tall, spreading to 9 inches across.

FOLIAGE: 4 inches long, basal, gray-green, evergreen, in a rosette.

FLOWERS: Many colored, in umbels during spring.

P. denticulata (den-tik-ū-lā´tà) Finely toothed, the leaves.

COMMON NAME: Himalayan primrose.

NATIVE HABITAT: Himalayas.

HARDINESS: Zones 4 and 5.

SIZE: 10 inches tall, spreading to 12 inches across.

FOLIAGE: Oblong to oblanceolate, soft-green, to 6 inches long in a basal rosette.

FLOWERS: In shades of purple to pinkish-purple, with yellow or white eye; 1/2 inch in diameter, in dense heads in early spring.

P. japonica (jà-pon´i-kà) Of Japan.

Figure 3–162 *Primula × polyantha* (0.5 life-size)

COMMON NAME: Japanese primrose.

NATIVE HABITAT: Japan.

HARDINESS: Zone 5.

SIZE: 18 to 24 inches tall, spreading to 2 feet across, varying directly with moisture.

LEAVES: Obovate-oblong to spatulate, in basal rosette, to 10 inches long, medium green.

FLOWERS: Rose-purple or white, 1 inch across, in several superimposed umbels on scapes, from mid-spring through early summer.

P. sieboldii (sē-bold´i-i) After Siebold, German botanist and traveler.

COMMON NAME: Siebold primrose.

NATIVE HABITAT: Japan.

HARDINESS: Zone 4.

SIZE: 9 inches tall, spreading to 12 inches across.

LEAVES: About 4 to 8 inches long, ovate to oblong-ovate, margin lobed and toothed, light green.

FLOWERS: White, rose, or purple and about 1 1/2 inches across; borne in late spring to early summer.

SPECIAL CULTURE: Will withstand more sun than other species of primula.

Generally Applicable to All Species

CULTURE: Soil: Organically rich, well-drained soil is required for *P. polyantha* and *P. auricula*. They do best in slight to moderately acidic soils. *P. denticulata*, *P. japonica*, and *P. sieboldii* are acid-loving plants and grow best when pH is kept low.
Moisture: All require high levels of soil moisture, especially *P. japonica* and *P. sieboldii*. Plants should be planted in cool locations, sheltered from strong and drying winds.
Light: Light to moderate shade.

PATHOLOGY: Diseases: Bacterial leaf spot, anthracnose, leafspots, root rot, rusts, crown rot, leaf blight, viral mosaic, and spotted wilt; also aster yellows.
Pests: Aphids, corn root aphid, beetles, mealybugs, slugs, black vine weevil. Also bulb, stem, and southern root knot nematode. The two spotted mite is a problem, especially when grown in the sun.
Physiological: Iron-deficient chlorosis due to high soil pH is common.

MAINTENANCE: Lightly mulching with composted leaves on a yearly basis will maintain high levels of organic matter in the soil. Plants may be divided every couple of years to prevent overcrowding and increase floral display.

PROPAGATION: Division: Divide clumps in spring or fall.

Seed: Commercial supplies are available. Stratify moist for 1 month at about 35 °F; then sow in early spring. Keep medium temperature around 70 °F and expect germination in 3 to 5 weeks. Light is necessary for germination, thus seed should not be topdressed.

***P*ulmonaria** (pul-mo-nā´ri-à) From Latin, *pulmo*, pertaining to the lungs, one species having been regarded as a remedy for diseases of the lungs, or, as likely, because the leaves appear to look like diseased lungs. Commonly called lungwort, this genus is composed of about 12 species of hairy, spring flowering, rhizomatous, herbaceous perennials. They are native to Europe and Asia.

FAMILY: BORAGINACEAE: Borage family.

HABIT: Herbaceous, low, often clump forming, ground covers.

RATE: Moderate; space plants from pint- to quart-sized containers 12 to 18 inches apart.

LANDSCAPE VALUE: Often used as edging along shady paths or in a perennial border. Effective under shrubs and deciduous trees for facing, and as a general cover in moderate-sized areas. Valued both for attractive foliage and flowers. No foot traffic.

*   *   *

P. angustifolia (an-gus-ti-fō´li-à) Narrow leaved. Sometimes listed as *P. azurea*.

COMMON NAME: Cowslip lungwort.

NATIVE HABITAT: Europe.

HARDINESS: Zones 2 or 3 to 9.

SIZE: 8 to 12 inches tall, spreading 12 to 18 inches across.

FOLIAGE: Alternate, evergreen, simple; basal leaves long petioled, linear-lanceolate to oblong lanceolate; the few stem leaves are linear-lanceolate to lanceolate-elliptic; each entire, to 6 inches long, pubescent dark green.

INFLORESCENCE: Nodding terminal, forked cymes on 12 inch stems; flowers with tubular-campanulate calyx, 5 parted, pink corolla (in bud) matures to blue or violet, effective in early to midsummer.

FRUIT: Nutlets, not ornamentally significant.

SELECTED CULTIVARS AND VARIETIES: 'Alba' Flowers white.
'April Opal' Flowers intense blue.

'Azurea' Flowers sky blue.
'Munstead Blue' Flowers dark blue.
'Rubra' Flowers red to deep reddish-violet.

* * *

P. saccharata (sak-a-rā′ta) Sugared, possibly in reference to the white-powdered leaves (Figure 3–163).

COMMON NAME: Bethlehem sage, spotted mary, Virgin Mary's milkdrops.

NATIVE HABITAT: Europe.

HARDINESS: Zones 2 or 3 to 9.

SIZE: 9 to 12 inches tall, spreading to 2 feet across.

FOLIAGE: Alternate, simple, evergreen, profusely spotted white on dark green; basal leaves elliptic, narrowing to petiole, acuminate; stem leaves ovate-oblong, petioled or sessile; about 6 inches long, pubescent.

INFLORESCENCE: Terminal forked cymes; flowers 5 parted, calyx tubular, corolla 5 lobed with narrow tube and colored pink to rose in bud, opening blue to violet in early to midspring.

FRUIT: Similar to *P. angustifolia.*

SELECTED CULTIVARS AND VARIETIES: 'Alba' With white flowers.

'Margery Fish' With foliage heavily marbled silvery-white on dark green.

'Mrs. Moon' With large pink buds that open to showy gentain-blue.

OTHER SPECIES: *P. mollis* Foliage, as the name would suggest, is soft, in fact velvety to the touch; reaches 18 inches tall by 2 feet across with dark blue flowers that fade to tones of red and purple.

P. officinalis Roughly hairy, white spotted leaves; basal leaves ovate with cordate base and petioled; stem leaves ovate auriculate-cordate and sessile; flowers rose-violet to blue, sometimes reddish, blooming in early spring.

P. rubra Flowers coral-red, foliage pale green. This species is noteworthy because it blooms in early to midwinter. The cultivar 'Bowles Variety' has mottled foliage while 'Albororollata' has white flowers.

Generally Applicable to All Species

CULTURE: Soil: Best in organically rich, well-drained, loamy soils, but adaptable to most soils with good drainage.

Moisture: Soil should be kept moist and the location should be such that it stays relatively cool. Hot conditions cause the foliage to desiccate along the margins.

Light: Moderate to dense shade.

PATHOLOGY: No serious diseases or pests reported.

MAINTENANCE: Little or no maintenance is required. Vigor might be increased with occasional thinnings if plantings become overcrowded. Top-dressing with a light application of rotted leaf mold on an annual basis is good for the soil.

PROPAGATION: Division: Divide plants in late summer or fall.

Seed: Sow ripe seed in late summer.

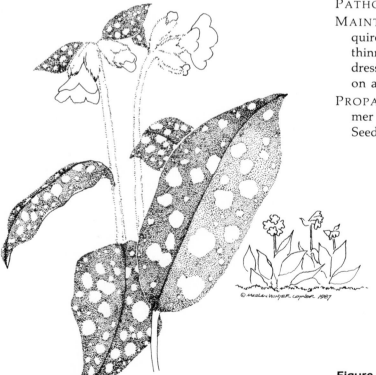

Figure 3–163 *Pulmonaria saccharata* 'Mrs. Moon' (0.5 life-size)

Figure 3-164 *Pyracantha koidzumii* 'Santa Cruz' (life-size)

++++++++++++++++++

*P*yracantha (pi-rä-kan´thä) From Greek, *pyr*, fire, and *akanthos*, a thorn, likely in reference to the brilliant red (firelike) berries and the spiny thorns. Not surprisingly, the common name of this genus is firethorn. It is composed of about 6 species of broadleaved, evergreen, usually thorny shrubs. They are native to southeastern Europe and Asia.

FAMILY: ROSACEAE: Rose family.

P. koidzumii (kōd-zum´ i-i) Meaning unknown (Figure 3-164).

COMMON NAME: Formosa firethorn.

NATIVE HABITAT: Japan.

HARDINESS: Zones 8 to 10.

DISCUSSION: The species is not itself useful as a ground cover, as it may reach 12 feet tall. Its ground covering cultivars are generally described as follows:

HABIT: Low growing, prostrate, mounding, woody shrubs.

RATE: Moderate to relatively fast; space plants from 1 to 2 gallon sized containers 3 to 4 feet apart.

LANDSCAPE VALUE: Good to use as low, informal hedges, or as rough general cover on rockery slopes of moderate to large size. Sometimes acceptable for foundation planting or as a specimen in rock garden. No foot traffic.

FOLIAGE: Semievergreen, alternate, simple, oblance-olate, 1 to 3 inches long, 1/2 to 3/4 inch across, entire, slightly serrate terminally, apex emarginate, shiny dark green, glabrous or sparsely pubescent on both sides.

INFLORESCENCE: Small corymbose racemes; flowers small, to 1/4 inch across, 5 parted, with two fertile ovules; corolla 5 petaled, white, effective early to midspring, fragrant but rather unpleasant smelling.

STEMS: Reddish and pubescent in youth, later purplish and glabrous.

FRUIT: Globose, 1/4 inch diameter red pomes that are profuse and very showy. Supposedly edible, but after tasting one it is hard to imagine why anything would include these in its diet.

OF SPECIAL INTEREST: Despite their bitter tasting fruit, pyracantha are quite valuable for attracting wildlife. Raccoons and gray squirrels are said to savor the fruits, as do quail and winter song birds. Additionally, the thorny branches offer good protection to birds who build their nests among the branches.

SELECTED CULTIVARS AND VARIETIES: 'Santa Cruz' Prostrate selection with dark green leaves; reported to be very resistant to scab; reaching 2 to 3 feet high.

'Walderi Prostrata' Prostrate selection with large fruit; reaching 2 to 3 feet tall.

HYBRIDS: *P.* × *'Red Elf'* Reported to be hardy in zones 7 to 10; dwarf mounding habit; foliage dark green; fruit bright red; reaching 2 feet tall and 3 to 4 feet across.

P. × *'Ruby Mound'* Reported hardy in zones 7 to 10; dwarf selection to about 1 1/2 feet tall and 3 feet across, with mounding habit.

Generally Applicable to the Cultivars and Hybrids

CULTURE: Adaptable to wide range of well-drained soils with pH from 7.0 to more acidic.

Moisture: Generally quite tolerant to drought. Plants are actually best sited in areas where air movement is good, as it lessens the incidence of disease. Occasional thorough watering in summer is beneficial.

Light: Full sun; protection from winter sun may be necessary in some states.

PATHOLOGY: Diseases: Bacterial fire blight, scab, twig blight, canker, root rot.
Pests: Apple aphid, hawthorn lace bug, scales (greedy, olive, and calico), mites, leaf rollers.

MAINTENANCE: Periodically prune out upright growing shoots that may arise. A light shearing on an annual basis will maintain compact habit.

PROPAGATION: Cuttings: Selected clones need to be propagated vegetatively and this is the most practical method. Both softwood and hardwood stem cuttings root readily. A 3000 ppm IBA treatment is helpful.

R*anunculus* (rȧ-nung´kū-lus) From Latin, *rana*, a frog, some species inhabiting marshy places where frogs abound. Commonly called frogwort—(really buttercup), this genus is composed of about 250 species of herbs that are native to many areas throughout northern temperate areas of the world.

FAMILY: RANUNCULACEAE: Buttercup family.

R. repens (rē´penz) Creeping (Figure 3–165).

COMMON NAME: Yellow gowan, butter daisy, creeping buttercup, creeping crowfoot, sitfast, creeping yellow gowan.

NATIVE HABITAT: Europe and Asia.

HARDINESS: Zones 3 to 9.

HABIT: Low, herbaceous, stoloniferous ground cover.

SIZE: 8 to 24 inches tall, spreading indefinitely.

RATE: Relatively fast; space plants from 3 to 4 inch diameter containers 12 to 18 inches apart.

LANDSCAPE VALUE: Often used as a general cover for moderate- to larger-sized areas. Spreading nature may be controlled easily by using boundaries such as walks, edging, or foundation to stop spread. Should not be planted with less vigorous plants of the same habit as it will quickly overrun them. No foot traffic.

FOLIAGE: Deciduous, alternate, primarily basal or arising from runners, 3 lobed, lobes to 2 inches long and wide, cut into 3 coarsely toothed segments, mostly long petioled, dark green above, lighter below, pubescent both sides.

STEMS: Hairy, rooting as they touch.

FLOWERS: Solitary, borne on erect branched stalks that are 1 to 2 feet tall, perfect, 5 parted; corolla 1/2 to 1 inch across, bright yellow, petals are 1/4 to 1/2 inch long with a small nectary at the base; effective midspring through late summer.

FRUIT: Achenes, not ornamentally significant.

SELECTED CULTIVARS AND VARIETIES: 'Pleniflorus' ('FlorePleno') Showy double-flowering selection.

CULTURE: Soil: Best in well-drained, rich loam, but adaptive to most soils. A pH that is either acid or alkaline, but near the neutral point, is best.

Figure 3–165 *Ranunculus repens* (0.5 life-size)

Moisture: Not notably tolerant to drought; soil should be kept slightly moist with regular watering.
Light: Full sun to moderate shade.

PATHOLOGY: Diseases: Leaf spots, mildews, root rots, rusts, viral curley-top, aster yellows.
Pests: No serious pests reported.

MAINTENANCE: Clip off flowers prior to seed ripening if self-sowing has become troublesome. Cut back runners as they outgrow their bounds.

PROPAGATION: Division: Divide plants in spring or fall.
Seed: Collect seed when ripe, store dry and cool, then sow the following spring.

Rhus (roos) Ancient Greek name for this genus. The sumacs, as species of this genus are commonly called, number about 150. Mostly they are erect, primarily dioecious vines, shrubs, and trees with milky sap. They are native to various temperate and subtropical regions.

FAMILY: ANACARDIACEAE: Cashew or sumac family.

R. aromatica (a-rō-mat´ik-à) Aromatic (Figure 3–166).

COMMON NAME: Fragrant sumac, sweet scented sumac, lemon sumac, polecat bush.

NATIVE HABITAT: Eastern North America, Vermont and Ontario to Minnesota, south to Florida and Louisiana.

HARDINESS: Zones 3 to 9.

HABIT: Low, variable, dense, suckering woody shrub.

SIZE: 1 1/2 to 6 feet tall, spreading from 6 to 10 feet across.

RATE: Slow to moderate; space plants from 1 to 2 gallon sized containers 3 1/2 to 4 1/2 feet apart.

LANDSCAPE VALUE: Valuable as an erosion controlling cover for banks and slopes. A fine and tough general cover for use in large- or moderate-sized areas. Frequently used on highway slopes and in median sections. No foot traffic.

DISCUSSION: This species is sometimes used for ground cover, although its variability (from seed-grown material) is often considered a drawback. For this reason, vegetatively grown material is more frequently cultivated and the cultivar 'Grow Low' is most common.

FOLIAGE: Deciduous, alternate, compound (petioled), trifoliate; leaflets subsessile, ovate in outline; terminal leaflet 1 1/2 to 3 inches long, apex acute or acuminate, base cuneate, coarsely toothed margin; lateral leaflets similar but about one-half as big; pubescent turning glabrous above, sometimes glossy, medium blue-green, often turning brilliant yellow or scarlet in autumn. Best color occurs when growing in full sun on well-drained soils; fragrant when crushed.

INFLORESCENCE: Polygamous or dioecious; male inflorescences are persistent catkins to 1 inch long; female inflorescences are short terminal panicles; yellow in either case and born in spring before leaves unfold.

FRUIT: Red, hairy drupes on female plants if dioecious, about 1/4 inch across, effective late summer to early fall, often persisting into winter, but becoming brownish; may be eaten by wildlife.

STEMS: Slender, pubescent, brown.

SELECTED CULTIVARS AND VARIETIES: 'Gro-low' A fine, low, wide spreading selection with glossy foliage; to about 2 feet tall and spreading to 8 feet across.

Figure 3–166 *Rhus aromatica* 'Gro Low' (0.5 life-size)

CULTURE: Soil: Adaptable to most well-drained soils. Best growth is in infertile, dry, well-drained soils with a moderately acidic reaction.
Moisture: Established plants are well equipped to withstand drought. An occasional deep watering in summer may be to some benefit.
Light: Full sun to moderate shade.

PATHOLOGY: Diseases: Fusarium wilt, cankers, leaf spots, powdery mildew, rusts, root rot.
Pests: Aphids, mites, scales.

MAINTENANCE: Annual hard shearing in midspring will keep plantings dense and compact. Dead branches may require selective pruning.

PROPAGATION: Cuttings: Both softwood and hardwood cuttings root quite well. Treatment with 2000 ppm IBA/talc may be beneficial. Medium should be very porous, and mist discontinued immediately upon rooting.
Layering: Pin branches down in spring and divide the following spring.
Seed: Sometimes propagated by seed, which is collected in mid to late summer and not allowed to dry out. Fruit turning red is an indicator of ripeness. The seed should be soaked in concentrated sulfuric acid for 1 hour at room temperature, washed, then fall planted outdoors or stratified at 40°F for 3 months prior to sowing in spring. As stated before, variability is high, and thus seed propagation is not of primary importance.

✛✛✛✛✛✛✛✛✛✛✛✛✛✛✛✛✛

Rosa (rō´zȧ) The ancient Latin name for the rose, likely from Celtic, *rhod*, meaning red. Currently appointed the national flower of the United States, there are more than 100 species of these thorny, sometimes clinging or trailing shrubs. They are native to temperate regions of the northern hemisphere.

FAMILY: ROSACEAE: Rose family.

HABIT: When unsupported, many types of roses make good ground covers. Generally, they are low, woody, trailing shrubs. With support they will climb.

RATE: Usually moderate to relatively fast; spacing of 3 1/2 to 4 1/2 feet is usually adequate for plants from gallon-sized containers.

LANDSCAPE VALUE: Good for rapid cover in medium to large areas. Often very attractive on a gently sloping bank or above a stone wall or on a terrace where branches can trail over. Excellent showy floral display, and a good deterrent to trespassing. No foot traffic.

* * *

R. banksiae (banks´ē-ē) After Lady Dorothea Banks.
COMMON NAME: Lady Banks's Rose, Banksia rose.
NATIVE HABITAT: China.
HARDINESS: Zones 7 to 10.
SIZE: To 18 inches tall when unsupported, spreading 10 to 20 feet across.
OF SPECIAL INTEREST: Withstanding moderate exposure to salt spray.
FOLIAGE: Evergreen, compound, alternate, odd pinnate; 3 to 5 leaflets, rarely 7, elliptic-ovate to oblong-lanceolate, 1 to 2 1/2 inches long, apex acute or obtuse, margin serrulate, shiny dark green, glabrous except at base of midrib on underside; rachis pubescent.
STEMS: With few thorns, climbing with support.
INFLORESCENCE: Many flowered umbels or solitary on thornless stalks; flowers are about 1 inch across, 5 petaled, white or yellow, effective in spring, sometimes again in fall, slightly fragrant.
FRUIT: Fleshy hip (receptacle) containing hairy achenes, red, pea sized.
SELECTED CULTIVARS AND VARIETIES: 'Alba Plena' Double white flowers.
'Lutea' Double flowered and yellow.
'Lutescens' Single and yellow flowered.

* * *

R. wichuraiana (wi-shur-ā-an´ȧ) After Max E. Wichura, Prussian diplomat who collected this species (Figure 3–167).
COMMON NAME: Memorial rose.
NATIVE HABITAT: China, Korea, and Japan.
HARDINESS: Zone 5 or 6 to 10.
SIZE: 12 to 18 inches tall, spreading 6 to 15 feet across.
FOLIAGE: Semievergreen, alternate, odd pinnately compound, 7 to 9 leaflets, leaflets 2/5 to 1 inch long, suborbicular to broad ovate or obovate, apex usually obtuse, margin coarsely serrate, shiny dark green on both sides, glabrous.
STEMS: With few thorns, green, trailing, rooting, climbing with support.
INFLORESCENCE: Many flowered pyramidal corymb; flowers about 2 inches across, fragrant, 5 petaled, many stamens, effective in late spring to early or midsummer.

FRUIT: 1/2 inch long egg-shaped, red, receptacle (hip) with hairy achenes inside, effective in fall.

SELECTED CULTIVARS AND VARIETIES: 'Poterifolia' Lower growing, less invasive selection; reaches 3 to 5 inches tall.

OTHER SPECIES AND HYBRIDS: *R. brachteata* Leaves dark green, evergreen; leaflets odd pinnate, 5 to 9, narrow-obovate in outline; flowers solitary, white, 2 to 3 inches across, pleasantly fragrant (like lemon). The cultivar 'Mermaid' has creamy yellow flowers that are somewhat larger. Fruit of the species is a globose orange-red hip.

R. × 'Max Graf' A good, large-scale ground cover with pinkish non-fragrant, golden centered, 3 inch diameter flowers in summer.

R. × 'Red Cascade' Miniature rose with 1 inch wide, double, red flowers.

Generally Applicable to All Species and Hybrids

CULTURE: Soil: Adaptable to most well-drained soils, but best when amended with liberal amounts of organic matter. A pH range from 5.5 to 7.0 is preferred.

Moisture: Roses need ample watering throughout the summer months. Water thoroughly when needed to keep soil moist. Avoid frequent light waterings as diseases will follow. Watering in the morning is the preferred time.

Light: Best in full sun, tolerant to light shade.

PATHOLOGY: Diseases: Black spot, canker, mildew, rust, viral mosaic, viral streak.

Pests: Sawflies, aphids, leaf hoppers, thrips, Japanese beetles, rose bugs and chaffers, mites, slugs.

MAINTENANCE: Selectively prune out dead or diseased stems and upright growing shoots immediately after flowering. A light mulching of the soil on an annual basis moderates the moisture level and helps keep roots cool.

PROPAGATION: Cuttings: Softwood cuttings in summer or semihardwood in fall can be rooted in a well-drained medium. Treating with 3000 ppm IBA is helpful.

Layering: Divide rooted stem sections in spring.

Seed: Harvest ripe seed, stratify 3 to 6 months at 40 °F; or fall sow in a cold frame for natural stratification.

*R*osmarinus (ros-ma-rī´nus, or roz-) From Latin, *ros,* dew (spray), and *marinus,* sea. Often inhabiting sea cliffs. This genus, commonly called rosemary, is popular among herbalists. It is composed of three species of evergreen herbs and subshrubs that are native to the Mediterranean region.

FAMILY: LAMIACEAE (LABIATAE): Mint family.

R. officinalis (o-fis-i-nā´lis) Common (Figure 3–168).

COMMON NAME: Rosemary.

Figure 3–167 *Rosa wichuraiana* (life-size)

NATIVE HABITAT: Southern Europe and Asia Minor.

HARDINESS: Zones 6 to 9.

DISCUSSION: The species itself is seldom used as a ground cover. While the species has a shrubby habit that often spreads broadly and ranges from 2 to 4 feet tall (occasionally to 6 feet), it is the cultivars with more uniform and lower habit that are prized for their ability to cover the ground. The size and spread of these cultivars will be addressed on an individual basis.

NATIVE HABITAT: Southern Europe and Asia Minor.

HARDINESS: Zones 6 to 9.

RATE: Moderate; space plants from quart- to gallon-sized containers 2 1/2 to 4 feet apart.

LANDSCAPE VALUE: Exceptional when allowed to cascade over retaining walls and rock ledges. Nicely suited to use on slopes where their branching patterns are well displayed. Other uses include general cover in moderate to larger areas, walkway edging, dwarf hedging, and specimens in the rock garden. No foot traffic.

FOLIAGE: Evergreen, simple, opposite, linear, 1/2 to 1 1/2 inches long, apex obtuse, tomentose, leathery, gray-green, tomentose and lighter below, aromatic.

STEMS: Squarish, pubescent in youth, becoming woody with age.

INFLORESCENCE: Few flowered verticillasters that are arranged in short axillary racemes on previous year's growth; calyx campanulate, corolla with upper lip entire and 2 lobed, lower lip 3 lobed with concave middle lobe, about 5/16 inch long, pale blue, rarely pink or white. Effective in spring.

FRUIT: Of 4 glabrous nutlets, not ornamentally significant.

SELECTED CULTIVARS AND VARIETIES: 'Collingwood Ingram' Graceful selection with curving branches; to 2 feet tall and 4 feet across.
'Huntington Carpet' Very compact, dwarf, light green, mounding clone. Reaches about 1 foot high and spreads to 4 feet across. Flowers are darker blue than the species. This cultivar originated at the Huntington Botanic Gardens in San Marino, California.
'Lockwood de Forest' Semiprostrate mounding selection that reaches about 2 feet tall and spreads to 4, sometimes 6, feet across. Flowers are dark blue.
'Prostratus' (trailing rosemary) Prostrate habit with foliage and flowers that are typical of the species. Probably hardy northward only to zone 8.

Figure 3–168 *Rosmarinus officinalis* 'Prostratus' (life-size)

OF SPECIAL INTEREST: In ancient times rosemary was considered to have tremendous curative properties. They ranged from reversing baldness, loss of speech, and the process of aging, to preventing drowsiness, tightening loose teeth, and returning speech to the dumb. Presently, few believe that rosemary can do all of these; however, the rinsing of one's skin in water in which rosemary stems and leaves have been boiled is said to be invigorating. Foliage is also used as a seasoning.

CULTURE: Soil: Adaptable to most soils provided drainage is excellent. Tolerant to infertile soils. Best with pH of 6.0 to 7.5.
Moisture: Highly tolerant to drought, but occasional deep watering in summer is recommended.
Light: Full sun.

PATHOLOGY: Diseases: None serious
Pests: Aphids, mites.

MAINTENANCE: Lightly shear plants immediately following bloom. This will promote branching and keep plants dense.

PROPAGATION: Cuttings: Cuttings are easily rooted during the growing season.
Seed: Commercial supplies are readily obtained for the species. Germination takes place in about 3 weeks at 55°F.

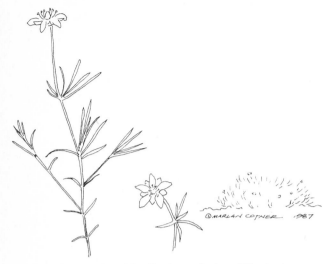

Figure 3-169 *Sagina subulata* (life-size)

+ + + + + + + + + + + + + + + + + +

*S**agina** (sȧ-jī´nȧ) Ancient name of spurry, which was originally regarded as a species of this genus, or possibly from Latin, *sagina* (fodder), sheep being fed with a related plant. Whatever the origin of the name, these pearlworts as they are commonly called, number about 25 species. They are small, herbaceous, often tufted annuals and perennials. They are native to various areas of the northern temperate zone and cool mountainous regions.

FAMILY: CARYOPHYLLACEAE: Pink family.

S. subulata (sub-ū-lā´tȧ) Awl shaped, the leaves (Figure 3-169).

COMMON NAME: Corsican pearlwort.

NATIVE HABITAT: Europe.

HARDINESS: Zones 5 to 10.

HABIT: Low, mat forming, mosslike, herbaceous ground cover.

SIZE: 2 to 4 inches tall, spreading indefinitely.

RATE: Moderate; space plants from 3 to 4 inch diameter containers 6 to 10 inches apart.

LANDSCAPE VALUE: Valuable as a filler between stepping stones and cracks in patio or walk. Tolerant of moderate foot traffic.

FOLIAGE: Evergreen, opposite, simple, often united at the base, subulate to linear-lanceolate, to about 1/4 inch long, aristate, glabrous and glandular dotted both sides, bright light green.

FLOWERS: Mostly solitary on stems to 5 inches tall, tiny, on thin pedicels about 1 inch long, 5 parted;

corolla sometimes apetalous, as long as sepals, white, about 3/16 inch across; stamens 10; effective in early to midsummer.

FRUIT: 4 or 5 valved capsule, not ornamentally significant.

SELECTED CULTIVARS AND VARIETIES: 'Aurea' Leaves light yellowish.

CULTURE: Soil: Organically rich loamy soil is necessary for good growth. Tolerant to other well-drained soils, but appearance is not as lush. Usually not successful in clay.
Moisture: Not notably tolerant to drought; soil should be kept moist with regular watering.
Light: Full sun to light shade.

PATHOLOGY: No serious diseases and pests reported.

MAINTENANCE: Occasionally, plantings will mound up when overcrowded. When mounding occurs, simply cut out the mounded portion and gently press the surrounding mat into the open area. Mow and bag clippings just after flowering to prevent seed spread.

PROPAGATION: Division: Plants are simply divided in spring or fall.
Seed: Commercial supplies are available. Germination takes 2 to 3 weeks at 65° to 75°F.

+ + + + + + + + + + + + + + + + + +

*S**alix** (sā´liks) Latin name for willow, likely from Celtic, *sal*, near, and *lis*, meaning water. Willows often inhabit banks of streams. Often referred to as willow or osier, this genus is made up of about 300 species of dioecious shrubs and trees, with some alpine species that are nearly herbaceous. They are native to cold and temperate regions throughout the northern hemisphere, with a few from the southern hemisphere.

FAMILY: SALICACEAE: Willow family.

OF SPECIAL INTEREST: The flowers of willows are valuable as they provide nutrition to hummingbirds and bees very early in the spring when little else is in bloom.

S. purpurea 'Nana' (pẽr-pū´rē-ȧ, nā´nȧ) Purple and dwarf, the stems and habit, respectively (Figure 3-170).

COMMON NAME: Dwarf arctic willow, dwarf purple osier willow.

NATIVE HABITAT: Europe and west to northeastern Asia (the species).

HARDINESS: Zones 4 to 9.

DISCUSSION: The species is not itself useful as a ground cover. The cultivar 'Nana' is described as follows:

HABIT: Low, woody, upright branched, suckering shrub.

SIZE: 3 to 4 feet tall, spreading 3 to 5 feet across.

RATE: Relatively fast; space plants from 1 to 2 gallon sized containers 2 1/2 to 3 feet apart.

LANDSCAPE VALUE: Works well as a low, formal, or informal hedge. Effective also as a general cover for large areas. No foot traffic.

FOLIAGE: Deciduous, opposite, simple, short petioled, primarily oblanceolate, 1 to 1 1/2 inches long, blue-green.

STEMS: Slender, erect, purplish to light grayish-brown, glabrous.

INFLORESCENCE: Catkins; male with yellow anthers; female gray; both appearing in early spring prior to leaves.

FRUIT: Small capsules, seeds covered with hair.

* * *

S. repens (rē'penz) Creeping.

COMMON NAME: Creeping Willow.

NATIVE HABITAT: Europe and Asia.

HARDINESS: Zones 4 to 9.

HABIT: Woody, low, generally prostrate, sometimes ascending, wide spreading, rhizomatous, clump forming ground cover.

SIZE: 2 to 3 feet tall, spreading 6 to 8 feet across.

RATE: Moderate; space plants from 1 to 2 gallon sized containers 2 1/2 to 3 1/2 feet apart.

LANDSCAPE VALUE: Useful as a general cover for moderate to large areas, or as a specimen in an ornamental shrub and tree border. Grows well along stream or pondside and in low marshy areas. No foot traffic.

FOLIAGE: Deciduous, simple, alternate, short petioled, ovate-elliptic, 3/4 to 2 inches long, apex recurved, pointed or rather obtuse; margins entire, glandular, slightly rolled under; pubescent in youth, becoming glabrous above, gray-green above, grayish white-green below.

STEMS: Procumbent, slender, pale, hairy in youth, becoming glabrous with age, brown with hints of red or purple.

INFLORESCENCE: Cylindrical catkins; male are a little over 1/2 inch long with showy white anthers; female longer to 1 1/2 inches; both borne in early spring prior to leaf emergence.

FRUIT: Small capsules, seeds covered with hair.

SELECTED CULTIVARS AND VARIETIES: 'Argentea' Silvery colored silky foliage, due to upper leaf surface remaining tomentose into maturity. Sometimes listed as 'Nitida.'

Figure 3–170 *Salix purpurea* 'Nana' (0.75 life-size)

* * *

S. tristis (tris´tis) Sad, bitter, dull, in reference to weeping habit.

COMMON NAME: Dwarf gray willow.

NATIVE HABITAT: Eastern North America from Maine to Minnesota and Montana, and south to northern Florida and northern Oklahoma.

HARDINESS: Zones 2 to 9.

HABIT: Low, woody, spreading, shrubby ground cover.

SIZE: 1 1/2 to 4 feet tall, spreading 3+ feet across.

RATE: Relatively slow; space plants from 1 to 2 gallon sized containers 2 to 3 feet apart.

LANDSCAPE VALUE: Good general cover for dry, open sites of moderate to large area. Also useful as a soil retainer on banks. No foot traffic.

FOLIAGE: Deciduous, alternate, simple, short petioled, crowded, narrowly ovate, apex blunt pointed, margins entire or slightly toothed, rolling downward slightly, gray-green above, gray and tomentose below, leathery.

STEMS: Gray-tomentose in youth, becoming glabrous with age.

INFLORESCENCE: Catkins, oval or rounded, appearing in early spring before leaf emergence.

FRUIT: Nonshowy capsules, seeds covered with hair.

OTHER SPECIES: *S. rosmarinifolia* Similar to *S. repens,* but with linear leaves that have 10 to 12 pairs of lateral veins, and catkins globose. Hardy in zones 4 to 9.

S. apoda Prostrate growing shrub with shiny green leaves above and pale green below. Male catkins are silvery, furry, over 1 inch long, and become bright yellow with maturity of the anthers. It is relatively drought tolerant and intolerant to high humidity.

S. uva-ursi (bearberry willow) Prostrate, mat forming species to 2 inches tall, with rather thick brown stems. Leaves are obovate, to 1 inch long, shiny green above, glaucous below; catkins reddish, appearing with the leaves; native from Labrador to Alaska and south to mountain summits of the northeast United States. Hardy in zones 1 to 4 and mild areas with cool winters. Commonly called bearberry willow.

Generally Applicable to All Species

CULTURE: Soil: For the most part, best growth is obtained in sandy, somwhat infertile soils. Growth becomes rank and leggy in rich soils. The pH is generally best either acid or alkaline within a point or so of the neutral mark.

Moisture: Generally best when soil is highly moist to saturated, the exception being *S. tristis,* which is better in dry conditions, and *S. rosmarinifolia,* which does not tolerate high humidity.

Light: All are at their best in full sun, but will tolerate light shade.

PATHOLOGY: Diseases: Bacterial twig blight, crown gall, leaf blight, cankers, gray scab, leaf spots, powdery mildew, rust, tar spot, witches'-broom. Pests: Aphids, willow leaf beetle, pinecone gall, willowbeaked gall midge, willow lace bug, willow flea beetle, mottled willow borer, poplar borer, satin moth, willow shoot sawfly, willow scurfy scale.

MAINTENANCE: Annual shearing of new growth in late spring will promote dense, compact habit. Sometimes shearing to within 6 inches of the ground every 2 years is practiced for the same reasons and to rejuvenate.

PROPAGATION: Cuttings: Cuttings, both softwood and hardwood, root very easily. Preformed root initials are present and one can expect high rates of success. Max Kawase of Ohio State University is credited with the extraction of a substance from willow that aids in the rooting process. To help root other hard-to-root plants, young willow stems can be cut into small pieces and steeped in water; the cuttings to be rooted are then allowed to stand in the extract for several hours. IBA applied thereafter is often more successful than the use of IBA alone.

*S*antolina (san-tō-lī´na) Meaning unknown. The genus is composed of about eight species of aromatic herbs, subshrubs, and shrubs that are native to the Mediterranean region.

FAMILY: ASTERACEAE (COMPOSITAE): Sunflower family.

LANDSCAPE VALUE: Excellent edging along walk or border. Fire resistance makes them suitable for foundation planting or as informal covers for facing fences. Sheared, they are fine as dwarf hedges. Can also be used to good effect as specimens in the rock garden. No foot traffic.

* * *

S. chamaecyparissus (kam-e-sip-ãr-i´sus) Ground cypress. Sometimes listed as *S. incana* (Figure 3–171).

COMMON NAME: Lavender cotton.

NATIVE HABITAT: Southern Europe and North Africa.

HARDINESS: Zones 6 to 9.

HABIT: Woody, low, spreading shrub.

SIZE: 1 1/2 to 2 feet tall, spreading 2 to 4 feet across.

RATE: Moderate to relatively fast; space plants from quart- to gallon-sized containers 1 1/2 to 2 1/2 feet apart.

FOLIAGE: Evergreen, alternate, pinnately divided into narrow segments, cylindrical, to 1 3/8 inch long, segments about 1/8 inch long, tomentose, silvery gray.

STEMS: To 2 feet, many branched, stiff.

INFLORESCENCE: Many solitary heads on long peduncles; heads to about 3/4 inch across, globular, regular; flowers with 4- to 5-lobed yellow corolla, effective in summer for about 1 month.

FRUIT: Achenes, not ornamentally significant.

SELECTED CULTIVARS AND VARIETIES: 'Nana' Dwarf selection to about 10 inches tall.

* * *

S. virens (vi´renz): Green, the leaves.

COMMON NAME: Green lavender-cotton.

NATIVE HABITAT: Mediterranean region.

HARDINESS: Zones 7 to 9.

HABIT: Woody, low, spreading shrub.

SIZE: 1 to 2 feet tall, spreading 2 to 4 feet across.

RATE: Moderate to relatively fast; space plants from quart- to gallon-sized containers 1 1/2 to 2 1/2 feet apart.

DISCUSSION: Similar to *S. chamaecyparissus* in habit; foliage 1 to 2 inches long, linear in outline, glabrous dark green; heads also globular, to 3/4 inches across and yellow, blooming mid to late spring.

Generally Applicable to Both Species

CULTURE: Soil: Adaptable to most any soil provided that drainage is excellent.
Moisture: Extremely tolerant to drought; an occasional deep watering in summer may be of some benefit.
Light: Full sun.

PATHOLOGY: No serious pathogens reported.

MAINTENANCE: Shear foliage halfway to the ground on an annual basis immediately after flowering. This will keep the plants dense, thus preventing woody stems (that can be unsightly) from showing.

PROPAGATION: Cuttings: Cuttings can be taken throughout the growing season; the medium should be very porous and kept relatively dry.
Seed: Collect seed when ripe, store dry and cool (40 °F), and sow the following spring.

*S*aponaria (sap-ō-nā´ri-à) From Latin, *sapo*, soap, the bruised leaves of *S. officinalis* producing a lather, and once used as a soap substitute. The roots, too, are reported to contain this juice that forms a lather. Soapworts, as they are commonly called, consist of about 30 species of usually coarse textured, annual, biennial, and perennial, erect or decumbent herbs. They are native to the Mediterranean region.

FAMILY: CARYOPHYLLACEAE: Chickweed or pink family.

Figure 3–171 *Santolina chamaecyparissus* (life-size)

S. ocymoides (ō-kim-oi´dez, or ō-sim-oi´dez) Ocimumlike (Figure 3–172).

COMMON NAME: Rock soapwort.

NATIVE HABITAT: Southern and central Europe.

HARDINESS: Zones 2 to 8.

HABIT: Low, mat forming, trailing but usually not strongly rooting, herbaceous ground cover.

SIZE: 4 to 8 inches tall, spreading to over 3 feet across.

RATE: Moderate to relatively fast; space plants from 2 1/2 to 3 1/2 inch diameter containers 12 to 18 inches apart.

LANDSCAPE VALUE: Exceptionally colorful flowers are like neon in the rock garden when used as a specimen. Fine as an edging for walks and garden paths. In a terrace or near the top of a retaining wall, their graceful stems look fantastic as they trail over. No foot traffic.

FOLIAGE: Evergreen, opposite, simple, small, to about 1 inch long and 3/8 inch across, spatulate or elliptic to ovate-lanceolate, margin entire, parallel veined, lower leaves short petioled, upper sessile, apex acute, dark green, medium to fine textured.

STEMS: Procumbent to ascending, many branched, reddish, with swollen nodes.

INFLORESCENCE: Broad, loose cymes; flowers tubular, 5 parted, calyx cylindrical and toothed; corolla purplish-pink, 5 petaled, petals narrowing to petiolelike bases, upper portion expanded and flattened, sometimes with shallow notches, effective for about 4 weeks in late spring and early summer, thereafter sporadically until fall.

FRUIT: 4-toothed capsules, not ornamentally significant.

SELECTED CULTIVARS AND VARIETIES: 'Alba' Flowers pure white.
'Rosea' Flowers bright rosy pinkish-red.
'Rubra' Flowers deep red.
'Splendens' Deep rose colored flowers that are larger than the species.

CULTURE: Adaptable to most well-drained soils. Tolerant to relatively infertile soils.
Moisture: Not notably tolerant to drought; soil should be kept slightly moist with occasional watering as needed.
Light: Full sun

OF SPECIAL INTEREST: This is one plant that should be grown in containers and not handled bare root, as it does not transplant easily.

PATHOLOGY: Diseases: None serious.
Pests: Aphids, easily controlled.

MAINTENANCE: Shear or mow plantings immediately after flowering to keep them dense, neat, and compact.

PROPAGATION: Cuttings: Take cuttings in late summer to early fall, treat with mild root inducing preparation, and give bottom heat of 70 °F. Pinch frequently in youth.
Division: Simply divide plants in spring or fall.
Seed: Commercial supplies are available. Germination takes 2 to 3 weeks at 60 °F.

Figure 3–172 *Saponaria ocymoides* (life-size)

S*arcococca* (sär-ko-kok´a) From Greek, *sarx*, flesh, and *kokkos*, a berry, the fruits being fleshy. The sweetboxes, as they are commonly referred to, are monoecious, evergreen shrubs that are native to western China, the Himalayas, and southeastern Asia.

FAMILY: BUXACEAE: Box family.

S. hookerana var. *humulis* (hook-ẽr-ā´na, hū-mū´lis) After Sir Joseph Hooker, and low growing, dwarf, respectively (Figure 3–173).

COMMON NAME: Dwarf Himalayan sarcococca, dwarf Himalayan sweet box.

NATIVE HABITAT: Western China.

HARDINESS: Zones 5 to 8; protection is needed in northern regions where snow cover is unreliable.

HABIT: Low, stoloniferous, spreading, woody shrub.

SIZE: 1 1/2 to 2 feet tall, spreading to 6 or more feet across.

RATE: Relatively slow; space plants from gallon-sized containers 3 to 4 feet apart.

LANDSCAPE VALUE: Fine for mass planting as a general cover in moderate- or large-sized areas. A good companion for rhododendrons and other ericaceous shrubs. Relatively tolerant of air pollution, but intolerant of foot traffic.

FOLIAGE: Evergreen, alternate, simple, lanceolate to narrowly elliptic or elliptic, 1 to 3 inches long by 1/2 to 3/4 inches across, margin entire, apex acute, base cuneate, glabrous, thin, leathery, shiny dark green, petiole short, being about 1/3 inch long.

INFLORESCENCE: Short 4-flowered axillary racemes, obscured somewhat by the foliage; flowers small, regular, apetalous; male flowers usually with 4 stamens, anthers cream colored or pinkish; female flowers below male, with 2 to 3 stigmas; overall appearing white, born in fall and winter, not very showy, but quite fragrant.

FRUIT: Black drupe.

CULTURE: Soil: Best growth in organically rich, well-drained, acidic soil.
Moisture: Soil should be kept slightly moist with occasional watering.
Light: Moderate to dense shade, except in cool summer areas where full sun and light shade conditions can also be tolerated.

PATHOLOGY: No serious diseases or pests reported.

MAINTENANCE: Annual light shearing helps keep plants dense and compact.

Figure 3–173 *Sarcococca hookeriana* var. *humulis* (0.75 life-size)

PROPAGATION: Cuttings: Cuttings taken in late fall and treated with preparation of 3000 ppm IBA/talc root readily.
Seed: Ripe seed is ready for germination and requires no stratification.

S*atureja* (sat-ū-rē´ya or sat-ū-rē´a) Sometimes spelled *Satureia*. The Old Latin name for savory, this genus name is probably derived from Arabic, *sattar*, a name applied to labiates in general. It is composed of about 30 species of herbaceous plants that are native to temperate and warm areas.

FAMILY: LAMIACEAE (LABIATEAE): Mint family.

LANDSCAPE VALUE: General covers for use in small- or moderate-sized areas where their spread can be contained. Withstanding very limited foot traffic.

* * *

S. douglasii (dug-las´i-ī) Formerly a species of the genus *Micromeria*. Named for Douglas (Figure 3–174).

COMMON NAME: Yerba Buena, the original name of San Francisco (after this plant).

NATIVE HABITAT: Western United States, Los Angeles County to British Columbia.

HARDINESS: Zone 4.

HABIT: Low, matlike, sometimes trailing, herbaceous ground cover.

SIZE: Usually around 2 inches tall, but varying somewhat, spreading to 3 feet across.

RATE: Moderate to relatively fast; space plants from quart-sized containers 10 to 16 inches apart.

FOLIAGE: Opposite, simple, petioled, ovate, to 1 1/4 inches long by 1 inch across, apex obtuse, margin crenate or crenate-serrate, glabrous, aromatic.

STEMS: Trailing, rooting, reaching 2 or more feet long.

FLOWERS: Solitary, axillary, tiny, irregular, on pedicils, calyx to 3/16 inch long, 5 toothed; corolla to 3/8 inch long, 2 lipped, upper lip erect, lower spreading and 3 lobed, white to purplish, effective early spring to late summer.

FRUIT: Of 4 globose nutlets, not ornamentally significant.

OF SPECIAL INTEREST: Dried leaves are reported useful in making tea.

*   *   *

S. glabella (glȧ-bel´ȧ) Meaning smooth.

NATIVE HABITAT: United States from Kentucky to Arkansas.

HARDINESS: Zone 6.

HABIT: Prostrate, mat forming, herbaceous ground cover.

SIZE: To 2 feet tall, spreading to 4 feet across.

RATE: Moderate to relatively fast; space plants from pint- to quart-sized containers 10 to 16 inches apart.

DISCUSSION: Leaves are oblanceolate, 3/4 to 1 1/2 inches long by 3/4 inch across, aromatic, dark green; attractive, small, purple flowers are born on thin stems to 3 inches long.

Generally Applicable to Both Species

CULTURE: Soil: Organically rich loam with good drainage is needed for best growth.
Moisture: Not notably tolerant to drought; soil should be kept moist with regular watering.
Light: Full sun in cool or humid areas such as along the coast; partial shade needed in warmer inland locations.

PATHOLOGY: No serious diseases or pests reported.

MAINTENANCE: Shear or mow in early spring to keep neat and compact.

PROPAGATION: Cuttings: Softwood cuttings are easily rooted early to midsummer.
Divison: Simply divide plants in spring or fall.
Seed: Seed germination is generally poor and is seldom practiced.

*S*axifraga (saks-if´rȧ-gȧ) From Latin, *saxum*, a rock or stone, and *frango*, to break; the application of the root words as they apply to this genus are disputed. It is likely that they refer to the plant's ability to exploit rocky terrain. The genus is commonly called saxifrage and consists of around 300 species of annual, biennial, and mostly perennial herbs of various habit. They are native to mountainous and rocky areas of temperate regions in Europe, Asia, North Africa, and North and South America.

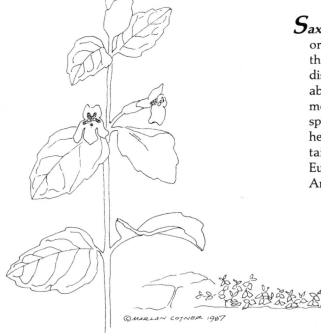

Figure 3–174 *Satureja douglasii* (life-size)

FAMILY: SAXIFRAGACEAE: Saxifrage family.

S. stolonifera (stō-lon-if´ẽr-á) With runners (Figure 3–175).

COMMON NAME: Strawberry saxifrage, mother-of-thousands, strawberry geranium, strawberry begonia, creeping sailor, beefsteak geranium.

NATIVE HABITAT: Eastern Asia.

HARDINESS: Zones 7 to 10.

HABIT: Low, stoloniferous, spreading, herbaceous ground cover.

SIZE: 6 to 8 inches tall, spreading indefinitely.

RATE: Relatively fast; space plants from pint- to quart-sized containers 12 to 18 inches apart.

LANDSCAPE VALUE: Good general cover for small- to moderate-sized areas. Fine companion for facing azaleas, rhododendrons, and other shade loving plants. Sometimes used as a rock garden specimen. No foot traffic.

FOLIAGE: Evergreen, mostly basal, stem leaves alternate, orbicular or cordate in outline, 2 to 4 inches long and wide, margin crenate to dentate and ciliate, conspicuous silvery gray venation above, gray-green and sparsely hairy over remainder of upper leaf surfaces; glabrous, warty and reddish below; glandular dotted both sides, petioles long with pink hairs. Overall very attractive.

INFLORESCENCE: Panicles on stems to 2 feet; flowers irregular, dainty, 5 petaled, two petals longer than other three, to 1 inch across, white, 10 stamened, effective spring and summer.

FRUIT: 2 beaked capsule, not ornamentally significant.

SELECTED CULTIVARS AND VARIETIES: 'Tricolor' Leaves variegated dark green, gray-green, and ivory-white, flushed with tones of rose; may be difficult to obtain.

CULTURE: Soil: Best in organically rich loam with good drainage. Adaptable to range of pH from 7.5 to more acidic.
Moisture: Not greatly tolerant of drought; soil should be kept moist with regular watering. Useful only in areas where summer temperatures are relatively cool. Plant in locations that are sheltered from strong winds.
Light: Moderate to dense shade.

PATHOLOGY: Diseases: Leaf spots, powdery mildew, rusts.
Pests: Slugs, snails, grape rootworm.

MAINTENANCE: Clip back plants as they outgrow their confines.

PROPAGATION: Division: Simply divide rooted plantlets that form on runners. If the plantlets have not yet rooted, place them in a porous medium and root as if they were cuttings.

OF SPECIAL INTEREST: Many other species of *Saxifraga* make excellent ground covers. For further reading, see the book *Saxifrages and Related Genera*, by Fritz Kohlein, Timber Press, 1984.

Figure 3–175 *Saxifraga stolonifera* (0.75 life-size)

✝✝✝✝✝✝✝✝✝✝✝✝✝✝✝✝✝

Sedum (sē´dum) From Latin, *sedate*, to assuage, in allusion to the healing properties of the houseleek to which the name was applied, as well as to the stonecrop, by Roman writers. Others feel that the name implies to sit (from Latin *sedere*, referring to the way some species grow upon rocks.

FAMILY: CRASSULACEAE: Stonecrop family.

The genus *Sedum*, commonly called orpine or stonecrop, is very large. It is composed of an estimated 500 or more species of fleshy, succculent, usually quite hardy perennial herbs and subshrubs. They are native to the northern temperate zone and the mountains of the tropics. Many sedums are of worth to the gardener who desires a reliable ground cover. A few of the more commonly cultivated species are addressed here. However, those wishing to study the genus in depth are encouraged to consult the books *An Account of the Genus Sedum*, by L. R. Praeger, and *Handbook of Cultivated Sedums*, by Ronald L. Evans (see bibliography). In general, sedums are well adapted to dry environments, and the reasons for this are not as obvious as a brief glance at their morphology might indicate. Initially, one will observe that the leaves and stems are usually rather fleshy, leathery to the touch, and covered with a waxy secretion. Certainly, these properties aid in the conservation of water. Moreover, the foliage is usually thick and rounded. This indicates a high volume to exposed surface area ratio, another important adaptation to help retain the precious little water that might be available in a native rocky environment. Important as these adaptations may seem, there is also a very important physiological adaptation that accounts for the ability of the sedums to conserve water. Recalling general botany, you will remember that the surfaces of plant leaves and, to a lesser extent, the stems are covered with specialized structures called stomates. Functioning in the process of transpiration and as a port of gas exchange, the stomates of a typical broadleaved dicotyledonous plant are open in the day and closed at night. This allows for cooling through transpiration and for gas exchange for the photosynthetic process. In several species (but not all) of the family Crassulaceae (and others), the stomates behave in just the opposite fashion. That is, the stomates remain closed during the day and open at night. They are able to do so because of a unique metabolism known as crassulacian acid metabolism (CAM). In simplified terms, during the process of CAM, carbon dioxide is taken into the plant and stored in the cells at night and then used in the photosynthetic process the following day. By doing so, the bulk of gas exchange occurs at night when temperatures are cool and water loss via transpiration is minimized.

General Description of Sedums for Ground Cover

LEAVES: Evergreen, semievergreen, or deciduous, usually alternate, often small and overlapping, sometimes whorled, usually sessile, fleshy, thick, in various colors ranging from yellow to green, blue, purple, and reddish; glabrous.

STEMS: Most often fleshy, sometimes woody at the base, branched or unbranched, upright, spreading or decumbent, often trailing and rooting.

INFLORESCENCE: Commonly terminal, usually a cyme of 2 or more cincinni (separate segments), flowers usually 5 parted, sepals separate or nearly so, lobes equal or unequal, sometimes spurred, corolla with petals separate or shortly united, frequently colored yellow, white, or shades of red or pink, outspreading stamens usually 10, less often 5, carpels usually separate or nearly so.

FRUIT: Follicles, which in most cases are of little ornamental significance.

*   *   *

S. acre (ahk´rē) Biting or sharp to the taste.

COMMON NAME: Goldmoss stonecrop, mossy stonecrop, golden carpet.

NATIVE HABITAT: Europe, Asia Minor, and North Africa.

HARDINESS: Zones 3 to 9.

HABIT: Very low, prostrate spreading, matlike, succulent cover.

SIZE: 1 to 2 inches tall, spreading indefinitely.

RATE: Moderate to relatively fast; sometimes invasive; space plants from 2 to 3 inch diameter containers 8 to 10 inches apart.

LANDSCAPE VALUE: General cover or specimen in the rock garden or small bed area if contained. Filler between cracks in stone retaining walls and as general covers for areas where they can be allowed to spread freely. Limited foot traffic.

DISCUSSION: A very common species that possesses very tiny evergreen, light green, triangular-ovoid (to 3/16 inch long), sessile, overlapping foliage. Flowers are bright yellow to 1/2 inch across in spring. Cultivars include 'Aureus,' with shoot tips

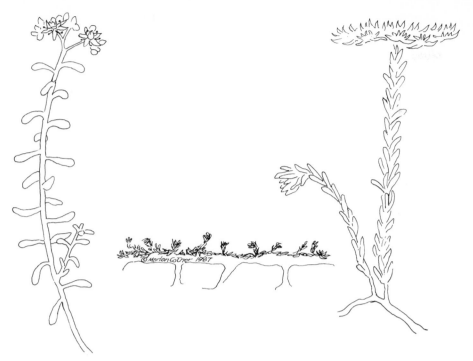

Figure 3–176 *Sedum album* 'Murale' (left) and *Sedum lydium* (right) (both life-size)

bright yellow, 'Elegans,' shoot tips silvery, 'Majus,' leaves and flowers somewhat larger, and 'Minus,' small in all respects.

*   *   *

S. album (al´bum) White, the flowers (Figure 3–176).

COMMON NAME: Worm grass, chubby fingers sedum, white flowered sedum.

NATIVE HABITAT: Europe and Siberia, West Asia, North Africa.

HARDINESS: Zone 3.

HABIT: Low, creeping, succulent ground cover.

SIZE: 2 to 4 inches high, spreading indefinitely.

RATE: Moderate to relatively fast; space plants 8 to 12 inches apart, from 2 to 3 inch diameter containers.

LANDSCAPE VALUE: Good filler between stones in retaining walls or barriers, and among large rocks for accent. Also acceptable as a general cover in moderate to larger bounded areas. Accepts limited foot traffic.

DISCUSSION: Another very commonly cultivated ground cover. Characterized by evergreen, sessile, flat, medium-green leaves that are linear-oblong to ovate or globose, not crowded, 1/8 to 5/8 inch long and nearly cylindrical in cross section. Flowers are

numerous, white, and borne in flat panicles, 1 to 2 inches across. Panicles are upon tall shoots that rise above the foliage in midsummer. The cultivar 'Murale' is probably more common than the species and is similar, with leaves somewhat larger and often tinged reddish purple. Flowers are often pinkish. 'Chloroticum' may be more difficult to obtain. It has vivid green leaves with white flowers. 'Brevifolium' has shorter leaves than the species.

*   *   *

S. anglicum (ang´gli-kum) Of England.

COMMON NAME: English sedum.

NATIVE HABITAT: Western Europe from Norway to Spain.

HARDINESS: Zone 3.

HABIT: Matlike, creeping, succulent ground cover.

SIZE: 2 inches tall, spreading indefinitely.

RATE: Moderate; space plants from 2 to 3 inch diameter containers 6 to 8 inches apart.

LANDSCAPE VALUE: Same as *S. brevifolium*.

DISCUSSION: Leaves are evergreen, sessile, crowded, elliptic, to 3/16 inch long, nearly cylindrical in cross section, green, often tinged red. Flowers are white, about 1 inch across, effective in midsummer.

* * *

S. brevifolium (brev-i-fō´lē-um) Short leaved.

COMMON NAME: Shortleaf stonecrop.

NATIVE HABITAT: Mediterranean area.

HARDINESS: Zone 5.

HABIT: Low, creeping, succulent cover.

SIZE: 2 to 3 inches tall, spreading indefinitely.

RATE: Relatively fast; space plants 8 to 12 inches apart from 2 to 3 inch diameter containers.

LANDSCAPE VALUE: General cover or specimen for rock garden or small bed area.

DISCUSSION: Leaves are evergreen, opposite or alternate, ovoid or nearly globose, 1/8 inch long, mealy gray, often tinged reddish, in 4 close vertical rows. Flowers are few, white, to about 5/16 inch across, borne in midsummer.

* * *

S. cauticolum (kâ-ti´kol-um) Named for its habit of growing on cliffs.

NATIVE HABITAT: Mountains of Japan, cliffs of south coast of Yezo, Japan.

HARDINESS: Zone 3.

HABIT: Low, creeping, succulent, mat forming ground cover.

SIZE: 3 inches tall, spreading indefinitely.

RATE: Moderate, space plants 6 to 10 inches apart from 3 to 4 inch diameter containers.

LANDSCAPE VALUE: Same as *S. brevifolium*, but unique among sedums in that it withstands relatively moist soils.

DISCUSSION: Characterized by having more or less deciduous foliage that is much like the more common species *S. seiboldii*, but short petioled rather than sessile; flowers are in a looser inflorescence, rosy-red, effective early to midautumn.

* * *

S. confusum (con-fūs´um) Meaning unclear.

NATIVE HABITAT: Mexico.

HARDINESS: Zones 7 or 8 to 10.

HABIT: Low, shrubby, succulent ground cover.

SIZE: 6 to 12 inches tall by 6 to 12 inches across.

RATE: Moderate; space plants 6 to 10 inches apart from 3 to 4 inch diameter containers.

LANDSCAPE VALUE: General cover for smaller areas or as a rock garden specimen.

DISCUSSION: Leaves are evergreen, flat, shiny green, sessile, obovate, 5/8 to 1 1/2 inches long by

1/4 to 5/8 inch wide, larger toward base; stem woody below; flowers yellow, 1/2 to 5/8 inch across, effective mid to late spring.

* * *

S. dasyphyllum (das-i-fil´um) Thick leaved.

COMMON NAME: Leafy stonecrop.

NATIVE HABITAT: Europe and northern Africa.

HARDINESS: Zone 5 or 6.

HABIT: Low, mosslike, mat forming, succulent ground cover.

SIZE: 2 inches tall, spreading indefinitely.

RATE: Moderate to relatively fast; space plants from 3 to 4 inch diameter containers 10 to 12 inches apart.

LANDSCAPE VALUE: Excellent in a rock garden or between and on large rocks of a retaining barrier.

DISCUSSION: Tiny evergreen leaves that are uniquely glandular-pubescent, ovoid-obovoid, 1/8 to 3/16 inch long, gray-green. Flowers are about 1/4 inch across, pinkish outside, white inside, borne in early summer; overall effect is similar to fine textured moss.

* * *

S. ellacombianum (el-là-kō-mē-ā´num) After Canon Ellacombe.

COMMON NAME: Ellacombe's sedum.

NATIVE HABITAT: Japan.

HARDINESS: Zone 3.

HABIT: Rhizomatous, upright stemmed, succulent ground cover.

SIZE: 6 to 10 inches tall, spreading indefinitely.

RATE: Moderate; space plants from 2 to 3 inch diameter containers 6 to 10 inches apart.

LANDSCAPE VALUE: Good, reliable general cover for banks and slopes of moderate grade. Also useful for accent or facing in an ornamental bed with dark green or purple foliaged shrubs.

DISCUSSION: Characterized by deciduous foliage that is light green, to about 1 1/4 inches long, 1/2 inch wide, margin crenate toward apex, fleshy and crowded in an irregularly opposite manner. Stems are upright and somewhat woody at the base. Bright yellow flowers are borne in early summer in elevated, terminal, flat umbellate inflorescences that are exceptionally beautiful.

* * *

S. kamtschaticum (kam-chat´i-kum) Of Kamchatka, Siberia.

COMMON NAME: Kamchatka stonecrop.

NATIVE HABITAT: Northeastern Siberia to Korea and central China, Kamchatka, and Japan.

HARDINESS: Zone 3.

HABIT: Low, mounding, rhizomatous, somewhat decumbent, succulent ground cover.

SIZE: 3 to 4 inches tall, spreading indefinitely.

RATE: Relatively slow; space plants 6 to 8 inches apart from 2 to 3 inch diameter containers.

LANDSCAPE VALUE: Rock garden specimen or general cover for small- to medium-sized area.

DISCUSSION: An exceptional, shiny, dark green, evergreen leaved species. Leaves are sessile, obovate to spatulate, 1 1/2 to 2 inches long, toothed toward apex. Flowers are orange-yellow, about 3/4 inch across; follicles orange, darkening to rusty red; effective time of bloom is from early to late summer.

SELECTED CULTIVARS AND VARIETIES: 'Variegatum' There appears to be two distinct clones that are sold under this name. Each is dull green, variegated, with creamy colored margins. The variegation on one is weak (almost unnoticeable) while the other is distinct. In either case, rate of growth is quite slow.

* * *

S. lineare (lin´ē-ēr-ē) Narrow leaved.

NATIVE HABITAT: Japan.

HARDINESS: Zone 7.

HABIT: Dense, low, trailing, succulent ground cover.

SIZE: 6 to 10 inches tall, spreading indefinitely.

RATE: Moderate to relatively fast; space plants 8 to 12 inches apart from 3 to 4 inch size containers.

LANDSCAPE VALUE: General cover for confined smaller areas or as a filler in a rock ledge or rock garden.

DISCUSSION: Leaves are evergreen, pale green, sessile, 3/4 to 1 1/4 inch long, linear-lanceolate, flat, acutely pointed. Flowers are yellow, 5/8 inch in diameter, starlike, in rather lax umbels, effective in late spring. The cultivar 'Variegatum' has leaves that are edged in white.

* * *

S. lydium (lid´i-um) Of Lydia, West Asia.

NOTE: See Figure 3–176.

COMMON NAME: Lydian stonecrop.

NATIVE HABITAT: Asia Minor.

HARDINESS: Zone 3.

HABIT: Low, erect or decumbent, rooting, mosslike, succulent ground cover.

SIZE: 2 to 4 inches tall, spreading indefinitely.

RATE: Moderate; space plants 8 to 12 inches apart from 3 to 4 inch diameter containers.

LANDSCAPE VALUE: Exceptionally fine texture makes this plant very useful in small areas as a general cover, or it may be used to good effect as a specimen or filler in a rock garden. Tolerates limited foot traffic.

DISCUSSION: Leaves are minute, linear, 1/4 inch long, sessile, evergreen, bright green, crowded, cylindrical, sometimes tinged reddish, exceptionally soft in texture. Flowers are white, 1/4 inch across, in compact terminal flat cymes that are elevated well above the foliage on erect stems in summer.

* * *

S. middendorfianum (mid-den-dor-fē-ā´num) After A. T. von Middendorf, a plant collector who introduced many Siberian plants.

HABIT: Low, decumbent, succulent, rhizomatous ground cover.

COMMON NAME: Middendorf sedum.

NATIVE HABITAT: Eastern Siberia, Inner Mongolia, Manchuria, Northern Korea.

HARDINESS: Zone 3.

SIZE: To 4 inches tall, spreading indefinitely.

RATE: Moderate; space plants from 2 to 3 inch diameter containers 6 to 10 inches apart.

LANDSCAPE VALUE: Same as *S. kamtschaticum.*

DISCUSSION: A very attractive and worthwhile species with dark green, crowded, alternate, 3/4 inch long, 1/4 inch wide, deflexed, sessile leaves. Flowers are bright yellow and arranged in terminal, compact, cymose inflorescences in mid to late summer. Following bloom are showy carpels of greenish-yellow, which turn rusty red and contrast well with foliage. The variety *diffusum* is an inch taller, laxer in habit, with somewhat wider leaves, and is may be a little more vigorous.

* * *

S. moranense (mor-an´en-sē) From Real de Moran, Mexico.

COMMON NAME: Moran stonecrop.

NATIVE HABITAT: Southern Mexico.

HARDINESS: Zone 7.

HABIT: Low, mat forming, bushy, succulent cover.

SIZE: 3 to 4 inches tall, spreading indefinitely.

RATE: Moderate; space plants 8 to 12 inches apart from 3 to 4 inch diameter containers.

LANDSCAPE VALUE: Specimen in rock garden or as a general cover for rock garden.

Figure 3–177 *Sedum populifolium* (left) and *Sedum spectabile* (right) (both life-size)

DISCUSSION: Leaves are closely set, evergreen, ovate to ovate-lanceolate, sessile, bluntly pointed, tiny, about 1/8 inch long, green. Flowers are few, white, tinged red on back, terminal, sessile, effective spring and summer.

* * *

S. oaxacanum (wah-hah-kā´num) Name derived from the native habitat, Oaxaca, Mexico.

COMMON NAME: Oaxaca stonecrop.

NATIVE HABITAT: Mexico.

HARDINESS: Zone 6.

HABIT: Low, matlike, creeping, succulent ground cover.

SIZE: 1 to 2 inches tall, spreading indefinitely.

RATE: Moderate to relatively fast; space plants 8 to 14 inches apart from 3 to 4 inch diameter containers.

LANDSCAPE VALUE: Same as *S. acre.*

DISCUSSION: Leaves are evergreen, sessile, 1/4 inch long, flat, obovate, 1/8 inch wide, grayish-green. Flowers are grouped 1 to 4 terminally, yellow, 5/16 inch across, effective in spring.

* * *

S. populifolium (pop-ū-li-fō´lē-um) Leaves like a poplar (Figure 3–177).

COMMON NAME: Poplar-leaved sedum, Siberian sedum.

NATIVE HABITAT: Central Siberia.

HARDINESS: Zone 3.

HABIT: Woody based, small subshrublike, succulent ground cover.

SIZE: 6 to 14 inches tall, spreading to about 1 1/2 feet wide.

RATE: Relatively slow to moderate; space plants 8 to 12 inches apart from quart- to gallon-sized containers.

LANDSCAPE VALUE: General cover for small- to moderate-sized areas or as a specimen for rock gardens.

DISCUSSION: Leaves are deciduous, light green, succulent, ovate or oblong in outline, coarsely and irregularly toothed, 1/2 to 1 inch long, petioled; branches are semiwoody; flowers pink or white, to 3/8 inch across, effective in summer.

* * *

S. reflexum (rē-flex´um) Leaves bent back, recurved.

COMMON NAME: Jenny stonecrop, spruce stonecrop.

NATIVE HABITAT: West, north, and central Europe.

HARDINESS: Zone 3.

HABIT: Creeping, carpet forming, prostrate to decumbent succulent ground cover.

SIZE: 4 to 10 inches high, spreading indefinitely.

RATE: Moderate; space plants 8 to 12 inches apart from 3 to 4 inch diameter containers.

LANDSCAPE VALUE: Unusual color of the foliage makes this an outstanding rock garden specimen. Also useful as a general ground cover for small- to moderate-size areas.

DISCUSSION: Thick fleshy stems trail along the ground. Leaves are more or less evergreen, crowded, 1/2 inch long, sessile, linear, acutely pointed, nearly terete, ascending or recurved, attractive bluish-green. Flowers are numerous, yellow, in flat terminal cymes; cymes are 1 to 1 1/2 inches across, elevated to about 12 inches on stiff, erect, leafy shoots, effective early to midsummer.

* * *

S. rubrotinctum (rūb-rō-tink´tum) Red-tinged, the leaves.

COMMON NAME: Christmas cheer, pork and beans.

NATIVE HABITAT: Guatemala.

HARDINESS: Zones 9 to 10.

HABIT: Low, creeping, succulent ground cover.

SIZE: 6 to 8 inches tall, spreading indefinitely.

RATE: Relatively fast; space plants 10 to 16 inches apart from 2 to 3 inch diameter containers.

LANDSCAPE VALUE: Same as *S. lineare.*

DISCUSSION: An excellent and reliable species. Leaves are sessile, evergreen, 1/2 to 3/4 inch long by 1/4 inch across, clublike, shiny green, often with varying amounts of coppery-red, especially if grown in full sun; flowers are yellow, 1/2 inch across, in terminal cymes, effective spring and summer.

* * *

S. rupestre (roo-pes´trē) Rock breaking.

NATIVE HABITAT: Portugal.

HARDINESS: Zone 7.

HABIT: Low, matlike, creeping, succulent ground cover.

SIZE: About 1 1/2 inches tall, spreading indefinitely.

RATE: Moderate to relatively fast; space plants from 3 to 4 inch diameter containers 10 to 16 inches apart.

LANDSCAPE VALUE: Excellent in rock garden or as a general cover for small- to medium-sized confined areas.

DISCUSSION: Evergreen leaves are linear to linear-oblanceolate, 5/8 inch long, 1/16 inch wide, acutely pointed, spurred, sessile, crowded, flat faced, blue-green turning purple in fall. Flowers are pale yellow, 1/2 inch across, arranged in umbellate cymes that are terminal on upright stems, effective in early summer.

* * *

S. sarmentosum (sär-men-tō´sum) Twiglike.

COMMON NAME: Stringy stonecrop.

NATIVE HABITAT: Northern China and Japan.

HARDINESS: Zone 3.

HABIT: Prostrate, trailing, succulent ground cover.

SIZE: Reaching 3 to 6 inches tall, spreading indefinitely.

RATE: Relatively fast; space plants from 3 to 4 inch diameter containers 12 to 16 inches apart.

LANDSCAPE VALUE: See *S. album.*

DISCUSSION: Thin, sprawling, rooting stems that lay relatively flat on the soil. Foliage is for the most part evergreen, ternate, about 3/4 inch long, sessile, entire, fleshy, light green and glabrous. Flowers are bright yellow in summer, arranged in flat topped, lax, terminal cymes that are held on leafy, decumbent shoots.

* * *

S. sexangulare (sex-ang-ū-lär´ē) Leaves in six rows.

COMMON NAME: Hexagon stonecrop.

NATIVE HABITAT: Europe.

HARDINESS: Zone 2 or 3.

HABIT: Low, broad, matlike, creeping, succulent ground cover.

SIZE: 2 to 4 inches high, spreading indefinitely.

RATE: Moderate; space plants 8 to 10 inches apart from 2 to 4 inch diameter containers.

LANDSCAPE VALUE: Same as *S. acre.*

DISCUSSION: Leaves are evergreen, crowded, sessile, commonly in 6 spiral rows, linear, 1/8 to 1/4 inch long, cylindrical, apex blunt, vibrant green. Flowers are canary yellow, 3/8 inch across, borne in terminal, flat cymes, effective early to midsummer.

* * *

S. sieboldii (sē-bōld´i-ī) After P.F. von Siebold.

COMMON NAME: Siebold stonecrop.

NATIVE HABITAT: Japan.

HARDINESS: Zone 2 or 3.

HABIT: Low, decumbent, clump forming, succulent ground cover.

SIZE: 3 to 8 inches tall, spreading about 12 inches wide.

RATE: Relatively slow; space plants about 6 inches apart from 2 to 3 inch diameter containers.

LANDSCAPE VALUE: Commonly used for edging in perennial border or along walks. Often grown in planters or as a specimen in rock garden. Tolerates light shade.

DISCUSSION: A fairly common species in cultivation. Leaves are thick, sessile, deciduous, whorled in threes, nearly orbicular in outline, 1/2 to 1 inch long, blue glaucous, often with a flush of reddish purple; margin colored brightly red, sinuate or bluntly toothed in the upper half. Flowers are pink, nearly 1/2 inch across, arranged in compact, terminal, flattish umbellate cymes, effective in midautumn. Cultivars include a variegated form, 'Medio Variegatum,' that has leaves blotched yellow, which may be difficult to obtain.

* * *

S. spathulifolium (spath-e-li-fō´lē-um) With spatula-shaped leaves.

NATIVE HABITAT: Western North America; British Columbia to California.

HARDINESS: Zone 5.

HABIT: Dense, low, matlike, succulent, spreading ground cover.

SIZE: 2 to 4 inches tall, spreading indefinitely.

RATE: Relatively slow; space plants 6 to 8 inches apart from 2 to 3 inch diameter containers.

LANDSCAPE VALUE: General cover in small, confined locations. Also good as rock garden specimen.

DISCUSSION: Leaves are sessile or nearly so, evergreen, arranged in rosettes, spatulate, 1/2 to 1 1/4 inches long, 1 to 1 1/2 inches across, blue-green, white below; apex abruptly pointed. Flowers are yellow, to 5/8 inch across, arranged in rather large terminal, flat topped cymes, effective mid to late spring. Cultivars include 'Cape Blanco' with appealing silvery (bloomy) foliage, and 'Purpureum' with deep purple to bluish-purple tinged foliage.

* * *

S. spectabile (spek-tab´il-lē) Showy, striking.

COMMON NAME: Showy stonecrop.

NATIVE HABITAT: Japan.

HARDINESS: Zone 3.

HABIT: Clump forming, neat, mounding, succulent ground cover.

SIZE: 12 to 18 inches tall, spreading to about 12 to 18 inches wide.

RATE: Relatively slow; space plants 6 to 10 inches apart from pint- or quart-sized containers.

LANDSCAPE VALUE: Excellent when used en masse as general covers for small- to moderate-sized areas. Also nice as rock garden specimens and combined with other sedums.

DISCUSSION: Foliage is gray-green, usually opposite or whorled in threes, subsessile, deciduous, obovate, to 3 inches long, 2 inches across, flattish, margin somewhat toothed. Flowers pink, 1/2 inch across, arranged in flat-topped, dense corymbs that may reach 3 inches across, showy in early to midautumn. There are many cultivars, some of which include 'Album' White flowered.
'Autumn Joy' Flowers salmon, tinged bronzy, effective late summer until frost.
'Brilliant' Flowers deeper pink than the species.
'Carmen' Flowers deep rosy pink.
'Meteor' Flowers deep rosy pink.
'Star Dust' Flowers ivory-white.
'Variegatum' Leaves variegated white and yellow.

* * *

S. spurium (spew´rē-um) False or doubtful, possibly in reference to its many false names.

COMMON NAME: Two-row stonecrop.

NATIVE HABITAT: Caucasus area of Asia Minor.

HARDINESS: Zone 3.

HABIT: Low, creeping, matlike, succulent ground cover.

SIZE: 2 to 6 inches tall, spreading indefinitely.

RATE: Moderate; some of the cultivars are relatively slow to spread. In general, spacing of 6 to 8 inches is adequate for plants from 2 to 4 inch diameter containers.

LANDSCAPE VALUE: Excellent general cover for small- or medium-sized areas that are surrounded with edging or pavement to limit spread. Unusual among the sedums in its tolerance of light shade; foliage color, however, is best in full sun.

DISCUSSION: A species that contains many widely cultivated cultivars. It is characterized by leaves that are deciduous, with the exception of crowded clusters at the ends of the stems, which persist, obovate-cuneate, 1/2 to 1 inch long, about 2/3 inch across, crenate-serrate margined toward apex, papillose-ciliate, dark green. Flowers are pink to purplish, sometimes varying to white, 1/2 to 3/4 inch in diameter, arranged in dense terminal cymes that are elevated about 2 inches above the foliage on leafy stems, effective mid to late summer. Cultivars include 'Album' Flowers white or cream. 'Bronze Carpet' Foliage larger and bronzy, flowers magenta.

Var. *coccineum* A name applied to many scarlet or reddish flowering varieties. Plants that have been propagated by seed and sold as the cultivar 'Dragon's Blood' should more properly be listed as var. *coccineum*.

'Dragon's Blood' ('Schorbusser Blut') Widely cultivated; foliage bronzy purplish becoming strikingly reddish in autumn; flowers are vibrant pinkish red. Relatively slow growing.

'Splendens' Likely a misnomer for var. *coccineum*.
'Tricolor' Foliage colored green, white, and red.

Generally Applicable to All Species

CULTURE: Soil: In general, sedums perform best in infertile, stony or sandy, porous soils. Good drainage is essential. Most species are adaptable to a range of pH from 5.0 to 7.5. High fertility causes abnormal growth and predisposes plants to fungal diseases.
Water: Extremely tolerant of drought; only infrequent watering during summer is needed for optimal growth. Generally, soil should be kept on the dry side.
Light: Full sun, unless otherwise noted.

PATHOLOGY: Generally healthy when cultural requirements are met. Crown rot, leaf blotch, leaf spot, stem rot, and rusts may sometimes be a problem.
Pests: Aphids, southern root knot, nematode, slugs, and weevils.

MAINTENANCE: Mowing to remove dead flowers after bloom is advisable as it makes plantings look neat and may prevent self-sowing. Mowing again in spring stimulates branching, thus encouraging thick growth.

PROPAGATION: Cuttings: Cuttings root readily for most species. Take cuttings from vegetative shoots only. Number and size of roots are usually increased by the use of 1000 to 2000 ppm IBA/talc, but chemical treatment is not a necessity for most species.
Division: Simply divide plants at any time of the year.
Seed: Many species are propagated from seed. Some come true, while others are variable, sometimes creating interesting textured and many colored mats. Commercial supplies are available and germination usually takes 2 to 3 weeks.

*S*empervivum (sem-pĕr-vī´vum) From Latin, *semper*, always, and *vivo*, alive, alluding to the tenacity of these plants.

FAMILY: CRASSULACEAE: Stonecrop family.

Commonly called houseleek, this genus is composed of about 40 species of succulent, reliable, long-lived, herbaceous perennials. They are native to Europe, Morocco, and western Asia. Most of the species make good ground covers, but not all of them are readily attainable. A general description of the species of *Semipervivum* is addressed here, with specific descriptions to follow.

HABIT: Low, matlike, stoloniferous, spreading, succulent covers.

SIZE: Foliage mass ranges from 1 to 4 inches tall; individual heights listed reflect the height of flowering stems.

RATE: Relatively slow; space plants 6 to 8 inches apart from pint- to quart-sized containers.

LANDSCAPE VALUE: Excellent and often very interesting rock garden specimens. Good on dry, gentle slopes or in small beds for use as general covers. Sometimes used to good effect in a border setting, but avoid planting near invasive growing sedums or other rampant growing plants as they will soon overrun the sempervivums. No foot traffic.

LEAVES: Evergreen, in dense basal rosettes, alternate, oblong to ovate, broad based, margin ciliate, often apex sharply pointed, frequently margin and apex in shades of pink to purple or brown, thick and fleshy, color varying from shades of green, blue, bronzy, to gray.

INFLORESCENCE: Compact cymes, born terminally on erect, leafy flowering stems; flowers 6 to 20 parted, sepals lanceolate, free, petals usually red, purple, yellow, or white, separate, margins often fringed with hairs, stamens numbering twice the number of petals.

FRUIT: Not of ornamental significance.

STEMS: Functioning to support inflorescences, erect, leafy, from 2 to 18 inches tall, arising from central mature rosette.

OF SPECIAL INTEREST: Upon senescence of the flowers, the rosette that the reproductive stem arose from dies and is replaced by offsets that it has produced. In Central Europe a superstition has been common that if a houseleek is planted on the roof the house will be protected from lightning and fire. In the 1900s, G. Klebs, a German, observed that *Sempervivum* would not flower if grown on a short day photoperiod. This is among the first observations of the phenomenon of photoperiodism. Later, Garner and Allard performed research using tobacco and other plants that enabled them to advance the theory of photoperiodism. Those wishing to obtain an excellent reference book on this genus should consult L. R. Praeger's book, *An Account of the Sempervivum Group* (see bibliography).

* * *

S. arachnoideum (ȧ-rak-noy-dē´um) From Latin, *araneosum*, cobwebbed, hairs on the foliage look like cobwebs (Figure 3–178).

COMMON NAME: Spiderweb houseleek, cobweb houseleek.

NATIVE HABITAT: Mountains of southern Europe.

HARDINESS: Zone 5.

SIZE: 4 inches tall, spreading indefinitely.

DISCUSSION: One of the most popular species of sempervivums. Rosettes are about 1 1/2 to 3 inches across; offsets crowded; leaves sessile, numbering about 50, oblong-oblanceolate, incurved, to 3/4 inch long by 1/5 inch across, margins lightly hairy, pale green, sometimes tipped brown or red, covered with white hairs that extend from leaf tip to leaf tip; actually appearing to be cobweblike, thus the common name. Flowers are born midsummer, bright rose-red, in 5 to 15 flowered, compact, flat-

tish cymes. The variety *tomentosum* has flattish rosettes that are nearly concealed by cobweblike hairs.

* * *

S. arenarium (ãr-e-nā´rē-um) Of sandy places.

NATIVE HABITAT: Eastern Alps.

HARDINESS: Zone 5.

SIZE: To 9 inches tall, spreading indefinitely.

DISCUSSION: Rosettes are 1/8 to 3/4 inch wide, offsets are minute; leaves 60 to 80, lanceolate, broadest near the middle, upcurved, flat above, convex below, glabrous, often red-brown tipped, margins ciliate. Flowers are pale yellow, tinged reddish outside, blooming in late summer.

* * *

S. × barbulatum (barb-ū-lā´tum) Meaning unknown.

HYBRID ORIGIN: *S. arachnoideum × S. montanum.*

HARDINESS: Zone 5.

SIZE: To 10 inches tall, spreading indefinitely.

DISCUSSION: A hybrid (sometimes listed as *S. × fimbriatum*) with variable morphology; rosettes are usually about 3/8 inch in diameter; flowers are on stems with color intermediate between the two parents.

* * *

S. × fauconnettii (faugh-con-ett´i-ī) Meaning unknown, probably a memorial.

HYBRID ORIGIN: *S. arachnoideum × S. tectorum.*

NATIVE HABITAT: Eastern France in the Jura Mountains.

HARDINESS: Zone 5.

SIZE: To 8 inches tall, spreading indefinitely.

Figure 3–178 *Sempervivum arachnoideum* (0.5 life-size)

DISCUSSION: Sometimes is listed as *S. pilioseum.* It is a hybrid similar to its parent *S. tectorum,* but its leaves are slightly hairy, rosettes denser and smaller, and inflorescences slightly larger.

*   *   *

S. montanum (mon-tā´num) Of the mountains.

NATIVE HABITAT: Alps.

HARDINESS: Zone 5.

SIZE: To 6 inches tall, spreading indefinitely.

DISCUSSION: A fine species with 1 to 1 1/2 inch diameter rosettes that are composed of 40 to 50 leaves. Leaves are broadly to narrowly oblanceolate, 1/2 to 1 inch long, apex acute, glandular pubescent with hairs longer toward the tip and margins, dark green, sometimes with tips darkened; flowers bluish-purple, 1 1/2 inches across, grouped 3 to 10, effective early to late summer. The variety *braunii* has flowers dull yellowish and rosettes that reach 2 inches across.

*   *   *

S. ruthenicum (ru-then´i-kum) Russian.

COMMON NAME: Russian sempervivum.

NATIVE HABITAT: Eastern Europe.

HARDINESS: Zone 5.

SIZE: To 12 inches tall, spreading indefinitely.

DISCUSSION: Rosettes are 1 1/2 to 2 1/2 inches across, leaves to 1 1/4 inch long, obovate, densely pubescent, green, sometimes with purple tips, older leaves flushed rose in upper portion; flowers pale yellow, 3/4 to 1 inch across, with purple center, blooming mid to late summer.

*   *   *

S. soboliferum (sob-ol-if´ēr-um) From Latin *suboles,* meaning to sprout, sucker, bear offspring.

COMMON NAME: Hen and chickens.

NATIVE HABITAT: Austria.

HARDINESS: Zone 5.

SIZE: To 9 inches tall, spreading indefinitely.

DISCUSSION: Rosettes are 1/2 to 1 1/4 inches across; leaves number 60 to 80, are glabrous, incurved, oblanceolate, to 1 inch long, broadest above the middle, light green, sometimes redtipped, outer leaves tinged brown. Flowers are greenish-yellow, 5/8 inch long, effective in summer.

*   *   *

S. tectorum (tek-tōr´um) Of roofs, the houseleek, from Latin.

COMMON NAME: Hen and chickens, common houseleek, old man and old woman, roof houseleek.

NATIVE HABITAT: Europe and Asia.

HARDINESS: Zone 4.

SIZE: To 12 inches tall, spreading indefinitely.

DISCUSSION: Probably the most frequently cultivated species of semipervivum. It is characterized by flattish, 3 to 4 inch diameter (sometimes larger) rosettes; leaves number 50 to 60, are glabrous, 1 1/2 to 3 inches long, green, frequently purple tipped; flowers are 3/4 to 1 inch across, purplish-red, effective in summer. Many cultivars and varieties exist, a few of which include Var. *alpinum* that has smaller rosettes (1 to 2 1/2 inches in diameter) with red leaf bases.

Var. *calcarium* Leaves blue-green, tipped reddish-brown; flowers pale red. 'Robustum' Leaves tipped in purple; rosettes to 6 inches across.

Generally Applicable to All Species

CULTURE: Soil: Adaptable to most soils, excellent drainage is a necessity. Tolerant of infertile soils. Moisture: Highly tolerant of drought, but benefited by occasional watering to moisten soil in hot, dry, summer months.
Light: Full sun to light shade; afternoon shade is desirable in very hot conditions.

PATHOLOGY: Diseases include rust, leaf rot, stem rot, root rot.

MAINTENANCE: Little or no special maintenance is required. Mother plants that have flowered die soon afterward. The dead rosettes are best removed so that offsets of their progeny can more readily fill the gaps.

PROPAGATION: Seed: Seed is sometimes used to propagate sempervivums. Commercial supplies may be obtained for some species, and a fair amount of variability should be expected.
Division: Simply divide small rosettes from around the base of the mother plant and transplant.

✛✛✛✛✛✛✛✛✛✛✛✛✛✛✛✛

𝑺oleirolia (sō-lā-rōl´i-à) After Joseph Francois Soleirol (1796–1863), who originally collected it in Corsica. A monospecific genus composed of a monoecious herb.

FAMILY: URTICACEAE: Nettle family.

S. soleirolii (sō-lā-rōl´i-ī) Same meaning as the genus (Figure 3–179).

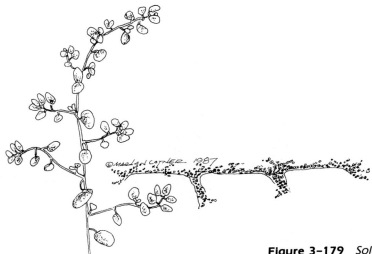

Figure 3-179 *Soleirolia soleirolii* (life-size)

COMMON NAME: Baby's tears, angel's tears, Irish moss, Corsican curse, Japanese moss, Pollyana vine, Corsican carpet plant, peace in the home.

NATIVE HABITAT: Corsica and Sardinia.

HARDINESS: Zone 10.

HABIT: Low, herbaceous, creeping, mosslike, ground cover.

SIZE: To 3 inches tall, spreading indefinitely.

RATE: Relatively fast to invasive; space plants 10 to 14 inches apart from 4 inch diameter containers.

LANDSCAPE VALUE: Excellent carpeting plant when used around rocks or stepping stones in naturalized settings. Fine for general cover as a companion with ferns or hostas in a woodland setting or border. Withstands infrequent foot traffic.

LEAVES: Evergreen, simple, alternate, nearly orbicular, about 1/4 inch long and wide, both sides sparsely hairy and glandular dotted, shiny light green, short petioled.

STEMS: Creeping, delicate, hairy, pinkish.

FLOWERS: Solitary, axillary, regular; male flowers with 4 stamens and 4 parted calyx; female flowers with 4 lobed tubular calyx; in either case flowers are very small, without petals, and of no ornamental significance.

FRUIT: Achene enclosed by calyx, not of ornamental significance.

SELECTED CULTIVARS AND VARIETIES: 'Aurea' A golden selection.

CULTURE: Soil: Best in organically rich, loamy soil. Moisture: Intolerant of drought; soil should be kept moist with regular watering. Intolerant of very high temperatures; thus planting in a cool location is necessary.

Light: Light to dense shade.

PATHOLOGY: No serious diseases or pests reported.

MAINTENANCE: Little or no unusual maintenance is required.

PROPAGATION: Cuttings root easily; just insert sections of stem into a well-drained medium and keep moist.

Division: Divide plants during cool seasons and keep moist until roots reestablish.

*S*piraea (spi-rē´à) Likely from Greek, *speira*, a wreath, or *speiraira*, a plant used in garlands. The genus is composed of about 100 species of deciduous shrubs. They are native to many areas throughout the northern hemisphere.

FAMILY: ROSACEAE: Rose family.

LANDSCAPE VALUE: Ground covering spireas are exceptional as specimens in a rock garden, border, or woody ornamental bed. They are also nice used as edging along a walkway or steps for accent or as dwarf hedge. No foot traffic.

S. japonica (jà-pon´i-kà) of Japan (Figure 3-180).

COMMON NAME: Japanese spirea.

NATIVE HABITAT: Japan.

HARDINESS: Zone 5.

DISCUSSION: The species itself is a rather stiffly erect shrub, which may reach 6 feet in height. It is seldom cultivated, as lower growing cultivars are usually favored. The species is characterized as follows:

LEAVES: Deciduous, simple, alternate, 1 to 3 inches

long, lanceolate-ovate, acute apex, cuneate base, short petioled, doubly incised-serrate margin, pale green or bluish below with veins often pubescent, medium green above.

STEMS: Stiffly erect, glabrous to pubescent in youth, few branched, sometimes angled.

INFLORESCENCE: Large terminal flat topped corymbs; flowers bisexual, small, 5 parted, varying from pink to deep pink or whitish, effective early to midsummer.

FRUIT: Dehiscent uninflated follicles, tiny, ornamentally insignificant.

HABIT: Ground cover types are low, often mounded, dense, shrubby ground covers.

SIZE: 12 to 30 inches tall by 3 to 6 feet across.

RATE: Slow to moderate; space plants 1 1/2 to 2 feet apart for the cultivar 'Alpina' and 2 1/2 to 3 feet apart for the cultivar 'Little Princess'; both are commonly available in gallon-sized containers.

SELECTED CULTIVARS AND VARIETIES: 'Alpina' ('Nana') Commonly called daphne spirea, this is a dwarf mounding selection that reaches 1 to 2 1/2 feet high. Leaves are medium green, dense, to about 1/2 inch long, and fine textured. Flowers are pink and showy in early summer.
'Little Princess' Also a dwarf selection with mounding habit, more robust than 'Alpina'; blooming rosy-red in early spring.

OTHER SPECIES: *S.* × *bumalda* 'Nywoods' (bew-mahl´dä) After Bumaldus.

COMMON NAME: Nywood's bumald spirea.

HYBRID ORIGIN: *S. albiflora* × *japonica*. The cultivar *S. bumalda* 'Nywoods' is similar to *S. japonica* 'Alpina' in many ways, but is generally more vigorous and has foliage that is deep blue-green.

HARDINESS: Zones 3 to 8.

HABIT: Low, dense, woody, spreading, shrubby ground cover.

RATE: Moderate; space plants 1 1/2 to 2 feet apart from gallon-sized containers.

Generally Applicable to All Selections

CULTURE: Soil: Adaptable to most well-drained soils. The soil pH is generally best in the range from 5.5 to 6.5.
Moisture: Fairly well adapted to drought once established. Occasional deep watering during summer months is beneficial.
Light: Full sun.

PATHOLOGY: Diseases: Fire blight, hairy root, leaf spots, powdery mildew, root rot.

Figure 3–180 *Spiraea japonica* 'Alpina' (life-size)

Pests: Spirea aphid, oblique banded leaf roller, scales, southern root knot nematode, saddled prominent caterpillar.

MAINTENANCE: A light shearing following bloom is commonly practiced and will keep plants looking dense and compact.

PROPAGATION: Cuttings: Spireas root easily from either softwood or hardwood cuttings. Treatment with 1000 to 3000 ppm IBA/talc will induce heavier and more vigorous root development, but is not essential.

*S*tachys (stā´kis) From Greek, *stachus*, a spike, alluding to the pointed inflorescences. This genus is composed of about 300 species of annual and perennial herbs and subshrubs. They are native to temperate and subtropical regions and tropical mountains throughout the world.

FAMILY: LAMIACEAE (LABIATAE): Mint family.

S. byzantina (bi-zan-tēn ´ä) Of Byzantium (now Istanbul); sometimes still listed as *S. lanata,* which means woolly, the foliage (Figure 3–181). In the past, this species has been listed as *S. olympica.*

COMMON NAME: Lamb's ears, woolly betony.

NATIVE HABITAT: Caucasus to Persia.

HARDINESS: Zone 4.

HABIT: Herbaceous ground cover.

SIZE: 12 to 18 inches tall when in bloom, otherwise 4 to 6 inches tall by 2 to 3 feet across.

RATE: Moderate; space 12 to 16 inches apart from 3 to 4 inch diameter containers.

LANDSCAPE VALUE: Commonly used as an edging along walk or in a border. En masse, lamb's ears makes an attractive general cover for small- or moderate-sized areas, especially on a slope. Unusual foliage color and texture make for contrast and add interest when used as a facer in the foreground of various green-foliaged specimens. No foot traffic.

LEAVES: Evergreen, opposite, simple, densely white-tomentose top and bottom, to 4 inches long, base cuneate to attenuate, lower leaves oblong spatulate, upper leaves elliptic. Foliage is the primary ornamental attribute of this species.

STEMS: Erect, white tomentose, squarish in cross section.

INFLORESCENCE: Many flowered terminal verticillasters on stout stems, spikelike; flowers sessile or short pedicelled; calyx to 1/2 inch long, 5 to 20 nerved, bristly toothed; corolla cylindrical, tubular, 2 lipped, upper 2 lobed, lower 3 lobed, pink or purple, not very showy, densely white-woolly, blooming from early to midsummer, frequently until frost in autumn.

FRUIT: Of 4 ovoid nutlets, not of ornamental significance.

SELECTED CULTIVARS AND VARIETIES: 'Silver Carpet,' Nonflowering selection, with thicker, more woolly foliage.

OTHER SPECIES: *S. grandiflora,* sometimes commonly called big betony or errantly *S. macrantha* or *S. betonica* 'Superba,' is a species notable for its 1 inch wide violet-purple flowers in early to midsummer. Flowers are born in 10 to 20 flowered verticillasters on stalks that may ascend 1 1/2 feet. Leaves are ovate to 2 1/2 inches long and coarsely crenate-serrate. They are dark, hairy, and arranged in rosettes.

HARDINESS: Zone 3.

CULTIVARS: 'Alba,' with white flowers.

CULTURE: Soil: Adaptable to most well-drained soils.
Moisture: Fairly tolerant of drought. Keeping soil only slightly moist with periodic watering is beneficial.
Light: Full sun.

PATHOLOGY: Diseases: Leaf gall, leaf spot, powdery mildew.

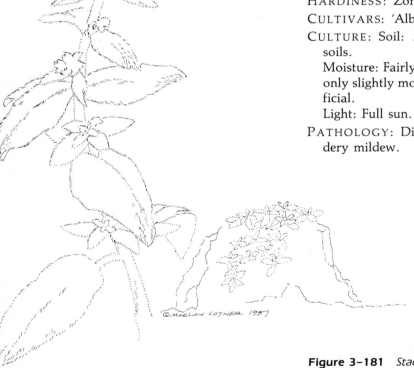

Figure 3–181 *Stachys byzantina* (life-size)

Pests: Slugs sometimes favor the foliage of stachys; nematodes.

PROPAGATION: Seed: Reported to propagate easily from seed. Collect ripe seed, store in a cool, dry environment through winter and sow in spring. Commercial supplies are available.

Division: Simply divide plants in spring or fall.

*S*tephanandra (stef-à-nan´drà) From Greek, *stephane*, a crown, and *andros*, a stamen, in reference to the form of the stamens, that is, crownlike. The genus contains four species of deciduous shrubs. They are native to temperate eastern Asia.

FAMILY: ROSACEAE: Rose family.

S. incisa 'Crispa' (in-si´sà) From Latin, *incis*, carve, cut into, referring to the leaf margin (Figure 3–182).

COMMON NAME: Dwarf cutleaf stephanandra, crisped stephanandra, lace shrub.

NATIVE HABITAT: The species is native to Japan and Korea, while the cultivar 'Crispa' originated in Denmark in 1930.

HARDINESS: Zones 3 to 8.

SIZE: 1 1/2 to 3 feet tall by 4 feet across.

HABIT: Low, mounded, woody, dense, suckering shrubby ground cover.

RATE: Moderate to relatively fast; space plants from gallon-sized containers about 3 feet apart.

LANDSCAPE VALUE: Commonly used en masse as a general cover for moderate-sized areas. Good on slopes and banks for erosion control provided moisture requirements can be met. Also makes a nice dwarf hedge. No foot traffic.

LEAVES: Simple, alternate, ovate, 1 1/2 to 2 1/2 inches long, long pointed, incisely lobed and serrate, lower incisions halfway to midrib, petioles 1/8 to 2/5 inch long, bright green above, veins pubescent below; fall color reddish purple or reddish orange.

STEMS: Terete, slender, brown, glabrous, zigzag.

INFLORESCENCE: Loose terminal panicles, 1 to 1 1/2 inches long; flowers small, sepals 5; petals 5, yellowish-white, not showy, borne late spring to early summer.

FRUIT: Follicles, not ornamentally significant.

CULTURE: Soil: Best in organically rich, well-drained, loamy soil. Soil pH is best in the range from 6.5 to more acidic conditions.

Moisture: Intolerant of prolonged drought; soil should be kept moist with regular watering.

Light: Full sun to light shade.

PATHOLOGY: No serious diseases or pests reported. I have had some trouble with slugs, but only on small plants.

PROPAGATION: Cuttings: Cuttings root readily (hardwood or softwood) throughout the year. Treatment with 3000 ppm IBA/talc is beneficial.

Division: Clumps, suckers with roots, and naturally or artificially layered branches may be dug and transplanted in the spring.

Figure 3–182 *Stephanandra incisa* 'Crispa' (life-size)

++++++++++++++++

Symphoricarpos (sim-fō-ri-kär´pos) From Greek, *symphoreo*, to accumulate, and *karpos*, a fruit, alluding to the clustered fruits. This genus, commonly called snowberry, is comprised of about 16 species of deciduous, variously hardy shrubs. They are native primarily to North America, with one from China.

FAMILY: CAPRIFOLIACEAE: Honeysuckle family.

LANDSCAPE VALUE: Large-scale ground covers for naturalized areas, sometimes used in a border or ornamental bed. Suckering forms are useful as bank stabilizers. No foot traffic.

OF SPECIAL INTEREST: The fruit of coralberries apparently is of little interest to songbirds; however, such game species as ruffed grouse and ringed-neck pheasant are reported to relish their fruit.

* * *

S. × chenaultii (sha-nolt´i-i) Memorial.

COMMON NAME: Chenault coralberry.

HYBRID ORIGIN: *S. orbiculatus × S. microphyllus*, 1910.

HARDINESS: Zones 4 to 8.

SIZE: 3 feet high by 4 feet across.

HABIT: Low, woody, spreading, shrubby cover.

RATE: Relatively fast; space plants 2 1/2 to 3 feet apart from gallon-sized containers.

LEAVES: Deciduous, opposite, simple, elliptic, to 3/4 inch long, pubescent below, soft medium green, sometimes lobed in youth.

STEMS: To 3 feet, erect and arching.

INFLORESCENCE: Terminal spikes and clusters along stems; flowers with 4 to 5 calyx lobes, corolla campanulate, pink, born midsummer.

FRUIT: 1/4 inch diameter, 2-seeded berry, reddish pink on side facing the sun, white on opposite side.

SELECTED CULTIVARS AND VARIETIES: 'Hancock' Exceptional dwarf selection that reaches about 2 feet high and 10 or more feet across. Growth is relatively fast with suckering and rooting stems. Reportedly of Canadian origin. Space 4 to 5 feet apart.

* * *

S. orbiculatus (ôr-bik-ū-lā´tus) Orbicular (round), the leaves (Figure 3–183).

COMMON NAME: Buckbrush, coralberry, Indian current, turkeyberry.

NATIVE HABITAT: South central United States, introduced into the trade in 1727.

HARDINESS: Zones 2 or 3 to 6 or 7.

SIZE: 2 to 5 feet tall by 4 to 8 feet across.

HABIT: Low, dense, suckering, woody ground cover.

RATE: Relatively fast; space plants from gallon-sized containers 2 to 3 feet apart.

LEAVES: Deciduous, simple, opposite, elliptic or ovate, 3/5 to 1 2/5 inches long, apex obtuse or acutish, base rounded, petiole 1/8 inch long, margin entire, glaucous and hairy below, dull gray-green above, turning bluish-gray in fall.

STEMS: Pubescent in youth with grayish papery bark at maturity, pith continuous.

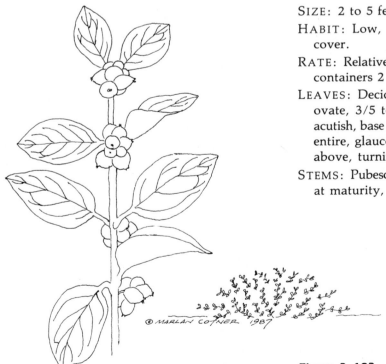

© MARLAN COTNER 1987

Figure 3–183 *Symphoricarpos orbiculatus* (life-size)

INFLORESCENCE: Dense axillary clusters or terminal spikes; flowers campanulate, 1/8 to 1/6 inch long, greenish-yellowish-white, sometimes tinged pinkish or purple, effective mid to late summer.

FRUIT: To 1/4 inch long berries, coral red to purplish, in clusters, effective in fall and persisting into winter.

SELECTED CULTIVARS AND VARIETIES: 'Leucocarpus' With whitish colored fruit.
'Variegatus' Foliage irregularly trimmed in yellow. Habit is rather lax.

Generally Applicable to Both Species

CULTURE: Soil: Tolerant of almost any soil. Adaptable to a wide range of pH.
Moisture: Relatively tolerant of drought once established; occasional thorough watering in summer months may be beneficial.
Light: Tolerant of range from full sun to moderate shade.

PATHOLOGY: Diseases: Anthracnose, berry rot, leaf spots, powdery mildews, rusts, stem gall.

Pests: Aphids, snowberry clearwing, glacial whitefly, San Jose scale.

MAINTENANCE: Pruning should be done in early spring.

PROPAGATION: Cuttings: Softwood, hardwood, and root cuttings are all viable means of propagation. Stem cuttings are helped with 1000 to 2000 ppm IBA/talc for *S. orbiculatus* and 7000 to 8000 ppm for *S* × *chenaultii*.

Tanacetum (tan-a-se´tum) Said to be derived from Greek, *athanatos*, immortal, in reference to the longevity of the flowers. This genus, commonly called tansy, is composed of between 30 and 50 species of annual and perennial herbs or subshrubs that are often aromatic. Primarily, they originated in Europe. Until very recently, the species *T. parthenium* was considered a species of the genus *Chrysanthemum*.

FAMILY: ASTERACEAE (COMPOSITAE): Sunflower family.

T. parthenium (pär-the´ni-um) According to Plutarch, the species name arose because the plant was used to save the life of a man who fell from the Parthenon during its construction (Figure 3–184).

COMMON NAME: Feverfew.

NATIVE HABITAT: Europe and Asia.

HARDINESS: Zones 6 to 10.

HABIT: Upright, compact, clump forming, herbaceous ground cover.

SIZE: 12 inches tall, spreading to 3 feet across.

RATE: Moderate; space plants from pint- to quart-sized containers 14 to 20 inches apart.

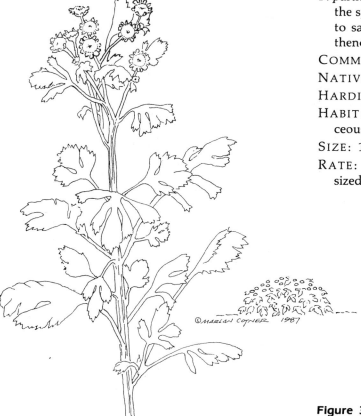

Figure 3–184 *Tanacetum parthenium* 'Golden Moss' (life-size)

LANDSCAPE VALUE: Primarily suited for covering small areas or for contrast as an accent plant in a border setting. No foot traffic.

FOLIAGE: Alternate, oblong to broadly ovate in outline, to 3 inches long, glabrous or lightly pubescent, pinnatifid, lower leaves with petioles, hairy below, medium dark green, aromatic.

STEMS: Erect, aromatic, branched.

INFLORESCENCE: Heads to 3/4 inch across, many arranged in corymbs; disk flowers yellow, ray flowers either white or lacking, borne in profusion in mid to late summer.

FRUIT: Achenes, not ornamentally significant.

OF SPECIAL INTEREST: The common name feverfew stems from the word febifuge, which alludes to the ability of the plant to reduce a fever. The whole plant was at one time used medicinally for fevers and nervous disorders. An extract was also reported to be an effective insect repellent when applied to the skin. Cut flowers are excellent.

SELECTED CULTIVARS AND VARIETIES: 'Aureum' Flowers white, single; young foliage yellowish green, then turning green later in the season.
'Crispum' With crinkled foliage.
'Golden Ball' Disc flowers bright yellow, ray flowers lacking.
'Golden Feather' With chartreuse foliage; height to 10 inches.
'Golden Moss' To 18 inches tall, very compact, with golden flowers.
'Silver Ball' Flowers double, colored white.

CULTURE: Soil: Well-drained, highly organic soil is preferred.
Moisture: Regular watering should be practiced during the heat of summer to maintain soil moisture.
Light: Full sun.

PATHOLOGY: Diseases: Bacterial and fungal leaf spots, stunting, wilting, rusts, aster yellows.
Pests: Aphids, bugs, beetles, mealybugs, whitefly, cutworm, mites.

MAINTENANCE: Shear foliage after blooming to neaten appearance and promote compact growth. Division may be needed every 3 to 4 years.

PROPAGATION: Cuttings: Stem cuttings can be taken in midsummer. They root well without the aid of a root inducing compound.
Division: Simple division is usually practiced in spring or fall.
Seed: Supposedly, seed will germinate in 2 to 4 weeks at 65° to 70°F.

Taxus (tak´sus) Latin name for a yew tree, probably from Greek, *taxon*, a bow, the resilient wood having been used for making bows. The genus is composed of eight species of dioecious, evergreen, widely cultivated shrubs and trees of varying hardiness. They are native to the northern hemisphere.

FAMILY: TAXACEAE: Yew family.

T. baccata (ba-kā´tá) Berried (Figure 3–185).

COMMON NAME: Berried yew, English yew.

NATIVE HABITAT: Europe, northern Africa, western Asia.

HARDINESS: Zones 5 or 6 to 7.

DISCUSSION: The species itself is a densely branched tree to 60 feet high. However, there are a couple of cultivars that have prostrate habits of growth and make excellent ground covers. They are not commonly cultivated but merit more attention. The characteristics of the ground covering types are as follows:

SIZE: 2 to 4 feet tall by 12 to 15 or more feet across.

HABIT: Low spreading, semiprostrate, woody shrubs.

RATE: Relatively slow; space plants about 4 to 5 feet apart from 2 or 3 gallon sized containers.

Figure 3–185 *Taxus bacatta* 'Repandens' (life-size)

LANDSCAPE VALUE: En masse these cultivars make fine cover in a larger setting. They are also used to good effect as specimens in a shrub bed or along walks and near building entrances. No foot traffic.

FOLIAGE: Evergreen, spirally arranged, simple, linear, 1/2 to 1 1/4 inches long, 1/8 to 3/16 inch wide, convex, shiny, dark (almost blackish) green above, margins recurved, midrib prominent, pale below with poorly defined stomatal lines, ending with horny tipped, acuminate apex.

STEMS: With brownish scales surrounding bases of branchlets, green in youth, turning brown with age.

INFLORESCENCE: Dioecious; male flowers in strobili, 6 to 14 stamened, globose, arising from leaf axils on lower side of previous years branchlets; female strobili are green, in leaf axils, solitary, opening in spring; neither the male nor female strobili are of ornamental significance.

FRUIT: 1/4 to 1/5 inch wide olive-brown seed, surrounded by scarlet aril.

SELECTED CULTIVARS AND VARIETIES: 'Cavendishii' Excellent, low growing female clone, less than 3 feet high, of wide spreading semiprostrate habit, with branch tips that frequently droop. 'Repandens' Low, wide spreading, fruiting selection, 2 to 4 feet high by 12 or more feet across. Hardier than the species, reliable in zone 5. A similar cultivar is reported, 'Repens Aurea' (leaves variegated yellow and green), but it is quite rare.

CULTURE: Soil: Best in sandy loam, with excellent drainage being essential. Adaptable to a wide range of pH, including alkaline and acidic conditions to as low as 5.5.
Moisture: Withstanding drought once established; occasional thorough watering during summer is beneficial. Keeping soil slightly moist is ideal.
Light: Full sun to moderate shade.

TRANSPLANTING: Nursery stock is usually available balled and burlapped or as recently potted field-grown plants. Establishment is relatively easy, but overwatering should be avoided, as the roots will quickly decay if they should become too wet.

PATHOLOGY: Diseases: Twig blight, needle blight, root rot, leaf drop, dieback.
Pests: Black vine weevil, strawberry root weevil, taxus mealybug, grape mealybug, taxus bud mite, ants, termites, nematodes.

MAINTENANCE: Annual shearing is recommended to keep plants dense and compact. This is usually done in spring or early summer.

PROPAGATION: Cuttings: Dormant cuttings are usually taken in the fall after a period of frost. They should be treated with a rooting compound of 8000 ppm IBA placed in well-drained medium and given bottom heat of 65° to 75° F.
Grafting: Side cleft grafting may be practiced, as some of the clones may be difficult to root. In such a case, seedlings of *T. baccata* are used as the understock, and grafting is performed in midwinter.

✦✦✦✦✦✦✦✦✦✦✦✦✦✦✦

Teucrium (tū´kri-um) Named after Teucher, a Trojan prince who first used one of the species in medicine. The genus is composed of about 300 species of shrubs and herbs. They are native to many regions, primarily the Mediterranean.

FAMILY: LAMIACEAE (LABIATAE): Mint family.

T. chamaedrys (kam-e´dris) Old name for germander, signifying on the ground (Figure 3–186).

COMMON NAME: Common or ground germander.

NATIVE HABITAT: Central and southern Europe.

HARDINESS: Zones 5 to 8.

Figure 3–186 *Teucrium chamaedrys* (life-size)

HABIT: Low, dense, rhizomatous, semiwoody, shrubby ground cover.

SIZE: 8 to 18 inches tall, spreading 1 1/2 to 2 feet across.

RATE: Moderate to relatively fast; space plants 12 to 18 inches apart from quart-sized containers.

LANDSCAPE VALUE: Often used to good effect as a soil binder or general cover for moderate or large, usually informal settings. A fine specimen in rock gardens or as a low hedge or walk edging. No foot traffic.

LEAVES: Semievergreen, opposite, simple, oblong to obovate-oblong, 1/2 to 3/4 inch long, margins of vegetative leaves deeply serrate to crenate-toothed, flower stalk leaves entire, pubescent, medium lustrous green, short petioled, slightly aromatic.

INFLORESCENCE: Loose to dense, upright terminal or axillary clusters of 2 to 6 flowers; flowers with calyx about 1/4 inch long, corolla to 5/8 inch long, purple to rose pink, spotted red and white, effective early summer to early fall.

FRUIT: Nutlets, not ornamentally significant.

SELECTED CULTIVARS AND VARIETIES: Var. *prostratum* Low selection to 8 inches high, heavy flowering.

CULTURE: Soil: Adaptable to most soils; good drainage a necessity.
Moisture: Relatively tolerant of drought; these plants take a good deal of heat and require only occasional thorough summer watering.
Light requirement: Full sun is best; withstands light shade.

PATHOLOGY: Diseases: Downy mildew, powdery mildew, leaf spots, rust.
Pests: Mites cause leaves to crinkle but are rarely serious.

MAINTENANCE: Annual shearing in early spring keeps plants dense and compact. Mulching in autumn will reduce winter damage to the stems.

PROPAGATION: Cuttings: Softwood cuttings are easily rooted when taken in early summer. Treatment with 1000 ppm IBA may increase the quantity of roots, but is not a necessity.
Division: Simply divide plants in spring or fall.

✛ ✛ ✛ ✛ ✛ ✛ ✛ ✛ ✛ ✛ ✛ ✛ ✛ ✛ ✛ ✛

T̲hymus (ty´mus, or thi´mus) Old Greek name used by Theophrastus either for this plant or for savory. This is a large genus of between 300 and 400 species of subshrubs and herbaceous perennials with aromatic foliage. Often they are of prostrate or creeping habit with squarish woody based stems. Species of *Thymus* are frequently used in flavoring, and the dried foliage is used in sachets and potpourri. They are native to Europe and Asia.

FAMILY: LAMIACEAE (LABIATAE): Mint family.

HABIT: Low, often decumbent, mat forming, woody based, subshrubby ground covers.

RATE: Relatively slow growing; space 6 to 10 inches apart from 3 to 4 inch diameter containers.

FOLIAGE: More or less evergreen, opposite, simple, entire, small.

FLOWERS: Solitary or combined in few or many flowered verticillasters, frequently crowded in a terminal head, calyx cylindrical to campanulate, 10 to 13 nerved, frequently 2 lipped, throat hairy; corolla tube straight, 2 lipped. Flowers of thymes are attractive to bees.

FRUIT: Nutlets, not ornamentally significant.

LANDSCAPE VALUE: Excellent aromatic fillers of patio cracks or between stepping stones. Sometimes used as a lawn substitute or general cover in small- or moderate-sized areas. Excellent as walkway edging where their fragrances can be appreciated. Tolerant of occasional foot traffic.

* * *

T. × *citriodorus* (kit-rē-ō´dōr-us or sit-) Citron scented (Figure 3–187).

COMMON NAME: Lemon thyme.

HYBRID ORIGIN: *T. pulegioides* × *T. vulgaris*.

HARDINESS: Zone 3.

DESCRIPTION: Highly branched, mostly erect stemmed; leaves narrowly rhombic-ovate to lanceolate, to 3/4 inches long, glabrous, scented like lemon.

SIZE: 4 to 8 inches tall, spreading indefinitely.

FLOWERS: Pale lilac, in 1/4 inch diameter oblong heads, effective in early summer.

SELECTED CULTIVARS AND VARIETIES: 'Aureus' Foliage gold and green variegated; somewhat lower growing.
'Argenteus' Foliage variegated silvery and green.

* * *

T. herba-barona (hẽr´ ba̅-ba̅-rō´na̅) Likely derived from French, meaning herb of nobility.

COMMON NAME: Caraway thyme.

NATIVE HABITAT: Corsica.

HARDINESS: Zone 4.

SIZE: 2 inches tall, spreading indefinitely.

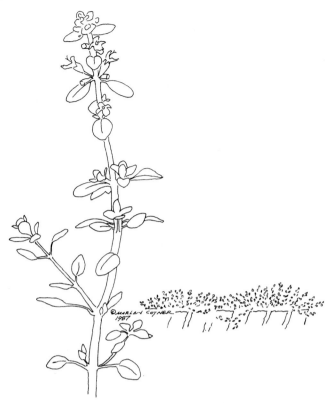

Figure 3–187 *Thymus × citriodorus* (life-size)

FOLIAGE: Ovate-lanceolate to 5/16 inch long, margin ciliate, apex acute, glabrous, dark green and glandular dotted above.

FLOWERS: Pink, in tiny heads in midsummer.

* * *

T. lanicaulis (lan-i-côl′ĭs) From Latin *lanis*, downy, and *caulis*, stalk of a plant, alluding to the hairs on the stems. Sometimes listed as *T. thracicus*.

COMMON NAME: Woolly-stem thyme.

NATIVE HABITAT: Balkans.

HARDINESS: Zone 3.

SIZE: 3 to 4 inches tall, spreading indefinitely.

FOLIAGE: Elliptic-spatulate, about 5/16 inch long, apex somewhat obtuse.

FLOWERS: Purple, in heads borne in early summer.

* * *

T. serpyllum (sẽr-pil′lum) From the old Greek name *kerpyllos*, for the wild thyme.

COMMON NAME: Wild thyme, creeping thyme, creeping red thyme, mother of thyme.

NATIVE HABITAT: Europe, western Asia, and northern Africa.

HARDINESS: Zone 3.

SIZE: 1 to 3 inches tall, spreading indefinitely.

FOLIAGE: Elliptic to oblong, 1/4 to 1/3 inch long, nearly sessile, dark green, resin dotted, pubescent above, margin ciliate, mintlike fragrance when crushed.

FLOWERS: Purple, in small heads, born late spring to early summer.

SELECTED CULTIVARS AND VARIETIES: 'Albus' Flowers white.
'Argenteus' Leaves variegated silvery white.
'Aureus' Leaves variegated golden-yellow. Var. *coccineus* With scarlet flowers.
'Lanuginosis' Foliage woolly-gray, pubescent.
'Roseus' With pink flowers.

* * *

T. vulgaris (vul-gā′ris) Common.

COMMON NAME: Common thyme, garden thyme.

NATIVE HABITAT: Southern Europe.

HARDINESS: Zone 5.

OF SPECIAL INTEREST: Common herb used as seasoning.

SIZE: 6 to 10 inches, spreading indefinitely.

FOLIAGE: Aromatic, linear to elliptic, 3/16 to 5/8 inch long, tomentose but margin not ciliate.

FLOWERS: Whitish to lilac, in dense, many flowered heads, borne in late spring.

OTHER SPECIES: There are many other species of thyme that are suitable for use as ground cover. Two species that may be readily available include *T. pseudolanuginosus* and *T. praecox* var. *arcticus*.

Generally Applicable to All Species

CULTURE: Soil: Best in well-drained, light textured soils of low fertility. A pH range of 6.5 to 7.5 is best. Other soil textures are tolerated, provided drainage is excellent.
Moisture: Well adapted to withstand drought; growth is best when soil is hot and somewhat dry.
Light: Full sun is best; light shade tolerated.

MAINTENANCE: Shear or mow plantings in early spring to keep them neat and compact.

PATHOLOGY: Diseases: Very susceptible to leaf blight during cold, rainy weather.
Pests: No serious pest problems.

PROPAGATION: Cuttings: Because of the small size of branchlets, thymes are not ordinarily propagated by cuttings; however, most cultivars are easily rooted. An effective manner to propagate thyme by cuttings is to lay the stems on top of the soil and let them root at many nodes instead of just a few that are poked into the soil if stuck in the conventional manner. The close monitoring of mist applications with this technique is essential as the stem, being removed from the soil, is not able to draw moisture out of it.

Division: Simply divide and replant in spring or fall.

Seed: Seed is available commercially for many species. Usually, germination ocurs in 2 or 3 weeks at 75 °F.

*T*iarella (ti-à-rel´à) From Greek, *tiara*, a turban (literally, a little turban), alluding to the shape of the seed pod. The genus is composed of six species of herbaceous, primarily perennial plants. They are rhizomatous, flower in spring, and are native to eastern and western North America, with one species from Asia.

FAMILY: SAXIFRAGACEAE: Saxifrage family.

T. cordifolia (kôr-di-fō´li-à) Heart shaped, the leaves (Figure 3–188).

COMMON NAME: Allegheny foamflower, foamflower, false miterwort.

NATIVE HABITAT: Moist woodlands throughout eastern Canada and the United States to the Carolinas and Tennessee, extending westward to Michigan.

HARDINESS: Zone 3.

SIZE: 6 to 12 inches high, spreading indefinitely.

HABIT: Low, herbaceous, compact, mounding, rhizomatous ground cover.

RATE: Moderate to relatively fast once established; space plants about 12 inches apart from pint- to quart-sized containers.

LANDSCAPE VALUE: Fine plant for general cover in naturalized small- or moderate-sized settings, or as specimen in a shaded section of the rock garden or perennial border. No foot traffic.

LEAVES: Basal, on downy 4 inch long petioles, ovate-cordate in outline, to 4 inches long, margin unevenly dentate, 5 to 7 lobed, apex blunt or pointed, veins prominent, rich green, bronzy in fall, scattered with pubescence on both sides.

INFLORESCENCE: 6 to 9 inch long racemes; flowers

5 petaled, small, white, stamens protruding, effective in midspring.

FRUIT: Capsules, not of ornamental significance.

SELECTED CULTIVARS AND VARIETIES: 'Major' Flowers rose or reddish.
'Marmorata' Foliage bronze to black-green, marbled with purple.
'Purpurea' Flowers purple.

CULTURE: Soil: Best in organically rich, acidic loam.
Moisture: Not notably tolerant of drought; soil should be kept moist with regular watering. Best in cool location.
Light: Light to dense shade.

PATHOLOGY: No serious diseases or pests reported.

MAINTENANCE: Lightly mulch soil in fall to supplement organic content in amended soils.

PROPAGATION: Division: Simply divide plants in spring or fall.
Seed: Collect mature seed and sow in late summer.

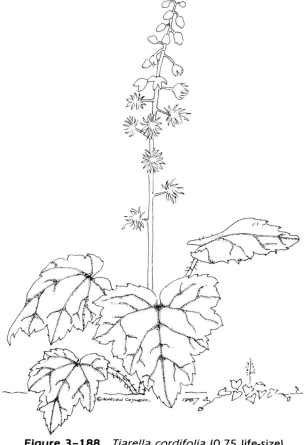

Figure 3–188 *Tiarella cordifolia* (0.75 life-size)

Figure 3–189   *Tolmiea menziesii* (life-size)

✛✛✛✛✛✛✛✛✛✛✛✛✛✛✛✛✛

Tolmiea (tol-mē´ǎ) After Tolmie, a surgeon of the Hudson Bay Company. This is a monospecific genus comprised of a herbaceous perennial that is native to western North America.

FAMILY: SAXIFRAGACEAE: Saxifrage family.

T. menziesii (men-zēz´ĭ-ī) After Archibald Menzies (1754–1842), naval surgeon and botanist who collected the species in western North America (Figure 3–189).

COMMON NAME: Piggy-back plant, youth-on-age, pick-a-back plant, thousand mothers.

NATIVE HABITAT: West coast, from Alaska to California.

HARDINESS: Zones 7 to 10.

HABIT: Low spreading, herbaceous ground cover.

SIZE: 12 to 24 inches high, spreading indefinitely.

RATE: Moderate; space plants 8 to 14 inches apart from 3 to 4 inch diameter containers.

LANDSCAPE VALUE: General cover, valued for foliage, in small- or medium-sized areas. No foot traffic.

LEAVES: Cordate in outline, to 4 inches across, hairy, light green, palmately veined, on long (to 8 inches) petioles that arise from a creeping rootstock; petioles with prominent stipules. New plantlets arise at the junction of leaf and petiole, rooting as they touch the ground.

INFLORESCENCES: Terminal raceme; flowers irregular, calyx tube cylindrical, greenish purple, to 3/8 inch long, lobes to 1/4 inch; overall rather inconspicuous and not of great ornamental significance.

FRUIT: Dehiscent capsule, not ornamentally significant.

CULTURE: Soil: Adaptable to most soils; tolerates wet soils.
Moisture: Intolerant to drought; soil should be kept moist with regular watering.
Light: Moderate to dense shade.

PATHOLOGY: Diseases: Powdery mildew.
Pests: Mealy bugs, mites.

MAINTENANCE: Little maintenance is required; simply clip back shoots as they climb out of their bounds.

PROPAGATION: Division: Separate new plantlets from the base of the leaves at the end of the petiole. Plant them anytime, being sure to keep them moist until roots become well developed. If they have not yet sprouted roots, they can be treated as cuttings and kept under mist for a few days until roots begin to form. Dipping them in 1000 ppm IBA will increase the number of roots.

Figure 3–190 *Trachelospermum jasminoides* (life-size)

✟✟✟✟✟✟✟✟✟✟✟✟✟✟✟✟✟

*T*rachelospermum (trā-ke-lō-spẽr´mum) From Greek, *trachelos*, throat, and *sperma*, seed, probably referring to the narrowness of the seeds. This genus is composed of about ten or more species of climbing (by twining) or clambering shrublike vines. Each has milky sap; they are native to tropical and subtropical eastern Asia. One species is native to North America.

FAMILY: APOCYNACEAE: Dogbane family.

HABIT: Ground cover types are low, creeping, woody vines or ground covers.

RATE: Moderate to relatively fast once established; space plants 1 1/2 to 3 feet apart from gallon-sized containers.

LANDSCAPE VALUE: Good general covers for moderate to large areas; can effectively grow atop tree roots. Often used to good effect as a bank cover. No foot traffic.

*   *   *

T. asiaticum (ā-shi-at´i-kum) From Asia.

COMMON NAME: Yellow star-jasmine.

NATIVE HABITAT: Japan and Korea, introduced into the trade in 1880.

HARDINESS: Zones 7 to 9.

SIZE: 10 to 14 inches tall when unsupported, spreading 8 to 10 feet across.

LEAVES: Evergreen, opposite, simple, elliptic to oblong or lanceolate, about 1 to 2 inches long, margin entire, leathery, dark and shiny green with whitish venation, dark purplish-red in winter along its northern range.

STEMS: Wiry, with holdfasts.

INFLORESCENCE: Loose thyrsus, terminal and lateral; flowers small, 5 parted, bisexual, calyx to 1/8 inch long, lobes shorter than narrow part of corolla tube, erect; corolla salverform, lobes oblong, stamens borne on corolla, yellowish-white, throat glabrous, effective midspring to early summer, fragrant, fewer flowers borne when grown as a ground cover than when climbing with support.

FRUIT: Paired cylindrical follicles, not of ornamental significance.

SELECTED CULTIVARS AND VARIETIES: 'Nana' ('Nortex') With leaves more lance shaped.

*   *   *

T. jasminoides (jas-min-oi´dēz) Jasminelike (Figure 3–190).

COMMON NAME: Confederate jasmine, Chinese star-jasmine.

NATIVE HABITAT: China.

HARDINESS: Zones 9 to 10.

SIZE: 10 to 16 inches tall unsupported, spreading 10 to 12 feet across.

LEAVES: Evergreen, opposite, simple, elliptic to oblong or oblanceolate, about 1 1/2 to 4 inches long by 1/2 to 1 inch wide, petiole about 1/4 inch long, apex acuminate, basally broadly cuneate, with darker venation.

STEMS: Wiry, with holdfasts.

INFLORESCENCES: Loose thyrsus, terminal and lateral flowers small, 5 parted, bisexual, calyx somewhat foliaceous, to 3/16 inch long, lobes longer than narrow part of corolla tube, strongly recurved; corolla salverform, white, to 1 inch in diameter, tube 5/16 inch long, lobes obliquely truncate at apex, stamens borne on middle of corolla tube, fragrant, effective midspring to early summer.

FRUIT: Paired cylindrical follicles, not of ornamental significance.

SELECTED CULTIVARS AND VARIETIES: 'Variegatum' Leaves variegated green and white, sometimes with a reddish tinge.

Generally Applicable to Both Species

CULTURE: Soil: Best in fertile, rich loam; pH in the range from 5.5 to 6.5 is best.
Moisture: Relatively intolerant of drought conditions; soil should be kept moist with regular watering.
Light: Full sun to moderate shade. Light to moderate shade is favored in warmer areas.

PATHOLOGY: Diseases: No serious fungal or bacterial diseases reported.
Physiological: Iron deficient chlorosis.
Pests: Mites, scales, greenhouse whitefly.

MAINTENANCE: Shear plantings in spring to maintain a low, neat appearance.

PROPAGATION: Cuttings: Softwood cuttings taken in late spring are generally easy to root and are benefitted with treatment of 1000 to 3000 ppm IBA.
Layering: A simple method of propagating these species: divide sections with rooted nodes in spring, winter, or fall.

*T*radescantia (trad-es-kan´shi-à) After John Tradescant, gardener to Charles I (1608 to 1622). This genus is composed of about 20 or more species of erect or trailing herbaceous perennials. They are native to North and South America.

FAMILY: COMMELINACEAE: Spiderwort family.

T. fluminensis (floo-min-en´sis) Of a river, from Latin, *flumineus* (Figure 3–191).

COMMON NAME: Wandering Jew, spiderwort.

NATIVE HABITAT: South America.

HARDINESS: Zones 8 to 10.

HABIT: Low, trailing, succulent groundcover.

SIZE: 2 to 4 inches tall, spreading indefinitely.

RATE: Moderate to relatively fast; space plants 1 to 1 1/2 feet apart from quart-sized containers.

LANDSCAPE VALUE: General cover for small- or medium-sized bounded areas. No foot traffic.

LEAVES: Semievergreen, alternate, simple, ovate-acuminate, to 1 5/8 inch long; 3/4 inch across, green above, deep violet below, glabrous, margin ciliate.

STEMS: Decumbent to erect, rooting at the nodes, soft and fleshy.

INFLORESCENCE: Of paired, sessile cincinni (coiled cymes); flowers to 1/4 inch long, 1/2 inch wide, white to pinkish, 3 petaled, effective summer through fall; reportedly individual flowers last only for 1 day.

FRUIT: 3-valved capsules, not of ornamental significance.

SELECTED CULTIVARS AND VARIETIES: 'Variegata' Leaves green and white.

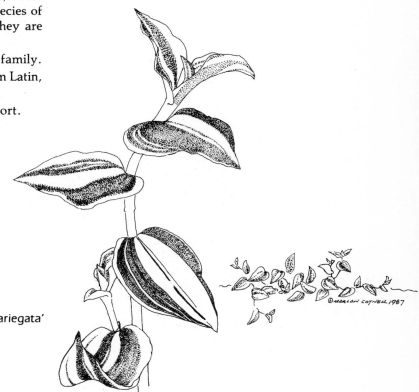

Figure 3–191 *Tradescantia fluminensis* 'Variegata' (life-size)

OTHER SPECIES: *T. albiflora* (Wandering Jew) Very similar to *T. fluminensis,* but the leaves are oblong to elliptic, to 2 to 3 inches long and 1 inch across. Color is green above and below.

T. virginiana (Virginia spiderwort, common spiderwort, widow's-tears).

HARDINESS: Zones 4 to 9.

DISCUSSION: This is relatively common as a herbaceous perennial and finds a good deal of use in the border, naturalized areas, or sometimes in ditches for erosion control. It is characterized by a rather wild appearance, with grasslike leaves to about 1 foot long and 1/2 inch to 1 inch wide. Flowers are violet-purple, deep blue, or rarely white. They are three parted and arranged in terminal umbels. Succession of bloom begins in late spring and continues until late summer, individual flowers lasting only for 1 day. Overall height may be from 1 to 3 feet; spreading indefinitely. Several cultivars have been ascribed to this species, but some are believed to be more properly classified as members of the hybrid *T. × andersoniana.* They include: 'Blue Stone' With dark blue flowers. 'Innocence' White flowered, clump forming plant to 24 inches tall.
'Iris Prichard' Flowers white, tinged with blue.
'James C. Weguelin' Flowers larger and pale blue.
'Pauline' Flowers large and pale pink.
'Purple Dome' Flowers bright rosy purple.
'Red Cloud' Flowers rosy red.
'Snowcap' Flowers snowy white.
'Zwanenburg' Flowers violet-purple.

Generally Applicable to All Species

CULTURE: Soils: Best in organically rich, loamy soil. Tolerant of poorly drained soils.
Moisture: Not notably tolerant of drought. Soil should be kept moderately moist with regular watering.
Light: Light to moderate shade; *T. virginiana* is best in full sun.

PATHOLOGY: Diseases: Leaf spots, rust, blight.
Pests: Greenhouse leaf tier, morning-glory leaf cutter, orange tortrix caterpillar, citrus mealybug, chaff scale, southern root knot nematode, aphids.

MAINTENANCE: Little or no unusual maintenance is required.

PROPAGATION: Division: Easily propagated by simple division (all species).
Cuttings: Cuttings root readily; no root inducing compound is needed, but treating with 1000 ppm IBA will stimulate development of a larger root system (exclusive of *T. virginiana*).

✛✛✛✛✛✛✛✛✛✛✛✛✛✛✛✛✛

*V*accinium (vak-sin´i-um) Ancient Latin name for the blueberry. The genus is composed of about 150 species of deciduous and evergreen shrubs, with a few small trees and vines. Native to the northern hemisphere, mostly from North America and Eastern Asia, the fruits of some species are valued for being edible. Fruits that we commonly recognize as blueberry, huckleberry, cranberry, and others are produced by various species of *Vaccinium.*

FAMILY: ERICACEAE: Heath family.

RATE: Relatively slow; space plants about 12 inches apart from quart- to gallon-sized containers.

LANDSCAPE VALUE: General covers for small areas, naturalized gardens, or as specimens where proper cultural conditions can be met. Sometimes they are even used in rock gardens should the rock be of a nature that does not cause the soil pH to rise excessively. No foot traffic.

*   *   *

V. angustifolium (an-gus-ti-fō´li-um) Narrow leaved (Figure 3–192).

COMMON NAME: Lowbush blueberry, low sweet blueberry, sweet-hurts, late sweet blueberry.

NATIVE HABITAT: Northeastern North America, Newfoundland to Saskatchewan and south to Illinois and Virginia. Introduced into the trade in 1772.

HARDINESS: Zones 2 to 6.

HABIT: Low growing, woody, often open, shrubby cover.

SIZE: 6 to 24 inches tall by 2 feet across.

LEAVES: Deciduous, simple, alternate, lanceolate, to 3/4 inches long, margin serrulate with bristle tipped teeth, shiny blue-green, glabrous, bronzy-scarlet and crimson in fall.

INFLORESCENCE: Small clusters of small urnlike flowers with 4 to 5 lobed calyxes; corolla bell shaped, to about 1/4 inch long, white, often tinged red, not very showy, borne early to midspring.

FRUIT: Small (1/4 to 1/2 inch diameter), edible, many-seeded berry, bluish-black and bloomy, very sweet and pleasant tasting.

OF SPECIAL INTEREST: This species is grown commercially for fruit production in the state of Maine.

*   *   *

V. vitis-idaea (vī´tis, i-dē´ä) Old generic name, meaning vine or grape of Mount Ida.

COMMON NAME: Mountain cranberry, cowberry, foxberry, lingonberry.

NATIVE HABITAT: Northeastern United States and Canada, also Northern Asia and Europe.

HARDINESS: Zones 2 to 6.

HABIT: Low, woody, rhizomatous, spreading ground cover.

SIZE: 6 to 12 inches tall, spreading to 2+ feet across.

LEAVES: Evergreen, alternate, simple, obovate-convex to 1 1/4 inch long, leathery, glossy dark green above, glandular dotted and paler below, turning metallic reddish-brown in winter.

INFLORESCENCE: Drooping racemes; flowers with 4 to 5 lobed calyxes, corolla campanulate, 1/4 inch long, white or pinkish, borne mid to late spring.

SELECTED CULTIVARS AND VARIETIES: Variety *majus* (Cowberry) Eurasian variety with leaves and fruit somewhat larger, seems to be more useful in southern range. The var. *minus* (mountain cranberry, lingonberry, rock cranberry) is an Arctic variety with smaller leaves and fruit. Slow growing, it is more useful in its northern range and is not as tolerant of hot and dry conditions. Height usually reaches around 4 to 6 inches and sometimes will reach 8 inches; spreading to about 10 inches across.

OTHER SPECIES: *V. crassifolium*, a procumbent growing species that is similar to *V. vitis-idaea*. Commonly called creeping blueberry, this native to southeastern Unites States (North Carolina to Gerogia) makes a nice, evergreen mat.

Generally Applicable to All Species

CULTURE: Soil: Organically rich, acidic soil with good drainage, pH in range of 4.5 to 5.5 is best. Moisture: Soil should be kept moist with regular watering for best growth. Light: Full sun to light shade.

PATHOLOGY: Diseases: Dieback, leaf spots, viral ring spot, and stunt. Pests: Stem gall (wasp), scales, azalea stem borer, forest tent caterpillar. Physiological: Iron deficient chlorosis, boron deficiency.

MAINTENANCE: In amended soils, annual incorporation or light topdressing of peat will help maintain organic content and pH. In areas of amended, naturally alkaline soils, soil should be tested periodically, and pH regulated accordingly.

PROPAGATION: Division: Simply divide plants in early spring before growth begins. Cuttings: Softwood and hardwood cuttings can both be rooted and rooting is enhanced with the use of a rooting preparation of 1000–3000 ppm IBA, Rhizomes of *V. angustifolium* can be sectioned into 3 or 4 inch sections and will form new shoots and roots. Seed: Ripe seed should be harvested in late summer, depulped, and stratified cold and moist for 3 months. Germination may take up to 6 months.

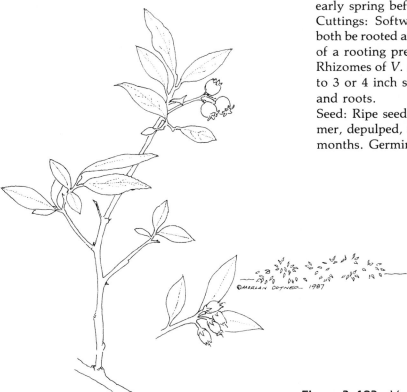

Figure 3–192 *Vaccinium angustifolium* (life-size)

✝✝✝✝✝✝✝✝✝✝✝✝✝✝✝✝✝

Vancouveria (van-kö-vē´ri-à) After Captain George Vancouver, English explorer (1758–1798). This genus is composed of three species of rhizomatous, herbaceous perennials. They are native to western North America.

FAMILY: BERBERIDACEAE: Barberry family.

HABIT: Dense, spreading, herbaceous, ground covers.

RATE: Moderate to relatively fast spreading; space plants 10 to 14 inches apart from pint- to quart-sized containers.

LANDSCAPE VALUE: Excellent general covers in native or naturalized settings. Good companion to taller clump forming ferns, rhododendrons, and other ericaceous plants. No foot traffic.

* * *

V. hexandra (heks-ăn´drà) Six-stamened (Figure 3–193).

COMMON NAME: American barrenwort.

NATIVE HABITAT: Pacific coastal forests of the United States.

HARDINESS: Zones 5 to 8.

SIZE: 12 to 18 inches high, spreading to 12 inches across.

LEAVES: Deciduous, basal, twice ternate; leaflets cordate in outline, somewhat 3 lobed, to 1 1/2 inch long, very thin, blue-green, nearly glabrous below.

INFLORESCENCE: Panicles on leafless scapes; flowers small, drooping, 1/2 inch long, corolla of 6 petals, narrow, white, effective mid to late spring.

FRUIT: Not of ornamental significance.

* * *

V. planipetala (plan-ni-pe´tā-là) Flat petaled.

COMMON NAME: Inside-out flower.

NATIVE HABITAT: Northern Oregon and California.

HARDINESS: Zones 6 to 8.

SIZE: 7 to 12 inches high, spreading to 12 inches across.

LEAVES: Evergreen, basal, two to three times ternate, sometimes with 5 leaflets to 1 1/2 inch long (infrequently somewhat larger), about 1 1/2 inch wide, margins cartilaginous, nearly glabrous below, light to medium green.

INFLORESCENCE: Panicles of 25 to 30 flowers on leafless scapes; flowers tiny, 6 petaled, white or tinged lavender, effective mid to late spring.

FRUIT: Not of ornamental significance.

Generally Applicable to Both Species

CULTURE: Soil: Best in organically rich, well-drained loam; relatively acidic pH; a range from approximately 4.5 to 6.0 is ideal.
Moisture: Not notably tolerant of drought. It is best to maintain soil in moist condition and plant in an area that is not subject to strong, drying winds.
Light: Moderate to dense shade.

PATHOLOGY: No serious diseases or insects reported.

MAINTENANCE: On an annual basis, lightly top-dress with compost in the fall to help maintain organic content of amended, light textured soils.

PROPAGATION: Division: Simply divide plants in spring or fall.

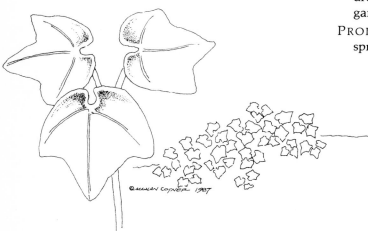

Figure 3–193 *Vancouveria hexandra* (life-size)

✝✝✝✝✝✝✝✝✝✝✝✝✝✝✝✝✝✝

***V**erbena* (vĕr-bē´nȧ) Ancient Latin name of the common European vervain, *V. officinalis*. Some authorities say that it is derived from Latin, *verbenae*, the sacred branches of olive, laurel, and myrtle. This genus, commonly called vervain, is composed of about 200 species of annual and perennial herbs and subshrubs. They are native to the tropics and subtropics of North and South America; often hairy, with erect, ducumbent, or prostrate habit.

FAMILY: VERBENACEAE: Vervain family.

HABIT: Ground cover types are low, mat forming, trailing, herbaceous ground covers.

RATE: Moderate; space plants 12 to 18 inches apart from 3 to 4 inch diameter containers.

LANDSCAPE VALUE: General low covers with good persistent, floral display for small- to medium-sized, bounded areas. They look striking trailing over an elevated planter. No foot traffic.

OF SPECIAL INTEREST: The flowers of verbena are attractive to butterflies.

*   *   *

V. bipinnatifida (bi-pin-ā´tif-i-dȧ) In reference to the leaf morphology; twice pinnate.

COMMON NAMES: Dakota verbena, Dakota vervain.

NATIVE HABITAT: Northern midwestern United States.

HARDINESS: Zone 3.

SIZE: To about 3 inches tall, spreading 1 1/2 to 2 1/2 feet across.

LEAVES: Opposite, triangular in outline, to 2 inches long, 3 parted or twice pinnate, segments oblong or linear.

INFLORESCENCE: Dense spikes; flowers with bracts shorter or of equal length as calyxes; corolla salverform, 1/2 inch across, lilac-purple, blooming late spring throughout the growing season until frost.

FRUIT: Of 4 nutlets, not of ornamental significance.

*   *   *

V. canadensis (can-ȧ-den´sis) Meaning unclear, as this is not a Canadian native; may have previously been classified as *Glandularia canadensis*, and the species name was retained when reclassified (Figure 3–194).

COMMON NAMES: Rose verbena, clump verbena, creeping vervain, rose vervain.

NATIVE HABITAT: Virginia to Florida, west to Iowa, Colorado, and Mexico (sandy and rocky barrens).

HARDINESS: Zones 5 to 10; mulching may be necessary in the northern range where snow cover is unreliable.

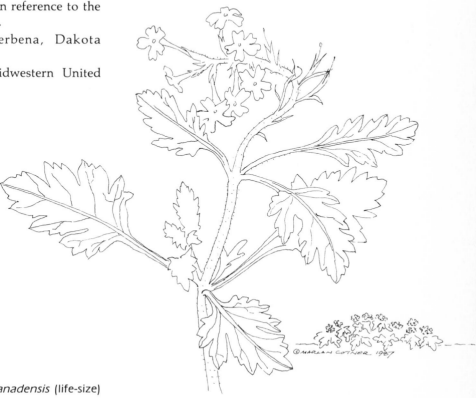

Figure 3–194 *Verbena canadensis* (life-size)

SIZE: 4 to 6 inches tall, spreading 2 or more feet across.

LEAVES: Opposite, simple, ovate or ovate-oblong, to 4 inches long, base truncate or broadly cuneate, margin sharply and irregularly toothed and cut, may be weakly to decidedly 3 cleft or lobed, stiffly hairy to nearly glabrous, shiny dark green.

STEMS: Branching, creeping, ascending or decumbent, rooting at the nodes, squarish in cross section.

INFLORESCENCE: Elongating, dense, terminal, short spikes; flowers are bisexual, numerous; bracts shorter than calyxes, calyx tubular, 5 parted; corolla salverform, to 5/8 inch across, with 5 unequal lobes, reddish-purple, lilac, rose or white, blooming late spring to early summer, then sporadically until frost.

FRUIT: Of 4 nutlets, not ornamentally significant.

SELECTED CULTIVARS AND VARIETIES: 'Candidissima' White flowered.
'Compacta' Dense, compact, lower growing selection.
'Rosea' More compact, with continually blooming rose-purple flowers.

OTHER SPECIES: *V. peruviana* (Peruvian verbena) A procumbent, tender, evergreen, dense, mat forming perennial. Native to Argentina and Brazil, its height reaches 4 to 6 inches with dark green, oblong-lanceolate, toothed leaves. Flowers are scarlet or crimson, in dense capitate spikes, effective throughout the summer. Frequently this species is grown as an annual in the north. Northern hardiness is limited to about zone 8.

V × *hybrida* (common garden verbena) This is a valuable hybrid with many cultivars. Derived from the cross *V. pulverulentum* × *V. sinuatum*, it is usually grown as an annual; habit is decumbent or creeping, often rooting along the way. Leaves are oblong to oblong-ovate, 2–4 inches long, toothed and somewhat lobed basally. Flowers of the various cultivars range from pink, red, yellow, blue, purple, or in combinations. Hardiness is probably limited to zones 9 and 10.

Generally Applicable to All Species

CULTURE: Soil: Adaptable to a wide range of well-drained soils. Tolerant of sandy soils and infertility. A pH slightly acidic to neutral (6.0 to 7.0) is preferred.
Moisture: Relatively tolerant of drought once established. Needs only occasional deep watering during the summer months thereafter.
Light: Full sun.

PATHOLOGY: Diseases: Bacterial wilt, flower blight, powdery mildew, stem rot, root rot.
Pests: Aphids, beetles, mites, northern root knot, fern nematodes, verbena leaf miner, verbena bud moth, greenhouse whitefly, caterpillars.

MAINTENANCE: Cut back trailing stems as they outgrow their bounds. Shear lightly in the spring and fall to keep plantings dense and compact.

PROPAGATION: Cuttings: Softwood cuttings should be taken in summer; they root readily. Application of 1000 ppm IBA is beneficial.
Division: Plants are readily propagated by dividing in spring, winter, or fall.
Seed: Seed germinates in 3 to 4 weeks at 70° to 85°F.

✦✦✦✦✦✦✦✦✦✦✦✦✦✦✦✦✦

*V*eronica (ve-ron´i-kȧ) Origin of the genus name is in question. Some believe it to be a Latin form of the Greek word, *betonika* (previously a small genus of herbs), and others refer to it as a derivation of Greek, *hiera eicon*, meaning sacred image, or from the Arabic term *viroo nikoo*, meaning beautiful remembrance. Others simply maintain that it is commemorative of Saint Veronica. At any rate, this genus, commonly called speedwell, is composed of about 250 or more species of annual and perennial herbs of varying habit. Some are clump forming with erect stems and slowly creeping rhizomes, while others trail with prostrate habit. They are native to the northern temperate zone. Several species are useful as ground covers. They are generally described next with more detailed descriptions to follow.

FAMILY: SCROPHULARIACEAE: Foxglove family.

RATE: Generally moderate to relatively fast; often invasive. Spacing of 8 to 12 inches is generally adequate for plants that come from 3 to 4 inch diameter containers.

LANDSCAPE VALUE: Good in rock gardens and for edging when spread is contained. The very prostrate forms make nice fillers for cracks in stepping stones and patio cracks. Using species of veronica as general cover is recommended only for moderate to large areas where they have room to spread without interfering with lawn areas, or small to moderate areas where their spread is contained. No foot traffic.

FOLIAGE: Deciduous to semievergreen in the north, frequently evergreen in areas of mild winter, lower stem leaves opposite, rarely whorled, occasionally becoming alternate higher on the stem, simple, margin entire or toothed, varying from broadly ovate to lanceolate.

STEMS: Branched or unbranched, creeping to upright depending on the species.

INFLORESCENCE: Axillary or terminal spikes, racemes, or corymbs, or sometimes solitary; flowers perfect, nearly to completely regular, calyx 4 or 5 parted, corolla usually spreading, 4 lobed (usually three rounded and one rounded and narrower), blue, purplish, pink or white, blooming spring and summer.

*   *   *

V. chamaedrys (kam-a´dris) On the ground and oak, the plant's habit and leaf shape, respectively.

COMMON NAME: Germander speedwell, angel's-eye, bird's-eye.

NATIVE HABITAT: Europe.

HARDINESS: Zones 3 to 10.

HABIT: Spreading, stoloniferous, compact, hairy, herbaceous ground cover.

SIZE: 12 to 18 inches tall.

DISCUSSION: *V. chamaedrys* has leaves that are 1/2 to 1 1/2 inches long, semievergreen, broadly margin toothed; flowers are in racemes to 6 inches, vibrant blue with light violet center, blooming late spring to early summer. The cultivar 'Alba' is white flowered.

*   *   *

V. incana (in-kā´nȧ) Hoary, hairy.

COMMON NAME: Woolly speedwell.

NATIVE HABITAT: Northern Asia.

HARDINESS: Zone 3.

HABIT: Herbaceous, dense, mat forming ground cover.

SIZE: 12 to 18 inches tall.

DISCUSSION: Stems and foliage are densely hairy, leaves are oblong or lanceolate, to 3 inches long, semievergreen, margin obtusely crenate, petioled, covered with silvery-gray soft hairs, overall effect of leaves is medium textured and soft; flowers are arranged in racemes to 6 inches, corolla bright blue, contrasting well with foliage, effective early to mid-summer. The cultivar 'Rosea' has flowers that are rose-pink.

*   *   *

V. latifolia 'Prostrata' (lat-i-fō´li-ȧ) Flat leaved.

COMMON NAME: Hungarian speedwell.

NATIVE HABITAT: Europe.

HARDINESS: Zone 3.

HABIT: Herbaceous, dense, hairy, matlike ground cover.

SIZE: 8 inches tall.

DISCUSSION: *V. latifolia* 'Prostrata' has semievergreen leaves that are lanceolate to ovate, margin toothed, dark green; flowers are blue to reddish, arranged in panacled racemes, effective late spring to early summer.

*   *   *

V. officinalis (ō-fis-i-nā´lis) Common.

COMMON NAME: Drug speedwell, gypsy weed, common speedwell, common veronica.

NATIVE HABITAT: Europe, Asia, and North America.

HARDINESS: Zone 3.

HABIT: Low, herbaceous, prostrate to decumbent, hairy, spreading ground cover.

SIZE: 12 inches tall.

DISCUSSION: Leaves are evergreen, ovate-elliptic to oblong, to 2 inches long, margin serrate; flowers arranged in many flowered axillary racemes, in-dividually the flowers are 1/2 inch wide, pale blue, effective from midspring to midsummer.

*   *   *

V. prostrata (pros-trā´ta) Prostrate, in reference to the habit of growth.

COMMON NAME: Harebell speedwell.

NATIVE HABITAT: Europe.

HARDINESS: Zone 5.

SIZE: 6 to 10 inches tall, spreading indefinitely.

HABIT: Herbaceous, dense, mat forming ground cover.

DISCUSSION: Leaves are semievergreen, linear to ovate, to 1 1/2 inch long, margin entire or sparse-ly toothed, dark green; flowers are arranged in terminal racemes, dark blue, effective late spring to early summer.

SELECTED CULTIVARS AND VARIETIES: 'Alba' With white flowers.
'Heavenly Blue' Flowers sapphire blue.
'Nana' Dwarf with lighter flowers.
'Rosea' With reddish-pink flowers.

*   *   *

V. repens (rē´penz) Creeping (Figure 3–195).

COMMON NAME: Creeping speedwell.

NATIVE HABITAT: Corsica and Spain.

HARDINESS: Zone 5.

HABIT: Low, herbaceous, dense, mat forming ground cover.

SIZE: 4 inches tall, spreading indefinitely.

DISCUSSION: Leaves are semievergreen, narrowly ovate, to 1/2 inch long, margin slightly crenate, glabrous, shiny, dark green; flowers are borne profusely in few flowered racemes, corolla pale blue, 1/4 inch across, effective and almost obscuring the foliage in late spring to early summer. The cultivar 'Alba' has white flowers, while 'Rosea' displays flowers that are reddish-pink.

Generally Applicable to All Species

CULTURE: Soil: Adaptable to most slightly to moderately acidic soils that have excellent drainage (essential to winter survival). Tolerant to and often better in infertile soils where spread is not rapid and habit is more compact.

Moisture: Relatively intolerant to extended drought; these shallowly rooted plants should be watered periodically to maintain enough moisture to keep them turgid.

Light: Full sun to light shade.

PATHOLOGY: Diseases: Downy mildew, leaf spot, leaf galls, root rot, leaf smut.

Pests: Checkerspot butterfly, Japanese weevil, southern root knot nematode.

MAINTENANCE: When fertilizing, be particularly careful to wash granules from foliage immediately, as leaves are exceedingly prone to desiccation when an osmotic imbalance exists. Shearing or mowing plantings after bloom neatens their appearance and rejuvenates growth.

PROPAGATION: Cuttings: Make softwood cuttings in either spring or summer.

Division: Plants should be divided spring or fall when dormant.

Seed: Seed should be collected when ripe, stored in a cool dry location, and planted the following spring. Germination usually takes 2 to 3 weeks at temperatures of 60° to 75°F.

***V**iburnum* (vi-bēr´num) Old Latin name for *V. lantana*, the wayfaring tree. The genus is composed of about 225 species of wide variation. Generally, the species are upright shrubs or small trees that are native to America, Europe, and Asia.

FAMILY: CAPRIFOLIACEAE: Honeysuckle family.

V. davidii (dā-vid´i-ī) After Abbé Armond David, French missionary and plant introducer and collector, in China (1826–1900) (Figure 3–196).

COMMON NAME: David viburnum.

NATIVE HABITAT: Western China, introduced into cultivation in 1904.

HARDINESS: Zones 7 to 9.

SIZE: 3 feet high by about 4 feet across.

HABIT: Low, broad, woody, mounding, shrubby ground cover.

RATE: Relatively slow; space plants 2 1/2 to 3 feet apart from gallon-sized containers.

LANDSCAPE VALUE: Excellent edging plant for along walks and for general cover in moderate- to large-size areas. A good companion for azaleas and rhododendrons. Widely cultivated in Europe, but not yet common in the United States. It does not tolerate extremes of hot, dry, or cold, and performs best in areas of moderate climate year round, such as in northern California and southern British Columbia. No foot traffic.

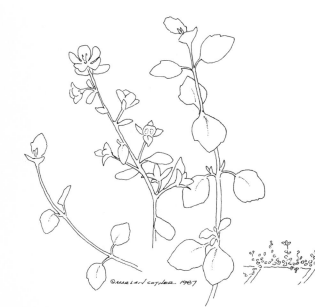

Figure 3–195 *Veronica repens* (life-size)

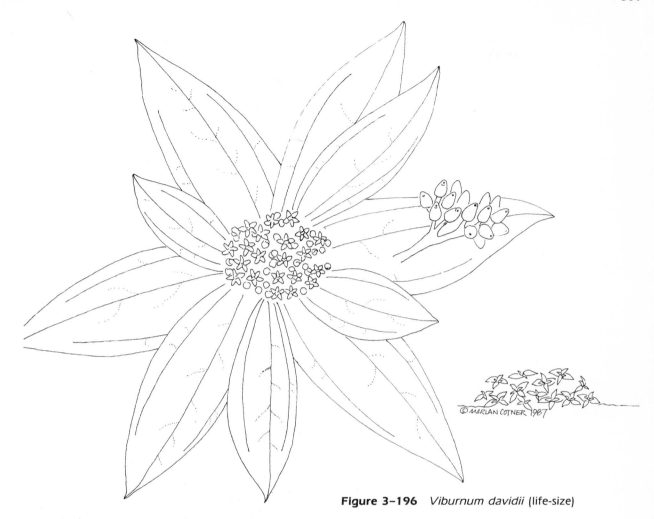

Figure 3–196 *Viburnum davidii* (life-size)

LEAVES: Evergreen, opposite, simple, narrowly ovate to somewhat obovate, 2 to 5 1/2 inches long, 1 to 2 1/2 inches wide, apex acuminate, base cuneate, margin sometimes slightly toothed toward apex, conspicuously depressed, 3 nerved, dark green above, pale below, glabrous with axillary tufts of hair below, petiole 1/4 to 1 inch long.

INFLORESCENCE: Peduncled, dense cymes to 3 inches across; flowers with 5 toothed calyxes, corolla 5 lobed, white, blooming in early summer.

FRUIT: 1/4 inch long, bright blue, 1-seeded oval drupes, not always produced as the plants appear to be functionally dioecious or require pollination from another clone. Excellent effect when present, late summer to midautumn.

SELECTED CULTIVARS AND VARIETIES: 'Jermyn's Globe' Compact and rounded selection.

CULTURE: Soil: Adaptable to most well-drained soils. A pH range of 5.0 to 6.5 is acceptable.
Moisture: Not notably tolerant of drought; best when soil is kept slightly moist with periodic watering in summer months.
Light: Light to moderate shade.

PATHOLOGY: Diseases: Bacterial leaf spot, bacterial crown gall, shoot blight, leaf spots, powdery mildew, rusts.
Pests: Aphids, beetles, thrips.

MAINTENANCE: Annual shearing in spring helps to keep plantings dense and attractive.

PROPAGATION: Cuttings: Softwood and hardwood cuttings are commonly rooted. Treatment with 8000 ppm IBA/talc is recommended.

✦✦✦✦✦✦✦✦✦✦✦✦✦✦✦✦✦

*V**inca* (ving´kȧ) Old Latin name, probably from *vincio*, to bind, alluding to the tough runners. This popular genus is composed of about 12 species of evergreen, trailing, vinelike subshrubs that are native to the Old World.

FAMILY: APOCYNACEAE: Dogbane family.

HABIT: Low, prostrate, creeping, nonclimbing, herbaceous, vinelike ground covers.

LANDSCAPE VALUE: Excellent general covers for moderate- or large-sized areas. *V. minor* is put to good use as edging along walkways, and both species have merits as foundation plantings. *V. minor* is one of the most popular ground covers for bank stabilization on large shady slopes and performs exceptionally well underneath trees and shrubs. Both species naturalize well and are exceptional in wooded settings, tolerating infrequent foot traffic.

*   *   *

V. major (mā´jôr) Greater; the greater periwinkle, meaning that the leaves are larger (Figure 3–197).

COMMON NAME: Big periwinkle, big leaf periwinkle, greater periwinkle.

NATIVE HABITAT: Southern Europe and Western Asia.

HARDINESS: Zones 7 to 9. I have seen the variegated form grown with success in protected sites with snow cover in zone 5. Top growth may die back in winter, but regrows quickly in spring.

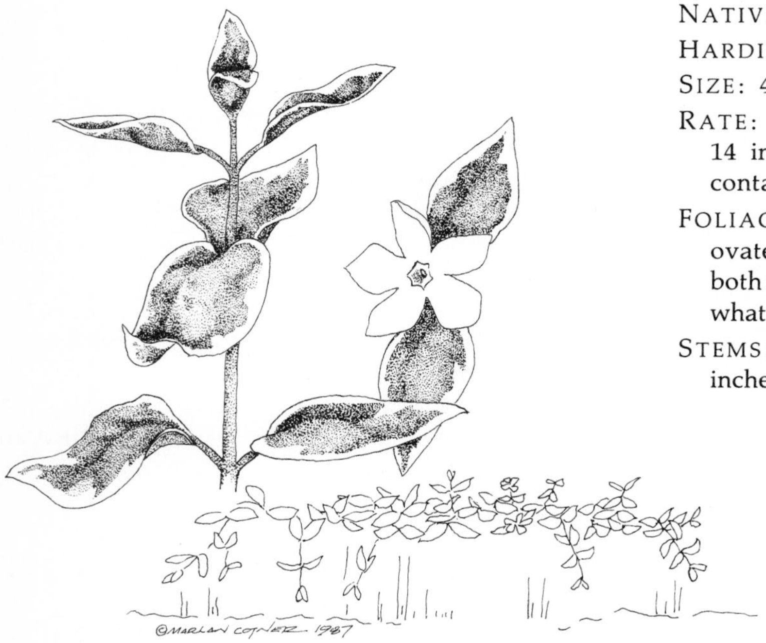

Figure 3–197 *Vinca major* 'Variegata' (0.75 life-size)

SIZE: 8 to 18 inches tall, spreading indefinitely.

RATE: Relatively fast; space plants from 3 to 4 inch diameter containers 10 to 16 inches apart.

FOLIAGE: Evergreen, opposite, simple, ovate, 1 1/2 to 3 inches long by 1 inch across, apex obtuse or acute, base truncate to subcordate, margin entire and ciliate; dark glossy green, glabrous, short petioled.

STEMS: Short and ascending, becoming trailing to several feet in length, rooting at the nodes.

FLOWERS: Solitary, axillary, 5 parted, perfect; corolla funnelform, lobes spreading, bright blue, 1 to 2 inches across, blooming primarily in early spring, then sporadically again in fall.

FRUIT: Of 2 follicles to about 2 inches in length, ornamentally insignificant.

SELECTED CULTIVARS AND VARIETIES: 'Gold Vein' Variegated golden yellow and green; yellow toward the inside surrounded by green. 'Variegata' ('Elegantissima') With leaves blotched and margined creamy white, no less vigorous than the species.

*   *   *

V. minor (mi´nôr) Lesser, the lesser periwinkle, presumably in reference to the diminutive morphology when compared to *V. major*.

COMMON NAME: Periwinkle, dwarf periwinkle, running myrtle, myrtle, lesser periwinkle.

NATIVE HABITAT: Europe and Western Asia.

HARDINESS: Zones 4 to 7 or 8.

SIZE: 4 to 6 inches tall, spreading indefinitely.

RATE: Moderate to relatively fast; space plants 8 to 14 inches apart from 2 to 3 1/2 inch diameter containers.

FOLIAGE: Evergreen, opposite, simple, oblong to ovate, apex acute or obtusely pointed, base acute, both ends attenuate, shiny dark green above, somewhat lighter below, petiole 1/2 to 1 1/4 inches long.

STEMS: Slender, trailing to usually fewer than 12 inches long, frequently rooting at the nodes.

FLOWERS: Solitary, axillary, perfect, 5 parted; corolla funnelform, lobes spreading, lilac-blue, nonfragrant, about 1/2 inch across, borne in early spring, then sparsely if at all through the remainder of the growing season; sometimes in number (but fewer than in spring) in fall.

FRUIT: Paired, erect 6 to 8 seeded cylindrical follicles, not of ornamental significance.

SELECTED CULTIVARS AND VARIETIES: 'Alba' Flowers white, leaves and stems smaller and less vigorous.

'Argenteo-Variegata' Leaves variegated with creamy white and pale green, flowers light blue, rather vigorous and attractive.

'Atropurpurea' Flowers deep purple.

'Aurea' New foliage golden yellow, becoming variegated green and gold in maturity.

'Aureo-Variegata' Leaves variegated with dull golden-yellow margins. The specimens that I have seen were not very attractive.

'Azurea Flore Pleno' Flowers double, sky blue.

'Bowlesii' ('Bowles's Variety') Flowers with somewhat more intense blue and larger than the species. Blooming more reliably and numerously in summer and fall after the primary flowering period in early spring. Leaves are somewhat more rounded and broader (sometimes folding under) than the species. Named for E. A. Bowles, sometimes this cultivar is called 'LaGrave' or 'Graveana'. It is more apt to stay in a clump than the species and spreads by crown division as well as trailing stems. I have also found it to be slightly more difficult to root from cuttings than the species.

'Gertrude Jekyll' Dainty, white flowered selection that trails lower to the ground than the species. Soil myst be kept relatively dry for success.

'Flore Pleno' Flowers are double, purplish-blue. Sometimes this cultivar is listed as 'Alpina' or 'Multiplex.'

'Ralph Shugert' I registered this sport of 'Bowlesii' in the autumn of 1987, and named it to commemorate the most enthusiastic and sharing horticulturist that I have had the pleasure of knowing. Characterized by foliage which is the same shape and size as 'Bowlesii', it has leaves that are colored deep glossy green, edged with a thin margin of white. Flowers are typical of the parent plant. Patent applied for.

Generally Applicable to Both Species

CULTURE: Soil: Best growth is attained in fertile, organically rich, well-drained, loamy soils. A pH of 4.5 to 7.5 is acceptable.

Moisture: Capable of withstanding summer drought, but best when soil moisture is maintained with occasional thorough watering. Protect from drying winter winds and sun if snow cover is unreliable in northern areas.

Light: Full sun to dense shade. In the south, best growth is attained in moderate to dense shade. If grown in full sun in the north, plants will perform well and hold their color provided that water is supplied regularly.

PATHOLOGY: Diseases: Blight, canker, dieback, leaf spots, root rot.

Pests: Aphids, nematodes.

MAINTENANCE: Mowing with a rotary mower in the spring before new growth begins will result in rejuvenation and thickened plantings.

PROPAGATION: Cuttings: Cuttings may be rooted in early summer with fair results. Usually they are started in small but deep plugs and then are transferred, three or four together into 2 1/4 inch pots. Division: Simply divide plants in late spring through winter. I have observed difficulty in establishing divisions that are dug in early spring when plants are actively growing.

Tissue culture: In vitro methods have been successful, and may increase the availability of select cultivars.

Viola (vi´ō-là) The ancient Latin name for *violet*. The genus is enormous and contains about 500 species. Widely distributed throughout the temperate regions, many species display an odd flowering characteristic whereby two types of flowers are formed. Typically flowers produced in spring are showy, 5 petaled, and sterile, while others in the summer are cleistogamous (flowers that do not open but are self-pollinated in the bud), apetalous, nonshowy, but heavily fruiting.

FAMILY: VIOLACEAE: Violet family.

HABIT: Ground cover types are low, tufted, herbaceous and often stoloniferous.

RATE: Moderate to relatively fast; space plants from 3 to 4 inch diameter containers 8 to 12 inches apart.

LANDSCAPE VALUE: Commonly used for edging or as general ground covers. Usually it is best to contain them with an edging to control their spread, or plant them in a naturalized area where they have room to roam. No foot traffic.

FRUIT: 3 valved capsule, not ornamentally significant.

* * *

V. hederaceae (hed-ēr-ā´sē-ē) Ivy (Hedera) -like, the shape of the leaves (Figure 3–198).

COMMON NAME: Ivy-leaved violet, Australian violet.

NATIVE HABITAT: Australia.

HARDINESS: Zones 9 and 10.

SIZE: 1 to 4 inches tall, spreading indefinitely.

FOLIAGE: Evergreen, basal, alternate, long petioled, orbicular to reniform in outline, 1/2 to 1 1/2 inches across, margin entire or crenate, glabrous or pubescent, medium green, glandular dotted on both sides.

FLOWERS: Solitary, irregular, 5 petaled, four of which are paired and the lower one spurred, to 1/2 inch long by 3/4 inch across, colored purple with white at tips of petals, blooming from spring through summer.

* * *

V. odorata (ō-do-rā´tȧ) Sweet scented.

COMMON NAME: Sweet violet.

NATIVE HABITAT: Europe, Asia, and Africa.

HARDINESS: Zones 5 to 10.

SIZE: 6 to 8 inches tall, spreading indefinitely.

FOLIAGE: Evergreen, basal, cordate to ovate or reniform, 1 to 2 inches long by 1 to 2 inches across, pubescent and glandular dotted both sides, medium green, long petioled; margin obtusely toothed, ciliate.

FLOWERS: Two types; fertile flowers (cleistogamous) are solitary, borne in summer, nonshowy and

greenish, concealed by foliage, and produce many seeds. Sterile flowers are solitary, showy, to 3/4 inch across, fragrant, violet, rarely pink or white, 5 petaled, effective mid to late spring.

FRUIT: Dehiscent 3 valved capsule, not ornamentally significant.

SELECTED CULTIVARS AND VARIETIES: Long in cultivation, there are many cultivars in the trade. Some of the more common include: 'Charm' Flowers white.
'Double Russian' Small, with double purple flowers.
'Marie Louise' Flowers double, white to bluish-lavender.
'Parma' Flowers double.
'Red Giant' Flowers large, reddish-violet.
'Rosina' Flowers rose-pink.
'Royal Elk' Flowers violet.
'Royal Robe' Flowers large, dark purple, highly fragrant.
'White Czar' Flowers white with purple markings on yellow centers.

* * *

V. pedata (pe-dā´tȧ) Footed, the leaves are shaped like birds' feet.

COMMON NAME: Birdfoot violet.

NATIVE HABITAT: Eastern United States, west to Minnesota and south to Florida.

HARDINESS: Zone 4.

SIZE: 5 to 6 inches tall, spreading indefinitely.

SPECIAL CULTURE: Tolerant to full sun and sandy soils.

FOLIAGE: Attractively divided into 3 to 5 segments, each of which is 2 to 4 cleft, each division often with toothed apex, colored bluish green.

FLOWERS: Solitary, deep blue, to 1 inch across, effective early spring to early summer.

Generally Applicable to All Species

CULTURE: Soil: Best in organically rich, loamy, well-drained, acidic soils.
Moisture: Not tolerant of drought; soil should be kept moist with watering as needed throughout the year.

Figure 3–198 *Viola hederacea* (life-size)

Light: Light to dense shade. In hot summer areas, moderate to dense shade.

PATHOLOGY: Diseases: Anthracnose, crown rot, downy mildew, gray mold, leaf spots, powdery mildew, root rots, rusts, scab, smut.

Pests: Aphids, cutworms, violet gall midge, greenhouse leaf tier, violet sawflies, slugs, mites, and nematodes.

MAINTENANCE: Annually apply a light topdressing of compost if they are grown in amended soils.

PROPAGATION: Cuttings: Cuttings may be taken from new shoots in the fall.

Division: Plantlets on stolons with roots can be separated spring and fall. Crown also may be divided and transplanted in spring or fall.

Seed: Many species and varieties are propagated by seed. Commercial supplies are available for some. Germination in 10 to 20 days at 70°F can be expected for most species.

***W**aldsteinia* [wâld-(or vâld)-stīn´i-á] After Count Franz Adam Waldstein-Wartenberg, an Austrian botanist (1739–1823). This genus is composed of only a few species of hardy, strawberrylike, herbaceous perennials that spread by shallow, short rhizomes. They are native to various areas of the northern temperate zone.

FAMILY: ROSACEAE: Rose family.

HABIT: Low, tufted, herbaceous, stoloniferous and rhizomatous, spreading, matlike ground covers.

RATE: Relatively slow to moderate; space plants about 6 to 12 inches apart from pint- or quart-sized containers.

LANDSCAPE VALUE: Fine general ground covers for use in small- or medium-sized areas. Edging around beds is desirable to check underground spread. Also nice as a filler among large shrubs and small trees. Tolerant of limited foot traffic.

* * *

W. fragarioides (frá-gā´ri-oy-dez) Strawberrylike, the leaves (Figure 3–199).

COMMON NAME: Barren strawberry.

NATIVE HABITAT: Southern Canada, northern and eastern United States.

HARDINESS: Zones 4 to 7.

SIZE: 4 to 8 inches high, spreading indefinitely.

FOLIAGE: Palmately compound, evergreen, alternate, basal, long petioled; 3 leaflets, wedge shaped, 3 lobed, to 3 inches long, dentate or cuneate at the apex, occasionally incised, deep shiny green, becoming bronzy in winter.

INFLORESCENCE: 3 to 8 flowered corymbs that are born on scapes held above foliage, flowers to 3/4 inch across, corolla with 5 yellow petals, reportedly sterile, effective in late spring.

FRUIT: Of 2 to 6 achenes, not ornamentally significant.

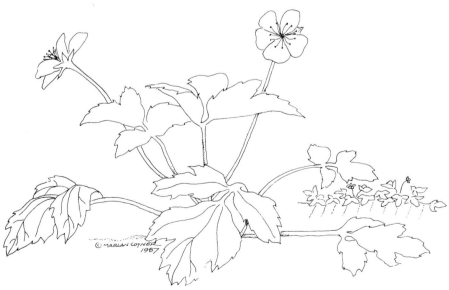

Figure 3–199 *Waldsteinia fragarioides* (0.75 life-size)

* * *

W. ternata (tĕr-nā´tà) In threes, the leaves (three leaflets).

COMMON NAME: Dry strawberry, yellow strawberry, T. K.'s special waldsteinia.

NATIVE HABITAT: Central Europe to Siberia, and Japan.

HARDINESS: Zone 4.

SIZE: 4 to 6 inches high, spreading indefinitely.

FOLIAGE: Evergreen, palmately compound, basal, alternate; leaflets wedge shaped, three, 1 to 2 inches long, irregularly toothed to lobed near apex, base cuneate, glossy green.

INFLORESCENCE: 3 to 8 flowered corymbs on bracted scapes to 8 inches high. Flowers are perfect, about 1/2 inch in diameter (to 3/4 inch), 5 sepals alternate with 5 bractlets, corolla is 5 petaled, yellow; reportedly sterile, blooming late spring to early summer.

OTHER SPECIES: *W. geoides* Leaves dull, rough, more or less deeply 5 lobed, cordate-reniform in outline, 3 to 5 inches long; flowers number 5 to 9, arranged in corymbs, 3/4 inch across, yellow; not as dense growing and thus less useful as a ground cover.

Generally Applicable to All Species

CULTURE: Soil: Adaptable to most well-drained soils; pH range of 4 to 7 is acceptable.
Moisture: Not notably tolerant of drought; soil should be kept moist with regular watering.
Light: Full sun to light shade.

PATHOLOGY: I have encountered serious damage to the foliage of Waldsteinias due to slugs. Otherwise, relatively free of problems.

MAINTENANCE: An annual light dressing of peat or compost will help to maintain soil moisture retention. Fertilizer should be washed immediately from foliage as it is prone to desiccation if an osmotic imbalance exists.

PROPAGATION: Division: Simply divide rooted plantlets that are formed on stolons. Crowns of established plants can also effectively be divided.

✦✦✦✦✦✦✦✦✦✦✦✦✦✦✦✦✦✦✦

Xanthorhiza (zan-thō-ri´zà) From *xanthos*, yellow, and *rhizoma*, root; hence, yellow root, the roots being yellow. This is a monospecific genus that is native to eastern North America.

FAMILY: RANUNCULACEAE: Buttercup family.

X. simplicissima (sim-pli-kis´i-mà) Most simple, that is, unbranched (Figure 3–200).

COMMON NAME: Yellow-root, shrub yellow-root.

NATIVE HABITAT: Eastern United States from New York to Florida.

HARDINESS: Zones 3 to 8.

HABIT: Woody, suckering, dense, shrubby ground cover.

RATE: Moderate to relatively fast; space plants 1 1/2 to 2 feet apart from gallon-sized containers.

SIZE: 2 to 3 feet tall, spreading indefinitely.

LANDSCAPE VALUE: Exceptional as a ground cover along stream and pond banks or other wet areas. Good as edging along contained areas such as walkways or building foundations. Sometimes used to good effect underneath open growing trees. No foot traffic.

FOLIAGE: Deciduous, compound, alternate, clustered, pinnate or bipinnate; leaflets usually numbering 3 to 5, ovate to ovate-oblong, 1 1/2 to 2 3/4 inches long, margin incisely toothed or serrate, medium shiny green, yellow to reddish orange in fall.

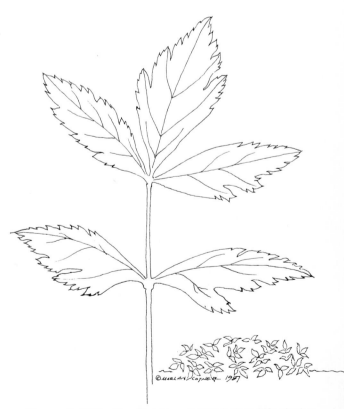

Figure 3–200 *Xanthorhiza simplicissima* (life-size)

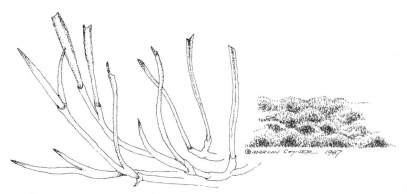

Figure 3–201 *Zoysia tenuifolia* (life-size)

STEMS: Outer bark yellowish-brown, inner bark yellow, branchlets pale greenish-gray; roots also yellow.

INFLORESCENCE: 2 to 4 inch long, many flowered racemes: flowers not very showy, tiny brownish-purple, star shaped, about 1/6 inch across, born before leaves in early spring.

FRUIT: Follicle, not ornamentally significant.

CULTURE: Soil: Adaptable to a wide range of soils; most luxuriant, however, in organically rich, moist, but well-drained soils. A pH range of 4.5 to 6.0 is generally best; tolerant of heavy clay soils.
Moisture: Although tolerating moderate drought conditions, soil moisture should be maintained with regular watering for lush growth. If grown in dry soil, growth rate is slowed and plantings may be less attractive. Its ability to withstand wet conditions, where other plants fail, is one of its main attributes.
Light: Full sun to dense shade.

MAINTENANCE: Shear plantings in early spring to promote lateral branching.

PATHOLOGY: No serious diseases or pests reported.

PROPAGATION: Cuttings: Cuttings from rhizome segments will form roots and shoots.
Division: Simply divide plants in early spring before vegetative growth begins.

Zoysia (zoi´si-à or zō-is´i-à) Named after Karl von Zoys, an Austrian botanist. This genus is composed of about five species of low growing perennial grasses of creeping habit. Many species are useful as lawn grasses in areas of warm climate.
FAMILY: POACEAE (GRAMINEAE): Grass family.

Z. tenuifolia (ten-ū-i-fō´li-à) From Latin, *tenuis*, slender, and *folia*, leaf, the leaves being narrow (Figure 3–201).

COMMON NAME: Mascarene grass, Korean grass, Korean velvet.

NATIVE HABITAT: Mascarene Islands.

HARDINESS: Zones 9 to 10.

HABIT: Low, stoloniferous, mounding, herbaceous, grasslike ground cover.

SIZE: 2 to 8 inches tall, spreading indefinitely.

RATE: Relatively slow; space divisions 6 to 10 inches apart.

LANDSCAPE VALUE: Very effective for filler in patio cracks, between stepping stones, or in a rock garden. Good also for general cover in small areas or as a lawn substitute where an informal effect is acceptable. Withstands moderate foot traffic.

FOLIAGE: Blades threadlike, acutely pointed, 1 to 2 inches long, strongly involute, bright green most of the year.

INFLORESCENCE: Florets arranged in narrow, compressed, spikelike racemes, not of ornamental significance.

FRUIT: Grain, not of ornamental significance.

CULTURE: Soil: Adaptable to most well-drained soils.
Moisture: Well adapted to withstand drought; only occasional watering is required throughout the summer months.
Light: Full sun to light shade.

PATHOLOGY: No serious diseases or pest problems reported.

MAINTENANCE: Little or no maintanance required. Unlike other zoysias, *Z. tenuifolia* does not take well to mowing.

PROPAGATION: Division: Simply divide into smaller sections in spring or fall.

Glossary

abscission: The separating of a leaf from its point of attachment.

achene: Dry, indehiscent, one-seeded fruit.

acuminate: A leaf apex where the sides are gradually concave and taper to a point.

acute: A leaf apex in which the sides are straight and taper to a point.

aerial rootlets: Roots produced above ground, usually arising from the stem.

aggregate flower: A flower that is crowded into a dense cluster.

aggregate fruit: A fruit formed by the coherence of separate carpels that were distinct in the flower. A common example is found in the mulberry.

alternate: An arrangement of leaves in which there is one leaf at each node, as compared to opposite or whorled.

ament: See *catkin.*

angiosperm: A vascular plant that has its seeds enclosed in an ovary.

annual: A plant that completes its life cycle in one season.

anther: The pollen-bearing divisions of a stamen (male reproductive structure).

apetalous: Without petals.

apex: The tip or terminal end, usually in reference to a leaf.

apical: Pertaining to the apex or tip.

appressed: Pressed closely against (usually in reference to the stem).

arching: Curving over gracefully.

aril: A fleshy or hairy outgrowth of the seed.

armed: Defended by thorns, spines, or barbs.

aromatic: Fragrant.

ascending: Curving upward.

attenuate: With a gradual, slender taper. Most often this term is used to refer to leaf apices, but sometimes it is used to describe leaf bases and other anatomical features.

auriculate: With earlike appendages.

awl-shaped: Tapering to a narrow stiff point; essentially similar to an awl used in leather working.

axil: Pertaining to the axis, usually in reference to the upper angle formed by the junction of leaf and stem.

axillary: Located in an axil.

bark: The outer protective tissue of the stems of woody plants.

basal: Referring to the base (crown) of a plant; leaves that arise from the crown are said to be basal.

berry: A fleshy, indehiscent, multiseeded fruit.

bifurcate: Forked, splitting into two branches.

bilabiate: Two-lipped, often referring to a corolla, especially for members of the mint family.

bipinnate: Twice pinnate.

bisexual: Stamens and pistil (male and female reproductive structures) present in a single flower.

biternate: Twice ternate; a structure that is ternate (usually a leaf), but whose primary divisions are also each ternate.

bract: A reduced leaf; see illustration of flower morphology, as bracts are usually associated with flowers.

branchlet: Smaller division of the branch.

bristle: A short stiff hair.

broadly-elliptic: Wider than elliptic, in reference to leaf shape.

broadly-ovate: Wider than ovate, in reference to leaf shape.

bud: A structure of immature tissues, which may develop into a leaf, a flower, or a new shoot.

bulb: A modified, fleshy, underground shoot with central axis surrounded by fleshy, scalelike, colorless leaves.

bulbil: Small bulbs arising around a parent bulb, or in the case of some ferns and succulents from the leaf margins or axils.

calyx: The outer and usually lowest set of perianth segments of a flower, usually green and leaflike.

campanulate: Bell shaped, usually used to describe the shape of a corolla.

capitate: Headlike, in a compact rounded structure, usually in reference to flower arrangements.

capsule: A dry, dehiscent fruit produced from two or more fused carpels.

carpel: One of the parts of the gynoecium. It is the structure that bears and encloses the ovules of flowering plants.

caryopsis: The typical fruit of members of the grass family.

catkin: A spikelike inflorescence that is modified for wind pollination, as seen in the birches.

caudate: With a taillike appendage.

cespitose: Growing in tufts, as with many species of grasses.

ciliate: The margin of a leaf that is fringed with hairs.

cladophyll: A flattened stem that has the form and function of a leaf.

cleft: Divided into segments from the middle.

cleistogamous: A closed, self-fertilized flower.

clone: Plants derived vegetatively from a single parent plant.

clustered: Crowded so as not to be distinctly opposite or alternate, in reference to leaf arrangement. Sometimes flowers, too, are said to be crowded.

compact: Densely arranged in a small amount of space.

complete flower: A flower that has a calyx, corolla, one or more pistils, and stamens.

compound leaf: A leaf formed by two or more leaflets.

compressed: Flattened from the sides.

concave: Curved, like the inner surface of a sphere, usually used in reference to leaf morphology.

cone: A coniferous fruit of woody, leathery, or fleshy scales, each with one or more seeds.

coniferous: Bearing cones.

confluent: Blending together, so that the separation of individual parts is unclear.

continuous pith: Solid, with no hollow points.

convex: Curved like the outer surface of a sphere, usually in reference to leaf morphology.

cordate: Heart shaped.

corm: A bulblike, solid, underground stem not differentiated into scales, as is a bulb.

corolla: Most often composed of petals; this is the inner whorl or floral envelope of an angiosperm flower.

corymb: A usually flat-topped, indeterminate inflorescence whose outer flowers open first.

creeping: Trailing along near the ground and rooting occasionally, usually in reference to stoloniferous plants.

crenate: Foliage with rounded teeth on the margins.

crenate-serrate: Combination of blunt and sharp teeth on the leaf margin.

crenulate: Having tiny rounded teeth on the leaf margin.

crown: The central point near the ground level of a plant from which new shoots arise.

culm: Stem of grasses, sedges, and reeds.

cultivar: Variety which arose in cultivation.

cultivated: Maintained by humankind.

cuneate: Wedge shaped, with rather straight sides, usually in reference to the morphology of leaf bases.

cuspidate: A leaf apex that is somewhat abruptly and concavely tapered into an elongated, acutely pointed tip.

cuticle: An outer layer of dead epidermal leaf or stem cells.

cyme: A predominately flat-topped, determinate inflorescence in which the outer flowers open last.

cymose: Like or arranged as a cyme.

deciduous: Defoliating, as leaves of these plants do (usually in fall), as opposed to being evergreen.

decumbent: Lying on the ground with tips ascending; a habit of growth.

defoliation: Falling off of leaves, from natural or induced means.

dehiscent: Splitting open, as do seed pods of dehiscent plants upon ripening.

deltoid: Triangular in outline.

dense: Crowded closely together.

dentate: A leaf margin with teeth whose apices are perpendicular to the margin.

depressed: Flattened or compressed; a term used primarily to describe morphological characteristics.

determinate: An inflorescence type with the terminal flowers opening first, causing the elongation of the axis to stop.

diffuse: Loosely or broad spreading, in reference to growth habit.

dimorphic: Having two forms, usually differentiated by maturity (adult versus juvenile).

dioecious: Having unisexual flowers, with individual sex confined to separate plants. In effect, male and female plants exist in these species.

disk flower: The flowers in the center of typical inflorescences of the composite family.

dissected: Divided in narrow segments.

distal: Located toward the apex of a plant. In effect, located away from the base.

divided: Separated into divisions; usually used to describe compound leaves.

dormant: Periods in which a plant is in a resting or nongrowing state.

dotted: Description of the surface of a leaf that has visible spots or hair glands.

double flower: A flower with greater than the normal number of petals.

doubly crenate, dentate, or serrate: A description of a leaf margin with small teeth of these types between the larger ones.

downy: Pubescent (or hairy) with fine hairs.

drooping: Hanging from the point of attachment.

drupaceous: Drupelike; a type of fruit morphology.

drupe: A fleshy indehiscent fruit whose seed is surrounded by a hard endocarp. An easily recognized example is a cherry.

drupelet: A small drupe.

dwarf: An atypically small plant when compared to the species or related plants.

ellipsoid: Football shaped.

elliptic-oblong: A shape that is a composite of those two forms; usually descriptive of leaves.

elliptical: Leaf shape with the outline of an ellipse.

elongate: Lengthened.

emarginate: With a shallow notch at the apex of a leaf.

entire: A leaf margin without teeth or crenations.

epidermis: The outer layer of cells on a leaf, stem, etc.

erect: Upright growth habit.

even-pinnate: A compound leaf in which leaflets are arranged so that the result is a lack of the terminal leaflet, since each is paired.

evergreen: Having foliage attached throughout the year.

exfoliate: Sloughing off in shreds, as bark often does.

fasciation: Fusion, usually of stems.

fascicle: A bundle, like the leaves of pines.

fertile: Capable of producing seed.

filament: The part of a stamen that functions as the stalk.

filamentous: Threadlike.

filiform: Threadlike in form.

fimbriate: Fringed as petals sometimes are.

flaccid: Limp.

flat: A horizontal habit of growth or outline of a plant.

fleshy: Pulpy or juicy at maturity, as fruits often are.

flexuous: Flexible.

floret: A tiny flower; usually applied to the flowers of members of the grass and composite family.

foliage: Leaves.

-foliate: -leaved.

follicle: A dry, dehiscent, many seeded fruit derived from a single carpel (simple ovary).

frond: Leaf; usually only applied to leaves of ferns.

fruit: The structure that develops from the ovary wall as the seeds mature.

fruticose: Shrubby in habit.

funnelform: Corolla shaped like a funnel or trumpet.

furrowed: With longitudinal grooves.

fusiform: Spindle shaped; tapering to each end from a broader mid-section; usually applied to seed shape or leaf outline.

gamopetalous: Petals that are joined to form a corolla of one piece.

glabrate: A plant which becomes glabrous with maturity.

glabrous: Not hairy.

gland: Oil-secreting organs, usually of leaves or stems.

glandular: With glands.

glandular-pubescent: With glands and hairs combined or mixed.

glaucescent: Slightly glaucous.

glaucous: Covered with a waxy, grayish-blue bloom or whitish material; in regard to color, bluish-green or greenish-blue.

globose: With a globelike or spherical shape.

globular: Circular.

glossy: Shining, reflecting light.

glutinous: Sticky.

grooved: With narrow furrows or depressions.

gymnosperm: A plant in which the seeds are exposed, as opposed to being enclosed in an ovary.

gynoecium: The female components (composed of a carpel or carpels) of an angiosperm flower.

habit: The general mode and form of growth a plant displays.

habitat: Location in which a plant grows.

hair: Fibrous outgrowth of the epidermis, usually of leaves or stems.

hairy: Pubescent, with long hairs.

hastate: With the shape of an arrowhead, the basal lobes pointed outward from the mid-rib; in reference to leaf shape.

head: A dense inflorescence typical of the sunflower family.

herb: A plant with soft or succulent, nonwoody stems.

herbaceous: Herblike.

hilum: The scar on a seed where it was previously attached.

hip: Fruit of the rose.

hirsute: With coarse, rough hairs.

hybrid: Plant which arose from cross-pollination between two or more different types of plants.

IBA: Indole-3-butyric acid, a commonly used chemical for inducing roots in stem and leaf cuttings.

imbricated: Overlapping.

imperfect flower: Flower lacking either stamens or pistils.

incised: Cut with irregular incisions, and intermediate between toothed and lobed in shape.

incomplete flower: A flower without calyx, corolla, stamens, or pistils.

indehiscent: Not opening automatically, usually in reference to a fruit capsule.

indeterminate: A type of inflorescence in which the terminal flowers open last.

indusium: A cover over the sori that contains the sporangia of a fern.

inferior: Beneath.

inflated: Bladderlike, usually referring to a membrane around a seed.

inflorescence: The term given to the arrangement of flowers on the axis.

internode: The part of a stem between two nodes.

interrupted: Not continuous.

introduced: A plant which is imported from another region or country for purposes of cultivation.

involucral: Of an involucre.

involucre: One or more whorls of bracts or leaves that are located underneath an inflorescence or flower.

involute: A longitudinal rolling upward; usually descriptive of leaf morphology.

irregular flower: A flower that can be cut into two equal halves at only one place.

juvenile: The early stage of plant growth, characterized by nonflowering, hairiness, vigorous growth, and frequently thorniness.

labiate: Lipped, as are projections of the corolla of plants in the mint family.

laciniate: Cut into narrow-pointed incisions or lobes.

lanate: Woolly.

lanceolate: Longer than wide, widest below the middle, and tapering to the apex; usually in reference to leaf shape.

lateral: Located on the side, as flowers frequently are.

lateral bud: A bud borne in the axil of a leaf.

leaflet: A foliar division of a compound leaf.

lenticel: A small corky projection or dot on young bark for the purpose of gaseous exchange between the inner plant tissues and the atmosphere.

lignified: Woody, hardened, with lignin present.

linear: Long and narrow, used primarily to describe leaf shapes.

lip: One of the divisions of an unequally divided corolla or calyx, commonly seen in the mint family.

loculicidal: See *dehiscent.*

loose: Irregularly and open in habit, not compact.

margin: The edge of a leaf.

marginal: Pertaining to the margin of a leaf or petal.

mature: A stage of growth characterized by flowering, fruiting, and often a reduced rate of growth.

mealy: Having a dry, granular appearance and texture.

meristem: Tissue that is capable of developing into specialized tissues.

monoecious: A plant with unisexual flowers, with both sexes on the same plant.

mossy: A matted growth habit that gives the impression of moss.

mucro: A sharp, short, abrupt point like a seed apex.

mucronate: Terminated by a mucro, as are some seeds.

multiple fruit: A fruit that is formed from the combination of several flowers into a single structure with a common axis.

mutation: A change in genetic composition.

needle: The narrow leaf typical of conifers.

nerve: A narrow rib or vein.

netted venation: Veins arranged so as to resemble a fish net.

nocturnal: Opening or appearing at night and closing during the day.

nodding: Drooping, bending over, often gracefully.

node: A joint of a stem.

nut: A dry, indehiscent, one-celled, single-seeded fruit with a hard mesocarp.

nutlet: A small nut.

obcordate: Leaf apex being cordate.

oblanceolate: Inversely lanceolate, usually in reference to leaf shape.

oblique: Lopsided.

oblong: Longer than broad, usually in reference to leaf shape.

oblong-lanceolate: A shape between oblong and lanceolate.

oblong-obovate: A shape between oblong and obovate.

obovate: Broadest toward the apex or end.

obovoid: Pear shaped, obovate in three dimensions.

obtuse: Rounded, or nearly so.

opposite: Grouped two per node, usually in reference to leaf arrangement.

ovary: The ovule-bearing structure of a pistil.

ovate: Egg shaped in outline, wider at the base than terminally.

ovate-oblong: A combination of these two forms.

ovoid: Three-dimensionally egg shaped.

ovule: The egg-containing part of an ovary that develops into a seed following fertilization.

palmate: Fanlike from a common point, like a hand; usually in reference to compound leaf arrangement.

palmatifid: Cut palmately, but not all the way down.

panicle: An indeterminate racemose inflorescence whose primary axis bears branches of pedicelled flowers.

paniculate: Bearing panicles.

papillate: With pimplelike outgrowths.

pappus: Modified calyxes of plants in the composite family, borne on the ovary and represented by awns, bristles, hairs, or scales. The pappus often adheres to the fruit.

parallel: Running side by side and not intersecting, as in veins of grass blades.

parallel venation: With veins extending in a predominantly parallel manner.

peduncle: The stalk of a flower cluster or a solitary flower.

peltate: A leaf with the petiole attached within the margin; such leaves are usually shield shaped.

perennial: A plant that lives for three or more years.

perfect flower: Having both stamens and pistils.

perfoliate: Situation when the leaf-blade surrounds the stem.

perianth: The term given to corolla and calyx collectively.

pericarp: A term used to designate the wall of a fruit.

persistent: Adhering for an extended period of time.

petal: A single unit of the corolla.

petaloid: A structure that resembles a petal.

petiole: Leaf-stem.

petiolule: Stem of a leaflet.

photoperiod: Time period during a 24-hour day in which a plant is exposed to light.

pilose: With soft, shaggy hairs.

pinna: The leaflets of a compound leaf, usually in reference to fern leaves.

pinnate: Compound leaf arrangement, with the leaflets or segments along each side of a common axis being featherlike in arrangement.

pinnatifid: Pinnately cleft or divided.

pinnule: The secondary leaflet of a pinna.

pistil: The unit of the gynoecium consisting of the ovary, stigma, and style.

pistillate: A flower without functional stamens, female.

pith: The central part of a stem that is usually corky.

plumose: Featherlike.

pod: A dry, dehiscent fruit (usually applied to legumes).

pollen: Microspores of seed plants that are contained within an anther.

polygamous: Bearing unisexual and heterosexual flowers on the same plant.

pome: A fleshy fruit that results from a compound ovary. Apples and pears are pomes.

prickle: A small, weak, spinelike projection.

procumbent: Trailing on the ground, but the stems not rooting.

prostrate: Lying flat on the ground.

proximal: Toward the base or crown.

puberulent: Minutely pubescent.

pubescent: Covered with soft, short hairs.

punctate: With dots, depressions, or pits.

pungent: Sharp and acid to the smell or taste.

quadrangular: Four angled.

raceme: A simple indeterminate racemose inflorescence with pedicelled flowers attached to a main axis.

racemose: Having flowers arranged in racemes.

rachilla: A secondary axis.

rachis: Main axis that bears leaflets or the primary axis of a fern frond.

radial: Originating and spreading from a common center.

radiate: Spreading from a common center.

ranked: Foliage arrangement in longitudinal planes around the stem.

ray: Usually used in reference to the marginal florets that surround the disc of tubular florets of the inflorescence of plants in the composite family.

recurved: Bent or curved backward.

reflexed: Bent sharply backward.

reniform: Kidney shaped.

revolute: Rolled toward the back or under, usually in reference to leaf margins.

rhizome: An underground stem with nodes, buds, or scalelike leaves. Roots do not have these.

rhombic: With four nearly equal sides, but angles unequal between them.

rhombic-ovate: Between diamond and egg shaped.

rhomboidal: With the shape of a rhomboid.

rib: Conspicuous vein or ridge.

rogue: A plant which varies from the norm for that species or variety in a manner such that it is undesirable; the act of removing and destroying such plants.

root: The descending anchor and absorptive apparatus of the plant; it has no nodes and is usually subterranean.

rootlet: A subdivision of a root.

rosette: A group of leaves radiating from a stem and often near the crown.

rugose: Wrinkled.

runner: A trailing shoot (stolon) that usually roots at the tip and may have a plantlet (or plantlets) associated with it.

sagittate: Shaped like an arrowhead, with basal lobes usually pointing backward; usually applied to leaf shapes.

salverform: Descriptive of a corolla with a slender tube and expanded flat ends that extend at right angles to the tube.

scabrous: Rough to the touch.

scale: A small dry bract or leaf, or a structure similar to such.

scape: A leafless, single or many flowered peduncle arising from the basal rosette of leaves or rhizome.

scapose: Bearing flowers on a scape.

scar: A mark left from a previous attachment, such as a leaf scar.

schizocarp: A dry, dehiscent fruit which splits into two segments.

seed: A fertilized, ripened ovule that usually contains an embryo and food supply.

segment: A division of a leaf or perianth that is divided, but is not itself compound.

semievergreen: Only part of the foliage fully held throughout the year, often dependent on climate.

sepal: Structure comprising the calyx; usually a green, leafy structure subtending the corolla.

serrate: Sawtoothed, with the teeth pointing forward; usually referring to leaf margin.

serrulate: Minutely serrate.

sessile: Without a stalk; usually referring to a leaf without a petiole.

seta: A bristlelike hair.

setaceous: Bristlelike.

setose: Covered with bristles.

shaggy: Covered with long, woolly hairs.

sheath: An elongated, tubular structure that surrounds a plant part.

shrub: A woody plant that is not treelike in size and often produces branches from the base.

silicle: The short fruit (dry and dehiscent) of some crucifers. It is usually less than one and a half times as long as wide.

silique: An elongated fruit of some crucifers, frequently three times (or more) as long as wide.

silky: With soft, appressed, fine hairs.

smooth: Not roughened or rugose, but smooth to the touch.

solitary: Occurring alone.

sori: Plural of sorus.

sorus: Group of sporangia (of ferns).

spadix: Club-shaped axis on which flowers are borne in plants of the Araceae.

spathe: A modified leaf surrounding or subtending an inflorescence.

spatulate: Spoon shaped.

spicate: With spikes.

spike: Basically an unbranched, simple, elongated, indeterminate inflorescence.

spikelet: A secondary spike.

spine: A sharp, pointed projection.

spinose: Endowed with spines.

spirally arranged: Arrangement of alternate leaves as one views them from atop.

sporangium: A spore-containing case.

spore: A simple, asexual, usually haploid reproductive body comprised of a single cell.

sporocarp: A structure containing sporangia or spores.

sporophyll: A spore-bearing leaf.

squamose: Covered with small scales.

stalk: Supporting structure of a leaf, fruit, or flower.

stamen: Male reproductive body that usually is made up of anther and filament.

staminate: An imperfect flower with only functional, male flower parts.

stellate: Starlike, usually in reference to arrangement of hairs.

stem: The primary aboveground axis of a plant.

sterile: Not capable of producing seed.

stigma: The part of the pistil that receives pollen.

stipe: A leafless stalk or petiole of a fern leaf.

stolon: A horizontal aboveground stem (runner) that roots at its tip and there produces a new plant.

stoloniferous: Bearing stolons.

stoma: Tiny pores in the epidermis, especially in the lower surface on the leaf or in stems where gas exchange takes place.

strigose: With straight, stiff, hairs or bristles.

strobilus: A cone-like reproductive structure.

subopposite: Pairs of leaves that are arranged at nearly the same level on the stem.

subulate: Awl shaped.

succulent: Thickened, fleshy, water-storing tissues that are often spongy to the touch.

sucker: A shoot arising from the roots or near the crown from beneath the surface of the ground.

suffrutescent: A plant that is woody basally but becomes herbaceous above distally.

ternate: In threes; three parted.

texture: The effect or impression that the surface or outward appearance of a structure portrays.

thyrse: A compact, paniclelike branching inflorescence in which the distal end is indeterminate and the lateral branches determinate.

tomentose: Densely hairy.

toothed: A leaf margin broken up into small, rather regular toothlike segments.

trailing: Prostrate but not rooting; procumbent.

trifoliate: Three leaved.

trifoliolate: Compound leaf of three leaflets.

tripinnate: With compound pinnules.

tuber: A short, thickened, usually underground storage stem.

tubercle: A tiny tuber.

tubular: With petals, sepals, or both joined into a tube.

tuft: A clump of hairs or leaves growing close together, originating from a common point, and spreading outward distally.

turgid: Swollen with water.

twining: A stem that winds spirally around a support.

umbel: An indeterminate inflorescence that usually is flat topped, with the pedicels and peduncles arising from a common point, such as a carrot inflorescence.

umbellate: With umbels.

unarmed: Without a defense such as spines, bristles, or thorns.

unilateral: One sided.

unisexual flowers: Of one sex.

urceolate: Urn shaped, usually descriptive of a corolla in the heath family.

valve: An individual part of a dehiscent fruit.

variegated: With a variety of color.

vein: A vascular structure of a leaf.

venation: Arrangement or organizational pattern of veins.

verticillaster: A spikelike inflorescence in which the flowers are arranged in pairs of dense sessile cymes which originate from leaf axils. Commonly this is witnessed among members of the mint family.

villous: With long, soft, unmatted hairs.

vine: A climbing or trailing plant with slender stems (usually).

weeping: Hanging pendantly.

whorl: An arrangement of three or more structures originating from a single node, such as leaves, petals, and calyx lobes.

whorled: Arranged in a whorl.

wilt: To lose turgor through a lack of water or inability to conduct water.

wing: Usually a dry, thin appendage of a seed.

Bibliography

ARTHURS, KATHRYN L., ed., *Lawns and Ground Covers.* Lane Publishing Co., Menlo Park, Calif., 1979.

BAILEY, L. H., *How Plants Get Their Names.* Dover Publications, Inc., New York, 1963.

———, and BAILEY, ETHEL Z., *Hortus Second.* Macmillan, Inc., New York, 1947.

BARON, MILTON, et al., *Handbook of Ground Covers and Vines.* Brooklyn Botanic Garden, Brooklyn, N.Y., 1982.

BLOOM, ALAN, *Alpines for Your Garden.* Floraprint U.S.A., Chicago, 1980.

———, *Perennials for Your Garden.* Floraprint U.S.A., Chicago, 1981.

BOLD, HAROLD C., *The Plant Kingdom.* Prentice-Hall, Inc., Englewood Cliffs, N.J. 1977.

BRICKELL, C. D., et al., *International Code of Nomenclature for Cultivated Plants.* Bohn, Scheltema and Holkema, Utrecht, Netherlands, 1980.

COBB, BOUTHTON, *A Field Guide to the Ferns and Their Related Families.* Houghton Mifflin, Boston, 1963.

CROCKETT, JAMES U., *Lawns and Ground Covers.* Time-Life Books, Inc., New York, 1972.

———, *Perennials.* Time-Life Books, Inc., New York, 1972.

DENNIS, JOHN V., *Beyond the Bird Feeder.* Alfred A. Knopf, Inc., New York, 1981.

———, *The Wildlife Gardener.* Alfred A. Knopf, Inc., New York, 1985.

DIAMOND, DON, and MacCASKEY, MICHAEL, *All about Ground Covers.* Ortho Books, San Francisco, 1982.

DIRR, MICHAEL A., *Manual of Woody Landscape Plants.* Stipes Publishing Company, Champaign, Ill., 1983.

———, and HEUSER, CHARLES, *The Reference Manual of Woody Plant Propagation.* Varsity Press, Athens, Ga., 1987.

DREW, JOHN K., *Pictorial Guide to Hardy Perennials.* Merchants Publishing Company, Kalamazoo, Mich., 1984.

ELLIOT, RODGER W., and JONES, DAVID L., *Encyclopaedia of Australian Plants Suitable for Cultivation.* Lothian Publishing Co. Ltd., Melbourne, Australia, 1982.

EVANS, RONALD L., *Handbook of Cultivated Sedums.* Science Reviews Ltd., Northwood, Middlesex, United Kingdom, 1985.

FISH, MARGERY, *Ground Cover Plants*. Faber and Faber, Boston, 1980.

FLINT, HARRISON L., *Landscape Plants for Eastern North America—Exclusive of Florida and the Immediate Gulf Coast*. John Wiley & Sons, Inc., New York, 1983.

FOLEY, DANIEL J., *Ground Covers for Easier Gardening*. Dover Publications, Inc., New York, 1961.

FOREST SERVICES, UNITED STATES DEPARTMENT OF AGRICULTURE, *Seeds of Woody Plants in the United States*. Agriculture Handbook No. 450, Washington, D.C., 1948.

GILES, F. A., et al., *Herbaceous Perennials*. Reston Publishing Company, Reston, Va., 1980.

GROUNDS, ROGER, *Ornamental Grasses*. Van Nostrand Reinhold Company, New York, 1981.

HAMILTON, BETTY R., et al., *Ground Covers for the Midwest*. Board of Trustees of the University of Illinois, Champaign, Ill., 1983.

HARTMAN, HUDSON T., and KESTER, DALE E., *Plant Propagation*. Prentice-Hall, Inc., Englewood Cliffs, N.J., 1983.

HILLIER, H. G., *Hillier's Manual of Trees and Shrubs*. Van Nostrand Reinhold Company, New York, 1981.

HOSHIZAK, BARBARA JOE, *Fern Grower's Manual*. Alfred A. Knopf, Inc., New York, 1975.

HOTTES, ALFRED C., *Climbers and Ground Covers*. A. T. Delamare Company, Inc., New York, 1947.

JOHNSON, A. P., *Plant Names Simplified*. Transatlantic Arts, Inc., Albuquerque, N.M., 1958.

LABADIE, EMILE L., *Ground Covers for the Landscape*. Sierra City Press, Sierra City, Calif., 1982.

LAMPE, KENNETH, and McCONN, MARY, *Handbook of Poisonous and Injurious Plants*, American Medical Association, Chicago, Ill., 1985.

MacCASKEY, MICHAEL, *Lawns and Ground Covers*. Horticultural Publishing Co., Inc., Tucson, Ariz., 1982.

MATHIAS, MILDRED E., *Flowering Plants in the Landscape*. University of California Press, Berkeley, Calif., 1982.

MICKEL, JOHN T., *Ferns and Fern Allies*. William C. Brown Company, Dubuque, Iowa, 1979.

MUNRO, WILLIAM, *A Monograph of the Bambusaceae*. Johnson Reprint Corporation, New York, 1966.

New Pronouncing Dictionary of Plants Names. Florist's Publishing Co., Chicago, 1964.

PIEROT, SUZANNE, *The Ivy Book*. Macmillan, Inc., New York 1974.

PIRONE, PASCAL P., *Diseases and Pests of Ornamental Plants*. John Wiley & Sons, New York, 1978.

PORTER, C. L., *Taxonomy of Flowering Plants*. W. H. Freeman and Company, San Francisco, 1967.

POTTER, CHARLES H., *Have You Tried Perennials?* Florists Publishing Company, Chicago, 1959.

PRAEGER, L. R., *An Account of the Genus Sedum as Found in Cultivation*. Stechert-Hafner Service Agency, Inc., New York, 1967.

———, *An Account of the Sempervivum Group*. Stechert-Hafner Service Agency, Inc., New York, 1967.

PROUDLEY, BRIAN, and PROUDLEY, VALERIE, *Heathers*. Blandford Press, Dorset, United Kingdom, 1983.

REILLY, ANN, *Park's Success with Seeds*. George W. Park Seed, Co., Inc., Greenwood, S.C., 1978.

ROSE, PETER Q., *Ivies*. Blandford Press, Dorset, United Kingdom, 1980.

STAFF OF LIBERTY HYDE BAILEY HORTORIUM, *Hortus Third*. Macmillan, Inc., New York, 1976.

STERN, WILLIAM T., *Botanical Latin*. David and Charles, Inc., North Pomfret, Pomfret, Vt., 1985.

STILL, STEVEN, *Herbaceous Ornamental Plants*. Stipes Publishing Company, Champaign, Ill., 1982.

SUNSET BOOKS. *New Western Garden Book*. Lane Publishing Co., Menlo Park, Calif., 1985.

Colloquial (Common) Name Index

Subject Index